ENGLISH PLACE-NAME SOCIETY. VOLUME XCIV
FOR 2017

GENERAL EDITOR

DAVID PARSONS

PRODUCTION EDITOR

PAUL CAVILL

THE PLACE-NAMES OF DORSET

PART V

THE SURVEY OF ENGLISH PLACE-NAMES
UNDERTAKEN WITH THE APPROVAL AND SUPPORT OF
THE ARTS AND HUMANITIES RESEARCH COUNCIL
AND
THE BRITISH ACADEMY

THE PLACE-NAMES OF

DORSET

BY

A. D. MILLS

PART FIVE

DISTRICTS, RIVERS, INTRODUCTION, ANALYSES, INDEX

NOTTINGHAM
ENGLISH PLACE-NAME SOCIETY
2020

Published by the English Place-Name Society,
School of English, University of Nottingham,
Nottingham NG7 2RD.

Registered Charity No. 257891

ISBN: 978-1-911640-02-8

Typeset by Paul Cavill
& Printed in Great Britain
by 4Word, Bristol.

CONTENTS

PREFACE

It gives me great pleasure and satisfaction to see this fifth volume of *The Place-names of Dorset* finally published, since it concludes the survey of the county's names I began in the early sixties and therefore completes what has turned out to be a lifetime's project. It has been a delight to work on the names of this splendid English county, and an enormous privilege to contribute to the English Place-Name Society's great survey of English place-names.

I would like to recall all the acknowledgements made long ago in the Preface to my first volume of the Dorset survey, and to once again thank all my colleagues past and present for all their help, advice and encouragement over the years. Above all I would once again like to thank Dr Paul Cavill, Production Editor for the English Place-Name Survey, for all his tremendous help and skill in bringing my work to publication.

Finally I would like to record my deep appreciation to my dear wife Solvejg: she has lived with this project throughout our marriage and without her unfailing love and wonderful patience it could not have been completed.

Monks Eleigh, October 2019 David Mills

DORSET

DORSET

(*mid*) *Dornsætum* l9 ASC (A) s.a. 837 [840], 845 [848], (*to* ~, *on*) *Dorsæton* 955 (14) S 1515, 997 ASWills, 1012 (12) *SherC* (S 1422), 1046 (12) *ib* (S 1474), 12 ASC (E) s.a. 845 [848]

(*on* ~, *mid*) *Dorsætum* m11 ASC (C) s.a. 978, 982, 12 ASC (E) s.a. 837 [840], (*on*) *Dorsætan* 1053–58, 1058–66 ASWrits, 12 ASC (E) s.a. 998, (*cum*) *Dorseton* e12 ASC (F) s.a. 845 [848]

Dorset' 891 (14) *Glast* (S 347(2)), 1195 P, 1196 Cur, 1212 Fees *et freq* to 1275 RH, *Dorset(h)*, *Dorscete* 11 France, *Dorsete* 1086 DB, 1197 P, 1225 Pat *et freq* to 1346 FA, *Dorseta* 1086 Exon, 1107 (1330) Ch, 1129 CartAnt, 1165–89 *Ilch*, 1190, 1194 P, *Dorsetha* 1129 CartAnt, 1190 (1332) Ch, *Dorsett(e)* 1154–5 RBE, 1201 ChancR *et freq* to 1401 *Marten*, *Dorsat'* 1174–5 P, 1191–2 AC, 1195 P, *Doreseta* 1178–9 ib, *Dorsetta* 1179–80 ib, 1196 ChancR, *Doresete* c.1190–1215 Laȝamon, *Dorsit* 1194 Cur, *Dorsettia* 1297 CoramR, *Dorcet* 1588 *AD*, *Dorcett(e)* 1595 *HarlCh*, 1596 *AD*

Thornsæta (sic) 893 (e11) Asser, *Torsete* (sic) c.1150 Wace (*Roman de Brut*)

Dorseteschire 940–6 (14) S 513, *-schyre* 955 (14) S 1515, *Dorsetasire* 1068–84 France, *-scira* 11 ib, *Dorseta scire* 12 SD, *Dorsets'* 1107–1120 CartAnt, *Dorsetscire* 1292 Lewes, *Dorset(t)shire* 1349–62 ib, 1608 *LymeR*, *Dorset(es)chire* 1359, 1360 Lewes

Dorchestre seten c. 1190–1215 Laȝamon

the shier of Dosshet 16 Hutch[3] (**1** 59)

Originally a folk-name, 'the Dorn people, the people of the region about Dorchester', from *Dorn* (a shortened form of *Dornwaraceaster* = Dorchester) and OE **sǣte** 'dwellers, settlers', *v.* Dorchester **1** 347–8. The early forms in *-sætum*, *-sæton*, *-seten* represent an OE dat. pl. *sǣtum* (*sǣte* is a pl. word). Forms with **scīr** 'shire, district' are relatively infrequent. Widén gives the local pronunciation as [ˈdɑːsət, ˈdɑːʀsət].

DISTRICT-NAMES

BLACKMOOR FOREST & VALE (**3** 274–5)
First recorded as *boscum de Blakemor*(*e*) 1212 Fees, named from the former manor of *Blakemor*(*e*), *v.* Blackmore Ctgs in Wootton G. par. **3** 268. For more early spellings and for references to the bounds of the Forest, *v.* full discussion **3** 274–5.

CRANBORNE CHASE (**2** 193–4)
First recorded as *bosc' de Craneburn'* 1234 Cl, named from Cranborne par. **2** 205. For more early spellings and for references to the bounds of the Chase and its jurisdiction, *v.* full discussion **2** 193–4.

ISLE OF PURBECK (**1** 1–3)
First recorded in the folk-name form *Purbicinga* 948 (15) *ShaftR* (S 534(1)), then as *Porbi*, *Porbiche* in 1086 DB. For more early spellings, etymology, and further details of extent and history, *v.* full discussion **1** 1–3.

MARSHWOOD VALE (**4** 466)
First recorded as *Merswodenaal* (sic for *-uaal*) 1319 FF, named from Marshwood par. **4** 400. For more early spellings and for a description of the extent, *v.* full discussion **4** 466.

RIVER-NAMES

The river-names included here are those of the principal rivers and streams in the county, together with a few minor ones of particular interest, most of the references being from local Dorset sources.

Stream-names that are now lost but which have become the names of settlements, e.g. Arfleet **1** 7, Blackwater **1** 139, Blashenwell **1** 8, Claywell **1** 44, Colesbrook **3** 19, Furzebrook **1** 91, Gorwell **4** 52, Holwell **1** 240, Honeybrook **2** 151, Knaveswell **1** 35, Lambrook **4** 317, Langbourne **2** 109, Litton Cheney **4** 23, Lowbrook **3** 181, Milborne **1** 282, 306, Northlake **1** 58, Palbroke **3** 210, Pink Lake Coppice **4** 54, Seawell **4** 6, Sherborne **3** 355, Shortlake **1** 214, Stanbrook **3** 286, Ulwell **1** 56, Winfrith **1** 174, Wynford **4** 117, are generally not included here since they are treated within their respective parishes.

Many other lost or unidentified stream-names are dealt with in the f.n. sections of the parishes in which they occur, e.g. *Alfricheswelle* **1** 30, *Clingerwell* **3** 242, *Cramborne* **1** 211, *Freshwater Lake* **2** 38, *Hlydanpol* **4** 299, *Holebroke* **3** 109, *Maynewater* **1** 338, *Melenbroc* **3** 236, *Mylenburnan* **3** 301, *Ruggewelle* **3** 253, *Sevewelle* **2** 242, *on seuen willes þry* **1** 32, *Smalebrok* **3** 109, *stanbrok* **3** 258, *Tadbrokes lake* **3** 288, *Todbere Lake* **3** 84. Many of these lost stream-names and others can of course be traced in the Elements section *infra* under the entries for **brōc**, **burna**, **lacu**, **rīð**, **wæter/water**, **wella**, etc.

ALLEN (affluent of R. Stour), *the Allen* (*or more commonly the Brook*) 1695 Hutch[3] (**3** 251). A back-formation from *Aleyn bridg*(*e*) 1535–43 Leland, *Aleynsbrydge* 1543 *Ct*, earlier *Aldewynebrigg'* 1268 *Ass*, later Canford Bridge in Wimborne Minster par., *v*. **2** 187 for further forms and discussion.

The Allen was originally called *Wimborne*, a name transferred to the pars. of Wimborne Minster and Wimborne St Giles. Early forms include (*on-*, *up andlang*) *winburnan* 946 (14) *Harl* (S 519), (*adune innan*) *winburnan* 956 (14) *ib* (S 609) (in bounds of Didlington in Chalbury par. **2** 134–5), (*on-*, *of*) *winburnan* 1033 (12) *SherC* (S 969), (*on* ~, *of*) *Winborne* n.d. (12) *ib* (f. 24v) (in bounds of Horton par. **2**

159), *aquam de Wymburn(')*, *-born(e)* 1280 *Ass*, QW, 1318 *Wim*, *aquam Wynbourn'* 1333 *Wim*, *Wymborn' Water* 1359 *Wim*, 17 *CecilMap*, *Winburn(e) River*, *-streame* 1535–43 Leland, *the water of Wimborne* 1593 Hutch[3]. According to local sources, the branch of the Allen passing under the westmost bridge in East Street in Wimborne Minster is known as the Win *q.v. infra.*

ASKER (affluent of R. Brit), a late back-formation from Askerswell par. (**4** 47). It is probably referred to as *the east river* 1784 *SalisT*. The original name of the stream was possibly *Loder*, *v.* discussion under Loders par. **4** 482.

ATRAM (possibly a lost stream-name), perhaps the original name of R. Simene *infra*, *v.* discussion under Atrim **4** 312–3.

AVON (lost), *on auenes broc* 955 (14) *ShaftR* (S 573), 'the stream called *Auen*', probably referring to the now unnamed stream rising near St Edward's Cottage (**1** 18, 77) and running into Corfe River. A common Brit r.n., from PrW ***aβon** 'river', *v.* **1** 30 and Ekwall RN 20–23.

AXE (rising near Chedington **4** 285 and giving name to Axminster and Axmouth in Devon), (*on* ~, *of*) *Axan* 1005 (12) S 910, *Axe* 1264 (15) *Forde*, 1268 *Ass*, *Axa* 13 (15) *Forde*, *Axe water* 1663 *DCMSurv*. An OE r.n. **Æsce* (metathesised **Æcse*) from Brit ***Escā** 'water', *v.* Ekwall RN 152.

BALLS LAKE (a channel in Poole Harbour), *v.* **1** 74.

BEAM BROC (lost), *on beam broc* 948 (15) *ShaftR* (S 534 (1)), *on beambroc* 955 (14) (S 573), 'brook with a tree-trunk across it', *v.* **bēam**, **brōc**, **1** 31, Ekwall RN 27. It refers to a now unnamed tributary of Corfe River, meeting at St Edward's Bridge (**1** 18).

BERE STREAM, 1984 Gazetteer, named from Bere Regis par. **1** 273. An earlier stream-name, *aqua de Bere* 1244 *Ass*, is named from Beer Hackett par. **3** 290 and may refer to the now unnamed stream that gave name to Trill **3** 291.

BIBBERN BROOK, *on bydeburnan* 933 (12) *SherC* (S 423) (in the Anglo-Saxon bounds of Stalbridge Weston **3** 283–4), *aqua de Biddeburn'* 1244 *Ass*, from OE **burna** 'stream', first el. probably OE **byden** either in its literal sense 'vessel, container' or in a transferred topographical sense 'hollow, small valley' (this preferred by Fägersten 37). The stream gives name to Bibbern Bridge and Fm in Stalbridge par. **3** 284. For an alternative view, that the name contains a pre-English r.n. borrowed as *Bȳd* found in R. Boyd Gl **1** 4, *v*. Ekwall RN 47.

BLACKWATER RIVER (affluent of R. Axe), *Black Water* 1769 *DROMap*, *Blackwater* 1811 OS, 1863 Hutch[3], from **black** and **water** and giving name to Blackwater Lane in Hawkchurch par. **4** 148.

Other instances of this name include: *Blak*(*e*)-, *Blacwater* 1449–59 *Weld*[1] (giving name to Blackwater Ford in Owermoigne par. **1** 139); *Black Water* 1811 OS in Canford M. par. **2** 2; *Black Water* 1811 OS (now Sleep Brook) in Alderholt par. **2** 196, 198); *Blackwater* 1824 Pigot, cf. *Blacke River* 1652 *ParlSurv*, in Cerne Abbas par. **4** 136. Cf. also Blackwater Fm in Bourton par. **3** 3 and Blackwater Bridge in Sturminster N. par. **3** 193.

BLECHENE (lost), *on Blechene* 956 (14) *ShaftR* (S 632) (in the Anglo-Saxon bounds of Corfe Castle), referring either to a spring or a stream, *v*. discussion under Blashenwell Fm **1** 8.

BLINCHES BROC (lost, affluent of Key Brook), *on blinches broc* 932 (15) *ShaftR* (S 419) (in the Anglo-Saxon bounds of Fontmell Magna), possibly a stream-name **Blinch*, *v*. discussion under Blynfield Fm in Cann par. **3** 90–2.

BOURNE, 1886 *DROMap*, *v*. **burna**. It gives name to Bourne Bottom and Valley in Branksome par. **2** 1, to Bourne in Kinson par. **2** 23, and of course to Bournemouth (formerly Ha).

There is another example of this r.n., recorded as *la Bourne* 1270 (1372) *ChrP*, an affluent of R. Piddle and rising in Cheselbourne **3** 202. It gives name to Bourne Fm in Piddlehinton **1** 310 *q.v.* for early forms and discussion.

BOW BROOK (affluent of the Cale), probably named from the fields called *Bowe Close* Eliz *LRMB*, *Gt & Lt Bow* 1846 *TA* in Kington M. par. **3** 43. The fields lie in a bend of the brook, cf. Ralph *atte Brouke* 1327 *SR*, *v*. **boga** in the sense 'river-bend'.

BRIDE, *aquam de Bridie*, *Bridye* 1288 *Ass*, a Celtic r.n. giving name to the pars. of Little and Long Bredy **4** 12, 50 (first recorded as *Bridian*

987 (13) KCD 656 (S 1217, Finberg 613), also to Burton Bradstock par. (*Bridetune* 11) **4** 477 and to Bredy Fm in that par., *loc. cit.* It is 'the gushing or surging stream' from an OWelsh ***Brydi** related to OCorn *bredion* 'to boil or well up', *v.* Ekwall RN 52. The stream rises at Bride Head Lake in Little Bredy **4** 13.

There is another instance of this r.n., recorded as *bridian beara lace* 1044 (m11) in the Anglo-Saxon bounds of Abbott's Wootton **4** 453.

BRIT, *The Ryver Birtt or Brit* 1535–43 Leland, *river Birt* 1774 Hutch[1], *river Bert*, *Brit or Birt* 1863 Hutch[3], a back-formation from Bridport **4** 353–4 (specifically from the form *Brit(e)port* found from the 13th to 16th centuries). The river was earlier called *Woth* (*v. infra*), cf. also the form *aquam de Kaynnes* 1288 *Ass* noted under Camesworth **4** 315 which may refer to this river, *v. Kaynnes infra*.

BROAD LOOE (a channel in Poole Harbour), *v.* **1** 75.

BROWNSEA ROAD (a channel in Poole Harbour), *v.* **1** 47.

BUDDEL (or BUDDLE), *the Buddel* 1535–43 Leland, *river Buddle* 1697 LymeR, an alternative name for R. Lim at its mouth, *v.* discussion **4** 396–7.

BURTON FRESHWATER (where R. Bride flows into the sea), named from Burton Bradstock par., *v.* **4** 478-9.

BYLE BROOK (lost, affluent of Corfe River), 1811 OS, *Bylebrook* (*rivulet called*) 1774 Hutch[1], *Byle-brook* 1861 Hutch[3], probably named from the family of Walter (*de*) *Bile* 1311 Hutch[3], 1332 SR, Walter *Byle* 1327 SR, William *Byle* 1546 *Ct*. The stream rises near Challow Fm (**1** 10) and falls into Corfe River at St Edward's Bridge (**1** 18). It is probably to be identified with *Swyneswell* 1381 (16) *Pitt*, *v.* **1** 32.

CALE (affluent of R. Stour), *Cawel* 956 (14) S 570, *aque … appellat' Caul* 15 *ShaftR*, *River Cale* 1811 OS, a pre-English r.n. of unknown origin and meaning, giving name to Cale Bridge in Fifehead Magdalen par. **3** 7, cf. Ekwall RN 63.

CAM (affluent of Caundle Brook), no early forms but possibly an old stream-name identical with R. Cam Gl **1** 4 which is probably from PrWelsh ***camm** (Brit **cambo-*) 'crooked', *v.* Ekwall RN 65. The stream has a winding course. Rye Water Fm (in a bend of the stream) and the f.ns. *Brook*, *Brook East*, *West of the Brook* (beside the stream), cf. also *atte Brouke* 1327, *Estbroke* 1516, all in Holnest par. **3** 338–9, refer to the Cam, as does *Week lake* 1569–74 *Map* (named from *Wyke Street Green* **3** 340).

CAREY (just possibly an earlier name for R. Piddle or Trent), *v.* discussion under Carey **1** 158, Keysworth **1** 159.

CAUNDLE BROOK (affluent of Lydden), named from Bishop's Caundle and Caundle Marsh (on one arm of the stream) and from Stourton Caundle (on another arm), *v.* **3** 276, 311, 316. *Swerebrok* m14 *Glast*, *Suerebrok* 1338–40 Glast, *v.* **swēora**, **brōc**, mentioned in the bounds of Sturminster Newton par. **3** 188, may refer to this stream.

CERNE (affluent of R. Frome), *aqua*(*m*) *de Cerne* 1244, 1268 *Ass*, *Cherne* (*brooke*) 1577 RN, a Celtic r.n. identical with R. Char, from PrWelsh ***carn** 'cairn, heap of stones', thus probably 'the rocky or stony stream' ('stony' perhaps referring to the greensand fragments found in the fields upstream, or to the piles of flints taken from the fields on the chalk lower down). It gives name to four Do pars., Cerne Abbas (**4** 134), Nether Cerne (**4** 139), Up Cerne (**3** 321) and Charminster (**1** 338). As noted **4** 134, the modern pronunciation with [s] instead of [tʃ] in the r.n. and in the three Cerne pars. has been assumed to be due to AN influence, whereas the expected [tʃ] occurs in Charminster (as well as in R. Char). The first el. of Charborough Ho in Morden par. **2** 59 may be another instance of the stream-name *Cerne*, *v.* discussion **2** 59.

CHAR, *ofer Cærn* 737 (12) BCS 158 (S 254 MS 1), *on cern* 737 (12) *Add* 15350 (S 254 MS 2), *ofer cern* 938 (12) BCS 727 (S 443), (*aqua*(*m*) *de*) *Cerne* c. 1235 Hutch[3], 1288 *Ass*, *amnis de Cernia* n.d. (15) *Forde*, *aquam voc' Churne* m13 (15) *ib*, *ripam de Cherne* l13 (15) *ib*, 1319 (15) *ib*, *aque de Cernemue* Ed 1 (15) *ib*, *la Char* 1342 Cl, *Charebroke*, *Chare Water* 1535–43 Leland, *the river Car or Chare* 1863 Hutch[3], a Celtic r.n. identical with R. Cerne *supra*. It gives name to Charmouth par. **4** 370 *q.v.* for further discussion of the name. The Char is referred to as *Whitchurch River* 1769 *DCMMap*, *v.* Whitchurch C. par. **4** 438.

CHECE (lost), *aquam que voca' chece* 1270 (1372) *ChrP* (in a rental of Fleet par. **4** 16), perhaps from OE ***ceacce** 'a lump, a hillock'. In any case it should be noted that Fleet shares a boundary with Chickerell par.

(**1** 204). Chickerell has been considered an obscure name, but it would seem that this lost stream-name (or the word behind it) may provide a clue as to its origin.

CHELWELL WATER (lost or local), 1822 *EnclA*, a small tributary of R. Stour near Lake Fm (named from this stream) in Canford M. par. **2** 3, *v.* **wella**, first el. perhaps **cegel** 'pole, post' or **ceole** 'throat, channel'.

CHID, a late back-formation from Chideock par. **4** 375 and an alternative name for R. Winniford *infra*, cf. *aquam de Cidioch*, ~ ~ *Chidioc*, *pontem aque de Cydioch* n.d. 13 *Weld*[2] and *Chidiokes Mowthe* **4** 384.

CHISELBURNE (lost), the stream that gives name to Cheselbourne par. **3** 202 *q.v.* for early forms of the r.n.

CHIVRICK'S BROOK, *rivulum de Chiner-*, *Chyneyate*, *Sinergate*, *Senergate* (for *Chiuer-*, *Chyue-*, *Siuer-*, *Seuer-*) 1338–40 Glast, *rivulum de Chiueyate*, *-iate* m14 *Glast* (all in the bounds of Sturminster Newton par. **3** 188), earlier references being *ouer cinerget* (for *ciuerget*) *mor*, *anlang cuterget* (sic) *mores*, *on þane stream anlang streames* 944 (15) *ShaftR* (S 502) (in the Anglo-Saxon bounds of Hinton St Mary par. **3** 159), and *on smeryate* (for *siueryate*), *by seuerȝate* 968 (14) Glast (S 764) (in the Anglo-Saxon bounds of Sturminster Newton). The *TA* f.n. Chiverick Md in Hinton St Mary **3** 161 (*Cheveret*, *Cheverytt*, *Chyverett* Eliz *LRMB*, *Cheverett* 1584 *Weld*[1]) also preserves the name, which is probably 'the beetle-infested gap or opening', from OE **cefer** and **geat**.

CLAY LAKE (lost, affluent of R. Crane), *v.* **2** 257–8.

COLEBROK (lost), *aquam de Colebrok'* 1288 *Ass*, associated with Mosterton par. **4** 308, no doubt 'cool brook' from OE **cōl**[2] and **brōc**.

COLLYER BROOK (the stream at Fontmell Magna **3** 103), cf. (bridge called) *Collier's* 1791 Boswell, no doubt from the surname *Collier*. It rises at Springhead (probably *Willeshed* 1505 *Glyn*), *v.* **3** 107.

COMBE STREAM (lost, affluent of R. Yeo), named from Coombe in Castleton par. **3** 302 and identical with Newell Water *infra*.

COMBESBROK (lost, affluent of R. Stour), recorded in the Anglo-Saxon and medieval bounds of Sturminster Newton as *on Conibrok'* (for *Combrok'*) 968 (14) *Glast* (S 764), *Combesbrok'* m14 *Glast*, *v*. Coombs **3** 191.

COOKWELL BROOK, *aque de Cokewyle* 1546 *PlR*, named from *Coguill Hylle* 1543 *ib*, *Cookwell* 1584 *Hen*[2], *a spring called Cook well* 1939 Kelly, and giving name to *Cuckwell Bridge* 1791 Boswell (all these forms associated with either Okeford Fitzpaine par. **3** 178 or Shillingstone par. **2** 238), *v*. **well(a)**, first el. uncertain but cf. the lost stream-name *Cuce 'the lively one' adduced for *on ~*, *of cucan dene*, *on cucan* 956 (14) *Harl* (S 609) in the Anglo-Saxon bounds of Uddens in Holt par. **2** 152, 158, apparently referring to a now unnamed stream forming the east boundary of the Anglo-Saxon estate.

CORFE RIVER, 1773 Bayly, 1811 OS, *the Corfe* 1774 Hutch[1], named from Corfe Castle par. **1** 5, earlier called *Wych*, *v*. Wych Channel and Lake *infra*. Another earlier name for (part of) the Corfe River may have been *Alflede* for which *v*. Arfleet Mill **1** 7–8.

CORRY BROOK (affluent of R. Axe), giving name to Corrie Fm and Ho, Lr Corry in Stockland par. **4** 418 and to Coryton Park De 641, a r.n. *Cori* of unknown meaning and origin, *v*. De 3, Ekwall RN 97. There is another example (now lost) of this r.n. just east of here, at Corscombe, which is 'valley of the **Cori*', *v*. discussion of this par. name **4** 288.

COWARD'S LAKE (small stream flowing into West Fleet), *v*. **4** 7.

COYORNE (lost), *aquam de Coyorne in villa de Cliffton*, *~ Clyfton* 1288 *Ass*, a stream in Clifton Maybank par. **4** 191, of unknown origin and meaning. It is perhaps an early name for the now unnamed tributary of R. Yeo forming the W bdy of the par.

CRANE, a late back-formation from Cranborne par. **2** 205 (*Creneburne* 1086 DB). The stream is referred to as *aquam de Craneborn'*, *-burn'* 1280 *Ass*, *the water of Cranborne* 1620 Hutch[3], *Cramborne River* 17 *CecilMap*, cf. *Crane Meare* e17 *ib*. An earlier name for the Crane may have been *Humber*, cf. the *TA* field Humber Md in Edmondsham par. **2** 220 which lies alongside R. Crane. This stream becomes Moors River *infra* in its lower course before joining R. Stour.

CRANEBURN (lost), *Craneburn* 1275 RH, *Cramborne Watter* 1535 *Ct*, *Cranbourne-lake* 1627 Hutch[3], 'stream frequented by cranes', from **cran** and **burna**, a stream in Gillingham par. **3** 32 identical with that giving name to Cranborne par. *v.* Crane *supra*.

CREEKMORE LAKE (a channel in Poole Harbour), *v.* **2** 5.

CRICHEL BROOK (affluent of R. Allen), 1952–70 RCHM, named from Long and More Crichel, the small stream that flows through the Crichels to join R. Allen, *v.* **2** 275.

CUCE (lost), *on ~*, *of cucan dene*, *on cucan* 956 (14) *Harl* (S 609) in the Anglo-Saxon bounds of Uddens **2** 152, *v.* discussion under Cookwell Brook *supra* and **2** 158.

CULMELAKE (lost), *Culmelake* 13 (15) *Forde*, 1485 *Pole*, a stream-name in Broadwindsor par., probably of Celtic origin and identical with R. Culm De 4, *v.* **4** 270 for further forms and discussion.

DARKNOLL BROOK (affluent of R. Stour), named from Darknoll Fm in Okeford Fitzpaine par. **3** 181. Its earlier name was *Tril* recorded in the Anglo-Saxon and medieval bounds of Sturminster Newton par. **3** 189 as *on trildoune vppe tril* 968 (14) *Glast* (S 764), *Trul* (p) 1268 *Ass*, *riuulum de Tril* m14 *Glast*. *Trildoune* is probably to be identified with Kingsdown **3** 191. For two (or three) more examples of this lost stream-name in Do and for further discussion *v.* *Tril infra*.

DEAD WELL WATER (lost or local, affluent of R. Frome), 1841 *TAMap*, in Stinsford par. **1** 371.

DEANSBROOK (affluent of R. Yeo), 1939 Kelly, in Chetnole par. **4** 189.

DEVIL'S BROOK (affluent of R. Piddle), 1870 Hutch[3], *Devilles broke* 1535–43 Leland, earlier *on deuelisc*, *of dune anlang streames*, *up of streame* 859 ?for 870 (15) *ShaftR* (S 334), *on deflisch*, *of dune anlang streames*, *up on streame* ?870 (15) *ib* (S 342), *on deulisc stream*, *of þane streame*, *to deulisc streame* 942 (15) *ib* (S 485(1)), *on deuelisc stream*, *anlang streames* 1019 (15) *ib* (S 955(1)) (all forms from the Anglo-Saxon bounds of Cheselbourne par. **3** 202). This stream-name is identical in origin with the following name Divelish (*q.v.* for further discussion), and gives name to Dewlish par. **1** 303. The modern stream-name is due to folk etymology, the early forms of the name having been associated with the words *devil* or *devilish*. A short stretch of the stream at Hilton is called Mash Water *q.v. infra*.

DIVELISH (affluent of R. Stour), *on deuelisch*, *of defelich* 968 (14) *Glast* (S 764) (in the Anglo-Saxon bounds of Sturminster Newton **3** 188), *Douelich* 1307 (15) *ForReg*, *Develysh'* e14 *GlastE*, *riuulum de dovelissh'* m14 *Glast*, *Defelisch*, *Defoeliosch*, *Dovelynch* 1338–40 Glast, *Diuelishe flu'* 1575 Saxton, cf. *Joh' Richeman de Devvelisbrege* 1477 *Wim* (*v*. **brycg**). This is a Celtic r.n., identical in origin with Devil's Brook *supra*, from PrWelsh ***duβ**, ***dü** 'dark, black' and ***gles**, **gleis** 'stream', thus 'the dark-coloured stream', *v*. Ekwall RN 130–3. Both streams rise on Bulbarrow Hill **3** 233, Divelish on the N slope of the hill to flow N into the Stour, Devil's Brook on the S slope to flow S into the Piddle. The Divelish is referred to as *alor ride* 'alder stream' in the Anglo-Saxon bounds of Woolland **3** 233, and part of it where it joins R. Stour may have been called *Blackwater*, *v*. Blackwater Bridge in Sturminster Newton **3** 193 (note the semantic connection between the names Divelish and Blackwater).

EBBLAKE STREAM, in Verwood par., and giving name to Ebblake (*Abbe*(*s*)*lake* 1280) **2** 256.

ELSFORD, *Elsford River* 1821 *DuCo*, *River Elsford* 1939 Kelly, named from Elsford Bridge (?*Elford* 1351) in Ryme In. par. **4** 219. The stream, which forms the Do-So county bdy, is called *Gallaker River* 1623 *Ct*, *v*. Gallica Bridge in Ryme In. **4** 220.

EMPOOL WATER (lost or local, affluent of R. Frome), giving name to Empool Bottom and Heath in W. Knighton par. **1** 209.

ENGINE STREAM or BACK BROOK (lost or local), 1822 *EnclA*, near R. Stour in the bounds of Canford M. **2** 2, 8.

FAIR WATER (affluent of Blackwater River), giving name to its source at Fairwater Head in Hawkchurch par. **4** 149.

FERN BROOK (affluent of R. Lodden), 1811 OS, *The Fern* 1970 RCHM, giving name to Fernbrook Fm (*Fernbroc*, *Farnbroc'* 1251) in Motcombe **3** 48 and to the *TA* f.n. Furnbrook Grd in Gillingham **3** 26.

FILLEY BROOK (affluent of Bow Brook), *aqua de Folebrock'*, ~ ~ *Fulebrok'* 1288 *Ass*, *Filly Brook* 1838 *TAMap*, 'the foul or dirty brook', from **fūl** and **brōc**, giving name to Filley Brook Fm **3** 86. Lying close to the brook is Caggypole Fm in Buckhorn Weston par. **3** 86.

FONTMELL BROOK (affluent of R. Stour), *rivus ... Funtamel*, *xxx manentes de aquilone rivus nomine Funtamel* (probably '30 hides bounded on the north by the *Funtamel'*) 670–6 (15) *ShaftR* (S 1164), *funtamel þry akeres wid eastem þane broc*, *anlang funtmeales*, *of funtemel* 939 (15) *ib* (S 445), a Celtic r.n., 'the stream or spring by the bare hill' from PrWelsh ***funtǭn** and ***mę̄l**. The stream gives its name to Fontmell Magna par. **3** 103 and to Fontmell Parva in Ch. Okeford par. **3** 61. The 'bare hill' may refer to Fontmell Down **3** 106 near the source of the stream. Cf. also the *TA* f.n. Smallbrook in Fontmell Magna **3** 109: this is *smale broke* 956 (14) *ShaftR* (S 630) in the Anglo-Saxon bounds of Compton Abbas **3** 99, 'the narrow brook' with reference to Fontmell Brook near its source. The 7th-century charter cited is the oldest West Saxon charter to survive in a text considered authentic: it records a grant of 30 hides of land by the *Funtamel* stream, to abbot Bectun for religious purposes, by Cenred (father of King Ine and of Ingeld, ancestor of King Alfred).

FROME [fru:m], *Frome* 859 ?for 870 (15) *ShaftR* (S 334), 966 (15) *ib* (S 744), 12 SD, 1233 FF, 1268 *Ass*, 1280 Ch *et passim*, *Frume* ?870 (15) *ShaftR* (S 342), *Frauu* 894 Asser, *Fraw* a1118 (12) RN (Florence of Worcester), *Froma* 1212 FF, *Forme* 1305 Cl, *Froome* 1384 BM, 1773 Bayly, *Wroune* p. 1483 *Sheen*, *Fraw or Frome ryver* 1535–43 Leland, cf. *Fromemouþe* 843 for 934 (eME) ASCharters, *-muthe* 843 for 934 (17) BCS 739, *From(m)uðan* 12 ASC (E) (s.a. 998, 1015), *Fromewater* 1462 *Weld*[1]. This is a Celtic r.n., identical in origin with the other rivers so called in Gl, He and So, probably simply 'running water' from PrWelsh ***frǭm** (whence ModWelsh *ffrau* 'stream'), *v.* Ekwall RN 166–8. The forms *Frauu*, *Fraw* above represent the normal Welsh development. For the form *Fromemouþe*, *-muthe* 'the mouth or estuary of the Frome', *v.* Green Island in Corfe Castle par. **1** 13–14. The river rises at Evershot, *v.* St John's Well **4** 86. It gives name to four Do parishes: Chilfrome **4** 80, Frampton **4** 88, Frome St Quintin **4** 93 and Frome Vauchurch **4** 96. Along various stretches of its 36 mile course it was referred to as *the South Ryver* 1584 (at Wareham, in contrast to *the North River* = R. Piddle *infra*), *the greate Ryver*, *the greate streame* 1548 *DCMDeed* (also at Wareham), *The Prince's River* 1617 *Add* (*v.*

Prince's Bridge in Dorchester **1** 360), and *Moones Bridge River* 1650 *ParlSurv* (*v.* Mohun's Bridges also in Dorchester *loc. cit.*)

GARBROKE (lost), giving name to the f.ns. Garbrooks and Outer Garbrook in Woolland par. **3** 235, *v.* **gāra**, **brōc**.

GREEN ISLAND LAKE (a channel in Poole Harbour), *v.* **1** 14.

GUSSAGE BROOK (local, affluent of R. Allen), 1970 RCHM(Do) **5** 19, with reference to the stream which gives name to the pars. of Gussage St Michael **2** 143 and Gussage All Saints **2** 276 as well as to Gussage St Andrew in Six. Handley par. **3** 114. It rises at Gussage St Andrew and flows through the other two pars. to join R. Allen. For the etymology of this English r.n., *v.* Gussage All Saints **2** 277–8.

THE GWYLE, near Monkton Wyld in Wootton Fitzpaine par. **4** 459, from ModE **goyle** 'ravine'.

HLOSMOC (lost, affluent of R. Char), (*in*) *on hlosmoc* 737 (12) BCS 158 (S 254), *Add* 15350 (ib.), in the Anglo-Saxon bounds of Charmouth par., *v.* discussion **4** 374. Also referred to as *aquam sub via que venit de Charrereyston'* m13 (15) *Forde*, *v.* Catherston Leweston par. **4** 368.

HOLEBROOK WATER (affluent of R. Lydden), *on Holambrok'* 968 (14) Glast (S 764), in the Anglo-Saxon bounds of Sturminster Newton **3** 188, forming the par. bdy between Lydlinch and Sturminster Newton and giving name to Holebrook Fm and Green **3** 348 *q.v.* for discussion and etymology. The stream-name that gives name to Holbrook in Yetminster par. **4** 230 has the same origin.

HOLLWYSSHE WATER (lost, affluent of Corfe River), 1381 (16) *Pitt*, named from *Hollish* in Corfe Castle par. **1** 14.

HOLY STREAM (affluent of R. Frome), in East Stoke par. **1** 149.

HOOKE (6"), HOOK (1") (affluent of R. Frome), *river Hook(e) or Owke* 1774 Hutch[1], 1863 Hutch[3], a back-formation from Hooke par. **4** 56. The old name for the river is *Toller q.v. infra*. One arm of the river rises in the N of Hooke par. at Park Pond **4** 57, the other at Toller Whelme in Corscombe par. **4** 291.

HORTON BECK (affluent of R. Allen), 1774 Hutch[1] **2** 144, named from Horton par. **2** 159, *v.* **beck**.

HUMBER (lost, six examples), a pre-English r.n. of uncertain origin and meaning (*v.* Ekwall RN 201ff., Jackson 510–1, 519) that occurs six times in Do:

i) possibly an earlier name of R. Crane *supra*, and giving name to the TA f.n. Humber Mead in Edmondsham par. **2** 220
ii) a small stream (now unnamed) giving name to Humber Wood in Milton Abbas par. **3** 219
iii) a small stream (now unnamed) giving name to Humber Wood in Pulham par. **3** 265
iv) an affluent of R. Yeo (now unnamed) giving name to Hummer (*Humbre* 1091–1106) in Trent par. **3** 394
v) a small stream (now unnamed) giving name to the TA f.n. Humber in Leigh par. **4** 209
vi) an earlier name of Wriggle River *infra*, and giving name to Heniford Fm and Mill in Melbury Bubb par. **4** 211.

IWERNE (affluent of R Stour), *on iwern broc* 958 (15) *ShaftR* (S 656) in the Anglo-Saxon bounds of Thorton in Marnhull par. **3** 171, *aquam que vocat' Hywerne* c. 1270 *Seymer*, *aquam de Iwerne* 1288 *Ass*, a Celtic r.n., possibly 'yew-tree river', from **ïw** 'yew' with suffix **-ern** (Ekwall RN 222), alternatively from a British goddess-name ***Īwernā** 'she who moves in a customary course' (*Celtic Voices* 294, 352), but cf. also Tengstrand MN 90–1. It gives name to the pars. of Iwerne Courtney **3** 37, Iwerne Minster **3** 123, and Steepleton Iwerne **2** 114.

JORDAN, a back-formation from Jordan Hill in Preston par. **1** 231 *q.v.* for discussion and etymology.

KAYNNES (lost), *aquam de Kaynnes in villa de Netherbur'* 1288 *Ass*, probably an elliptical reference to Camesworth in Netherbury par. **4** 315 and therefore also to R. *Woth* (Brit).

KEY BROOK, *cagbroc up anlang cagbroges* 939 (15) *ShaftR* (S 445), in the Anglo-Saxon bounds of West Orchard par. **3** 133, and giving name to *Keybrooke gate* 1701 *Glyn* in Fontmell Magna par. **3** 111. Etymology and meaning uncertain, though the first el. (allowing for the spelling from a 15th cent. copy of the charter) probably represents OE **cǣg** 'a key', also possibly used in a transferred sense for 'a key-shaped feature', alternatively perhaps an OE **cǣg** 'stone' (VEPN **2** 155). There is a R. Key in Wilts (PN W 8) but that seems to be a modern back-formation and therefore not useful for comparative purposes.

KNIGHTON BROOK (lost), 1853 *EnclA*, named from Knighton in Beer Hackett par. **3** 290.

KNOLLTON STREAME (lost), 16 *Midd* (map), named from Knowlton in Woodlands par. **2** 285.

KOKENHULE LAKE (lost, affluent of Corfe River), 14 Mansel, named from Cocknowle in Church Knowle par. **1** 89.

THE LAKE (local, R.J.S.), a stream in Langton Matravers par. **1** 33 flowing E towards Swanage, the N arm being known as The Spring (earlier *Knaveswell*, *v.* Knaveswell Fm **1** 35–6), the S arm as Pucklake, *v.* Putlake Fm **1** 39.

LAKE MILL STREAM (lost), 1822 *EnclA*, named from Lake Fm in Canford Magna par. **2** 3.

LANDSORE (lost), *Rivulum qui vocatur Landsore* c. 1270 *AddCh*, associated with Landchard Lane in Shillingstone par. **2** 240.

LECHER WATER (affluent of R. Parrett), *Lecher Water* (*Bottom*) 1838 *TA* (South Perrott), giving name to Lecher Bridge & Lane in South Perrott **4** 334. Perhaps from ME **leche** 'a stream, a bog', but early spellings are needed.

LIM, *occidentale ripam fluminis illius quod vulgo lim vocatum est* 774 (12) *SherC* (S 263), *on-*, *of Lym* 938 (14) BCS 728 (S 442), *the water of Lyme* 1322 Pat, *the Lime Water* 1577 RN, a Brit r.n. from the root **līm-* with senses referring to water, *v.* Ekwall RN 274, Forsberg 117. It gives name to Lyme Regis **4** 392 and to Uplyme De 649.

LISCBROC (lost), *lisebroke* 942 (15) *ShaftR* (S 485 (1)), *liscbroc* 1019 (15) *ib* (S 955 (1)), in the Anglo-Saxon bounds of Cheselbourne with reference to the stream at Lyscombe in Cheselbourne **3** 203 *q.v.* for etymology and discussion.

LODBOURN (lost), perhaps originally the name of (part of) Shreen Water *infra*, and giving name to Lodbourn in Gillingham par. **3** 15, identical with Lorteborn *infra*.

LODDEN (affluent of R. Stour), (*aquam de*) *Lyden*(*e*) 1234 Hutch[3], 1236 Cl, 1279 *For*, 1364 Hutch[3], (*aque de*) *Luden*(*e*) 1338–40 Glast, 1381 *GillCt*, *aquam de lydden'* 1438 *ib*, *the water of Lidden* 1568 Hutch[3], *Laden* 1774 Hutch[1], *Ledden River* 1811 OS, a Celtic r.n. identical with R. Lydden *infra*, 'the broad stream', from Brit ***litano-** 'broad' (PrWelsh ***lidan**), *v*. Ekwall RN 242. It gives name to Lodden Bridge (*Lydenebrigge* 1441 *GillCt*) in Gillingham par. **3** 21.

LODER (lost), possibly the original name of R. Asker *supra*, *v*. Loders par. **4** 482.

LORTEBORN (lost), 1412 *Weld*[1], a stream in Winfrith Newburgh par. **1** 184, 'the dirty stream', *v*. **lort(e)**, **burna**, identical in origin with the one that gives name to Lodbourn in Gillingham par. **3** 15 and Ludbourne Rd in Sherborne **3** 361, cf. also Lodburn in Powerstock par. **4** 65.

LUCKFORD LAKE (affluent of R. Frome), 1811 OS, *Luckford Lack* 1773 Bayly, the first el. of which may be an earlier name *Lugg* for this stream, *v*. discussion *s.n.* Luckford Coppice in East Stoke par. **1** 147.

LYDDEN or LIDDEN (affluent of R. Stour), *in Lydene* 740–56 (14) *Cott* (MS Faustina A.ii, f.25), Finberg (p. 156), *on Lideman bi lidenan* 968 (14) *Glast* (S 764), *aqua de Lidenne* 1244 *Ass*, *aqua de Lydene* 1268 *Ass*, *aquam de Lidelinch'*, *~ de Ludene*, *~ de Lydeneye* 1288 *ib*, *Lidenan* 1338–40 Glast, *Lydden flu'* 1575 Saxton, *River Lidden* 1811 OS, a Celtic r.n., 'the broad stream', identical with R. Lodden *supra*, *v*. Ekwall RN 242 (and also ForsbergCh 154 where it is noted that the form *Ludenham*(*e*) cited by Ekwall (who has it from Grundy) does not belong here). The river gives its name to Lydlinch par. **3** 346. The comment in Hutch[3] **4** 305 is an interesting example of folk etymology: 'a wide heavy sluggish river at the mouth, and from thence, by the common people, called the Leaden Water'. The R. Lydden is referred to as *neodbournen* in the Anglo-Saxon bounds of Buckland Newton, *v*. *infra*.

MAIN CHANNEL (a channel in Poole Harbour), 1811 OS, *Great Channell* 17 *DROMap*, *v.* **1** 76.

MANGERTON RIVER, named from Mangerton in Netherbury par. **4** 318. For the suggestion that *Power* may have been the original name of this river, *v.* Powerstock par. **4** 60, North Poorton par. **4** 336. A stream feeding into Mangerton River gives its name to Lodburn in Powerstock par. **4** 65.

MANSTON BROOK, named from Manston par. **3** 45.

MASH WATER, at Ansty in Hilton par. **3** 209. The form *Marsh Water* 1791 Boswell cited **3** 212 is almost certainly an error and the first el. is more likely to be the brewing term **mash**: the Blandford brewery of Hall and Woodhouse was started in Ansty in 1777 and was not moved to its present home in Blandford St Mary until c. 1900, with some brewing continued in Ansty until the 1940s (*Dorset Life* June 2012, p. 23).

MAYNEWATER (lost, affluent of R. Frome), (*aque de/vocat'*) (*le*) *Mayn*(*e*)*water* 1449 Weld[1] *et freq* to 1525 *AOMB*, also referred to as *aquam de Mayne Martel* 1288 *Ass*, named from Broadmayne par. **1** 337 and giving name to the *TA* f.n. Main Water Close in Moreton par. **1** 137, cf. also Watergates Ctg in Broadmayne par. **1** 338 and Blackwater Ford in Owermoigne par. **1** 139.

MIDDLEBERE LAKE (a channel in Poole Harbour), *Middlebeere Channel* 1811 OS, named from Middlebere in Arne par. **1** 72.

MIDDLEBROOK (lost, affluent of R. Char), giving name to Middlebrook Ctg, Middle Brook Fm **4** 207.

MIDDLE CHANNEL (a channel in Poole Harbour), *Little Channel*(*l*) 17 *DROMap*, 1811 OS.

MILBYRI WATER (lost, affluent of R. Yeo), 1535–43 Leland, rising in Melbury Park in Melbury Sampford par. **4** 100.

MILL STREAM, *of mylebroce*, *on myle broc* 1024 *Ilch* (S 961), in the Anglo-Saxon bounds of Portisham **4** 32. For other 'mill streams', cf. Milborne **1** 282, 306, *Mylenburnan* **3** 301, *Melenbroc* **3** 236, etc.

MOONES BRIDGE RIVER (lost), *the stream called* ~ ~ ~ 1650 *ParlSurv*, named from Gt and Lt Mohun's Bridges in Dorchester par. **1** 360.

MOORS RIVER (affluent of R. Stour), named from West Moors in West Parley par. **2** 232, cf. the river crossing at Wools Bridge in Verwood par. **2** 258, where *Wool-* may be from **wyll(a)** 'stream' with reference to this river. Moors River is R. Crane *supra* in its upper course.

MORKHAMS LAKE (small stream flowing into West Fleet), *v.* **4** 8.

NEODBOURNEN (lost), *on neodburnen endlang stremes* 941 (14) *Glast* (S 474), in the Anglo-Saxon bounds of Buckland Newton par. **3** 239 and with reference to R. Lydden *supra*, *v.* **burna**, first el. perhaps **nēd** 'need' in a sense such as 'tending to run dry', but Grundy 55.254 may be right in suggesting a scribal error for **hrēod** 'reed'.

NEWELL WATER (lost, affluent of R. Yeo), 1826 Gre, *Newelle, a litle burne* 1535–43 Leland, the stream giving name to the st.n. Newell in Sherborne par. **3** 362 *q.v.* for discussion and earlier spellings. It rises near Coombe Fm in Castleton par. **3** 302, hence the alternative name Combe Stream *supra*, cf. *Brook from New Well and Combe* 1802 Map. Fowler 12 believes that this stream may have been the original *scīraburna* that gave name to Sherborne par. **3** 355–6, but it may in fact rather have been the *lorteburna* that gave its name to the Sherborne st.n. Ludbourne Rd **3** 361, cf. the entry for *Lorteborn* stream *supra*.

NORTH CHANNEL (a channel in Poole Harbour), 1811 OS, *v.* **1** 163.

NORTH HAVEN LAKE (a channel in Poole Harbour), named from North Haven Point in Parkstone par. **2** 38.

NORTH WINTERBORNE, *v.* Winterborne *infra*.

OWER LAKE (a channel in Poole Harbour), 1811 OS, named from Ower in Corfe Castle par. **1** 16.

PARRETT, *Pedride* ?676–685 (13) S 1665, 794 (18) S 267, *Pedrit* 702 (13) S 244, *Pedrete* 708 (13) S 1176, *Pedredistrem* 725 (14) S 251, (*oþ*) *Pedridan* 9th cent. ASC (A) s.a. 658, (*æt*) *Pedridan muþan* ib s.a. 845 [848], *be eastan Pedredan* ib s.a. 894 [893], *Paret* 1243 Ass, *Perret* 1253 Ipm, *Pedrid* 1274 ib, of uncertain origin and meaning but

1253 Ipm, *Pedrid* 1274 ib, of uncertain origin and meaning but corresponding to OCorn **pedreda* possibly representing Brit **petru-rit-* which might mean 'fourfold stream', 'fourfold ford' or 'stream with four fords', *v.* Ekwall RN 320–1, Padel CoPNE 176, cf. further A. Breeze, 'The Rivers Pedredan and Parrett', *Trans. Worcs. Arch. Soc.* 25 (2016), 155ff., also R. Coates and A. J. Webb, 'Somerset's maritime place-names', in *A Maritime History of Somerset*, ed. A. J. Webb, vol. 3, Taunton (2017), 27–82. It rises near South Perrott par. **4** 332 which is of course named from the river and then flows NW into Somerset.

PIDDLE or TRENT, 1870 Hutch[3], *of ~, on pidelen stre*(*a*)*m*(*e*) 996 (15) *ShaftR* (S 744), *iuxta Pydelan* 987 (13) KCD 656 (S 1217), *aqua de Pidel* 1244 *Ass*, *aquam torrentem voc' Pudelwater* 1270 (1372) *ChrP*, *aquam de Pydele* 1288 *Ass*, *Trent ryver, alias Pyddildour* 1535–43 Leland, *Piddle flu'* 1575 Saxton, *River Trent or Piddle* 1811 OS. An English r.n. from OE ***pidele** 'marsh, fen', thus 'marshy stream', a diminutive of OE ***pide** 'marsh, fen'. The alternative name Trent, found from the 16th cent., is a back-formation due to a misunderstanding of the p.n. Piddletrenthide **4** 167 (in which *-trent-* represents OFr *trente* 'thirty'), cf. Trent Brook *infra* for a genuine occurrence of this British r.n. The name given for the river in a 12th cent. chronicle, *Terente* a1118 FW, is apparently unconnected, and may suggest that the lower Piddle originally had this name (Ekwall RN 416, cf. also D. Whitelock, *English Historical Documents* I 179). The stretch of the Piddle near Wareham is often referred to as (*the*) *North ryver* 1542 Hutch[3], 1586 *DCMDeed*, (*~ or River Trent*) 1774 Map, also *the north streame* 1551 *DCMDeed*, in contrast to *the South Ryver* = R. Frome *supra*. The Piddle rises at Acton Pancras **4** 122 and gives name to six Do pars (Piddletrenthide, Piddlehinton, Puddletown, Tolpuddle, Affpuddle, and Turner's Puddle), cf. also *Podeleford* in Bere Regis par. **1** 281.

PIMPERN (lost, affluent of R. Stour), (*to*) *pimpern*, (*of ~, to*) *pimpernwelle* 935 (15) *ShaftR* (S 429), in the Anglo-Saxon bounds of Tarrant Hinton **2** 119 and giving name to Pimperne par. **2** 110 *q.v.* for discussion of this difficult name. The name may be Celtic and is possibly an original r.n., but it could be a back-formation (Ekwall RN 326, cf. Coates CVEP 295 who notes the name's earliest occurrence is as a r.n.). Another name possibly analogous to Pimperne can be added to those adduced **2** 111, a lost stream-name in Stour Provost par. **3** 73: *curs' aque vocat' Pynperne Lake* 1529 *KCC* (no doubt to be associated with the f.n. Pimperne **3** 70). The name Langbourne in Langton Long

Blandford par. **2** 109 should also be noted: this lies on the *Pimpern* stream but there are no early spellings.

PLUSSHEBROKE (lost), c. 1325 *GlastE*, earlier *aque de Plys* 1268 *Ass*, named from Plush in Buckland Newton par. **3** 245, cf. also the *TA* f.n. Plush Brook Mead in Piddletrenthide par. **4** 172. The form *west over Plissh'* 891 (14) *Glast* (S 347 (2)) in the Anglo-Saxon bounds of Plush refers to the pool (OE ***plysc**, ***plisc**) which is the source of the stream.

POWER (lost), possibly an old r.n. and giving name to Powerstock par. **4** 59, North Poorton par. **4** 335 and South Poorton in Powerstock **4** 62, *v*. discussion **4** 59–60.

PRESTESWLLE (lost), *riuulu' de-* 113 *Sher*, in Yetminster par. **4** 235, *v*. **prēost**, **wella**.

PUCKSEY BROOK (affluent of R. Wey), giving name to the *TA* f.ns. Puxy Garden & Mead in Broadwey par. **1** 203, probably 'goblin haunted stream' from **pūca** and **ēa**. The stream-name *Pucklake* that gave name to Putlake **1** 39 is no doubt analogous.

RADDIPOLL (lost), *the River of* ~ 1663 *Weld*[1], named from Radipole par. **1** 239.

RAMSHORN LAKE (a channel in Poole Harbour), *Rams Horn* 1811 OS, *v*. **1** 24.

REDHORN LAKE (a channel in Poole Harbour), 1811 OS, named from Redhorn Quay in Studland par. **1** 49.

ROCK LEA RIVER (a channel in Lytchett Bay), 1811 OS, *Rock Lee River* 1826 Gre, named from Rock Lea Hill and Point in Hamworthy par. **2** 21.

RODLAKE (lost), *curs' aque apud* ~ 1414 *Weld*[1], in Winfrith Newburgh par. **1** 185, *v*. **hrēod**, **lacu**.

SCITERHUC (lost), *aqua*(*m*) *de Sciterhac*, *-huc* 13 (15) *Forde*, in Broadwindsor par. **4** 254, *v*. ***scitere** 'stream used as a sewer', **hōc**, **hūc** 'hook', hence often 'river bend'. For the first el. cf. Shitterton in Bere Regis par. **1** 276.

SELWAIE BROOK (lost), 1607 *DROMap*, probably to be associated with Silly Hill in Haselbury Bryan par. **2** 103, cf. R. Silley *infra*.

SHAG LOOE (a channel in Poole Harbour), *Shaggloo* 1811 OS, *v.* **1** 165.

SETE (lost, affluent of R. Lodden), Hy 3 (14) Cerne, *aquam de* ~ 1300 Hutch[3], *Suyte* 1247 (1270) *For*, *the water of Seate* 1568 Hutch[3], rising near Black Venn in East Stour par. **3** 68 and mentioned in the bounds of both the forest and the manor of Gillingham **3** 9, 13. Origin uncertain but perhaps transferred to the stream from a place named from OE **(ge)set** 'seat, dwelling, fold for animals'.

SHERFORD RIVER, 1774 Hutch[1], named from Sherford in Morden par. **2** 60, *q.v.* for suggestion that the river may originally have been called *Sher* (< OE **scīr**[2]) 'the bright one'.

SHREEN WATER (affluent of R. Stour), 1811 OS, *Shreen-water* 1774 Hutch[1], earlier *aqua(m) de Mere* 1279 *For*, 1300 Hutch[3], *aquam voc' Merewater* 1438 *GillCt*, *the River called Meare Water* 1609 *LRMB*, because it rises at Mere W. The origin of Shreen is obscure (could it be a garbled form of the word *serene*?). For other references to Shreen Water in Gillingham par., *v.* Huntingford **3** 14, Lodbourne **3** 15 and Colesbrook **3** 19.

SILLEY (affluent of R. Cerne), stream at Cerne Abbas **4** 134 noted by Penn 31 and Gazetteer 167, origin unknown but cf. *Selwaie Brook supra*.

SILVER (possibly a lost stream-name for an affluent of R. Winniford *infra*), perhaps giving name to Silverbridge Fm **4** 379, cf. Monksilver So (DEPN 329). But note also Silverdale in Castleton par. **3** 305 where the first el. seems to have a different origin.

SIMENE (affluent of R. Brit), a late back-formation from Symondsbury par. **4** 425–6, cf. the obscure name Atrim in Netherbury par. **4** 312 which may preserve the original name of the stream, note in particular the reference in 1861 Hutch[3] to 'the river Simene (better known as the river Atram)'.

SLEEP BROOK, named from Sleep Bottom in Alderholt par. **2** 198, earlier called *Black Water* 1811 OS, *v.* Blackwater **2** 196.

SLOD BROOK, *Sladbroke* 1577 HaRO, gives name to Slodbrook Fm and Lane in Gillingham par. **3** 23 and to the *TA* f.ns. Slade Brook in Silton par. **3** 67 (where one arm of the stream rises) and Slodbrook Common in East Stour par. **3** 70, 'stream in the valley', *v.* **slæd**, **brōc**.

SOUTH DEEP (a channel in Poole Harbour), *South Deeps* c. 1770 DCMMap, *v.* **1** 25.

SOUTH LAKE (a channel in Poole Harbour), *v.* **1** 77.

SOUTH WINTERBORNE, *v.* Winterborne *infra*.

SPRING (lost), *a Stream called the Spring* 1650 *ParlSurv*, *v.* **spring**, in Dorchester par. **1** 366.

STIRCHEL (affluent of R. Stour), *to stirchel an lang streames* 932 (15) *ShaftR* (S 419) (in the Anglo-Saxon bounds of Fontmell Magna par. **3** 103), *to stirthel*, *ouer stirtel* 939 (15) *ib* (S 445) (in the Anglo-Saxon bounds of West Orchard par. **2** 133), *on stirchel andlang streames* 956 (14) *ib* (S 630) (in the Anglo-Saxon bounds of Compton Abbas par. **3** 99), (*aquam de*) *Sturkel* 1244 *Ass*, 1280 QW, *Ass*, *aquam de Sterkel* 1280 QW, *Sterkhull'* 1395 (e15) *MiltRoll*, *Stirkell(')* 1474 *Wim*, 1620 Hutch[3], *Sterkele River* 17 *CecilMap*, *Sterke Hill* c. 1560, *Sterkehyll* (*aqu' vocat'* ~) 1576 *Glyn*, *Stirkehill* 1564 *ib*, *Sturkell Water* 1618 *Map*, possibly a diminutive of the r.n. Stour (into which it flows), from an OE **Stȳrcel* or the like (perhaps reduced from **Stȳrincel*), as suggested by Ekwall RN 382; the apparent loss of palatalisation is to be noted.

STOCK WATER, named from Stock Fm and Hill in Gillingham par. **3** 17.

STOKE WATER (affluent of R. Brit, a stream forming the par. bdy between Beaminster and Stoke Abbott), *Stokewater* 1651 *DCMCt*, *Stoke Water* 1702 *SxAS*, giving name to Stoke Water Fm & Ho **4** 342 and also transferred to the *TA* f.n. Stokewater in Beaminster **4** 249, earlier probably *aqua de Stokes* 1244 *Ass*, so named from Stoke Abbott par. **4** 338.

STONE ISLAND LAKE (a channel in Poole Harbour), *v.* **1** 50.

STONELAKE BROOK, giving name to Stonelake **4** 498, *v.* **stān**, **lacu**.

STOUR, *on ~*, *up anlang sture* 944 (15) *ShaftR* (S 502) (in the Anglo-Saxon bounds of Hinton St Mary **3** 159), *op of ~*, *on ~*, *vp on stoure* 968 (14) *Glast* (S 764) (in the Anglo-Saxon bounds of Sturminster Newton **3** 188), *Stures* 1242 Ch, 1244, 1268 *Ass*, 1276 RH, 1280 *Ass*, QW, 1288 *Ass*, 1364 Hutch[3], n.d. (1372) *ChrP*, 1620 Hutch[3], *Stur* 1242 Ch, *Le Esture* 1253 FF, *Sture* 1268 *Ass*, 1270 Hutch[3], 1280, 1288 *Ass*, n.d. (1372) *ChrP*, *Sturas* Ed 1 Hutch[3], *Stoure* 1274 (m14) *Glast*, 1300 Hutch[3], e.14 *GlastE*, 1324, 1367 Cl, 1404 *Marten*, 1431 *DLCt* (*le ~*), 1440 Pat, 1463 *MinAcct*, 1547 *DLMB*, 1501 *GillCt* (*~ Water*), 1547 *DLMB*, 17 *CecilMap* (*~ River*), *Staure* 1288 *Ass*, *Stoures* 1338–40 Glast, *Stowre* 1367 *Ilch*, 16 *Hen*[2], 1535 *Midd* (*the ~*), 1546 *Surv* (*the ~*), 1611 *DuCo* (*le ~*), 1653 *ParlSurv* (*the ~*), *le Stouer* 1529 *Ct*, *le Stower* 1620 *CH*, *the Stower* 1624 *Fort*, 1691 *Weld*[1]. An ancient r.n. (instanced for four other major rivers in England) of uncertain origin, possibly pre-Celtic, but perhaps rather an OE r.n. **Stūre* 'the strong one' from an OE adj **stūr* related to OE *styrian* 'to stir' and *storm* 'tumult' (all from a PrGerm root **stūr-* which seems to have had senses like 'strong, stiff, fierce'), *v.* Ekwall RN 378–382, Jackson 195 note 1, Coates CVEP 267, Watts CDEPN 582. The river gives name to the pars. of East & West Stour, Stourpaine, Stour Provost, Sturminster Marshall & Newton.

SWANEWYZ (lost), *aquam de ~*, 1268 *Ass*, a stream in Swanage par. **1** 52, perhaps that referred to in the name Northbrook or Norbrook **1** 58.

SYDLING WATER (affluent of R. Frome), *Sidling or Silling* 1535–43 Leland, named from Sydling St Nicholas par. **4** 174. The stream rises at Up Sydling **4** 177 (spring marked 6").

SYNDERFORD (affluent of R. Axe), named from Synderford in Thorncombe par. **4** 495.

TADNOLL BROOK (affluent of R. Frome), named from Tadnoll Dairy & Mill in Chaldon Herring par. **1** 110. For another stream associated with toads, cf. *Tadbrokes lake* in Stalbridge par. **3** 288.

TARRANT (affluent of R. Stour), *to terrente* 935 (15) *ShaftR* (S 429), *Tarent* 1237 Pat, 1242 Ch, *Tarrente* 1237 Pat, *Tharente* 1243 FF, *Tarente* 1253 FF, *Tarenta* n.d. (1372) *ChrP*, a British r.n. identical with Trent Brook *infra*, PrWelsh **Trihanton* (from Brit **Trisantonā*) metathesised as **Tirhanton*, 'river liable to floods', *v.* Ekwall RN 415–418, Jackson 257, 503, Watts CDEPN 600. The form *on terente dene*

956 (14) *ShaftR* (S 630), 'the Tarrant valley', refers to Stubhampton Bottom in Tarrant Gunville par. **2** 242, 248 where R. Tarrant rises, *v.* **3** 129. The river gives name to no less than eight different pars. along its course: Tarrant Crawford, ~ Gunville, ~ Hinton, ~ Keynston, ~ Launceston, ~ Monkton, ~ Rawston and ~ Rushton.

TEMPLE BROOK (affluent of R. Axe), giving name to Temple Brook Fm in Broadwindsor par. **4** 265.

TER(R)IG (lost, affluent of R. Allen), 1870 Hutch[3] **3** 389, 431: 'E of Hookes Wood, below Woodcotes, rises a small rivulet called by Holinshed the Terrig', *v.* Hookswood in Farnham par. **2** 222, Woodcutts in Sixpenny Handley par. **3** 116. Perhaps a 16th-century antiquarian coinage from Welsh **terig** 'ardent, rough' found in the Welsh r.n. (Afon) Terig, but cf. the British r.ns. Torridge and Tory Brook in Devon (PNDe 14, 15) from the antecedent of this word, *v.* Ekwall RN 413.

TOLLER (lost), the old name of R. Hooke *q.v. supra*, and giving name to the pars. of Toller Fratrum & Porcorum **4** 109, 110 and to Tollerford in Frome Vauchurch par. **4** 97. The river rises at Toller Whelme, *tollor æwylman* 1035 (12) *SherC* (S 975) in the Anglo-Saxon bounds of Corscombe par. **4** 287, 291, from OE **ǣwielm, ǣwylm** 'river-spring'. The form *tollor* possibly represents PrWelsh ***tull** (Co *toll*) 'a hole' with **duβr** (Co *dour*) 'water', thus 'the stream with hollows' or 'the stream in a deep valley'.

TRENT, a back-formation from Piddletrenthide **4** 167, *v.* R. Piddle *supra*.

TRENT BROOK (affluent of R. Yeo), giving name to Trent par. **3** 393, a British r.n. identical with Tarrant *supra*, *v.* Ekwall RN 415–418, Jackson 257, 503.

TRIL (a lost stream-name that occurs three (or four) times in Do):

i) an affluent of R. Stour, *Tril* being the earlier name of Darknoll Brook *supra*, recorded as *Tril* 968 (14) *Glast* (S 764);

ii) a second affluent of R. Stour, giving name to Trill Bridge (*Trul* 1268 *Ass* (p), *Tryllebrigge* 1492 AcctStAug) in Fifehead Magdalen par. **3** 7. The now unnamed stream joins R. Stour just below the bridge.

iii) an affluent of R. Yeo, a now unnamed stream giving name to Trill Fm & Ho (*æt Tril* 1002–12 ASWrits (S 1383)) in Beer Hackett par. **3** 291.
iv) another possible example may be the *TA* f.n. Trilcombe (*Trillecum* 1321 *MiltC*) in Sydling St Nicholas par. **4** 181, *Tril* perhaps with reference to Sydling Water *supra*.

This is an OE r.n. from ***tyrl** 'that which turns or rolls along', *v*. Ekwall RN 409, 418.

TWYFORD BROOK, named from Twyford ('double ford') in Compton Abbas par. **3** 100.

UDDENS WATER, named from Uddens Ho in Holt par. **2** 152. The suggestion that Uddens (*udding* 956 (14) *Harl* (S 609)) may originally have been the name of the stream should be noted. An alternative name for Uddens Water is Sedgebrook according to GazetteerSuppl 164, cf. William *Segbrouk'* 1332 SR (Melbury Osmond).

UPTON LAKE (a channel in Poole Harbour), in Canford Magna par. **2** 7, named from Upton in Corfe Mullen par. **2** 16.

VRTEBROK (lost), *aquam de Vrtebrok'* 1288 *Ass*, in Broadwindsor par. **4** 254, first el. probably OE **wyrt** 'plant', with **brōc**.

WANETINGE (lost), *aqua de Wanetinge*, ~ ~ *Wonetinge* 1244 *Ass*, possibly a stream near Bridport, identical with the stream-name *Waneting* that gives name to Wantage Brk 491, origin and meaning uncertain but most probably a derivative ***wanoting** of OE *wanian*, **wanotian* 'to wane, to decrease', thus referring to an intermittent stream, *v*. Ekwall RN 433.

WAREHAM CHANNEL (a channel in Poole Harbour), 1811 OS, ~ *Channell* 17 *DROMap*, named from Wareham par. **1** 166.

WAUSTICH (lost), *aquam que vocatur Waustich'* (sic) 1359 *Brid*, a stream near the Hospital of St John and East Bridge in Bridport par. **4** 355, 361, named from Walditch in Bothenhampton **4** 472.

WEST BROOK (affluent of R. Stour), giving name to Westbrook Fm in Gillingham **3** 24.

WESTHAY WATER, a stream named from Westhay Fm **4** 415.

WEST SPRING LAKE (a channel in Poole Harbour), in Arne par. **1** 78.

WEY, *aqua(m) de Waye* 1244, 1288 *Ass*, *the water of Weye* 1367 Pat, *Way ryver, that of sum ys caullid Wile* 1535–43 Leland. This is an ancient pre-English r.n., found also in Surrey, of uncertain origin. It may be a reflex of Brit * *u̯aisā* on the root **vis-*/**veis-* 'water', or from an IE **wegh-* 'move' seen in Lat *vehere* 'to carry' and OE *wǣg* 'a wave in the sea' (*v.* Ekwall RN 451, Jackson 45, Watts CDEPN 668). The meaning would probably have been simply 'flowing water'. The R. Wey rises at Upwey and flows S through Broadwey to Weymouth.

WHITE GROUND LAKE (a channel in Poole Harbour), named from *White Ground* 1811 OS, *v.* **1** 50.

WHITLEY LAKE (a channel in Poole Harbour), in Parkstone par. **2** 41.

WIMBOURNE (lost), the original name of R. Allen *supra*.

WIN (affluent of R. Frome), a back-formation from Winfrith Newburgh with reference to the stream which gave Winfrith its name ('the white or bright stream'), *v.* **1** 174–5.

Another stream-name Win (1998 GazetteerSuppl 199), applied to a branch of R. Allen *supra*, is also a back-formation, from Wimborne Minster par. **2** 183.

WINNIFORD, (*riuuli qui appellat'*) *Wynreford(')* 1258 (1356), 1356 *Weld*[2], the stream at Chideock, origin uncertain but possibly identical with the Celtic r.n. ('white or bright stream') that gave name to Winfrith Newburgh par **1** 175 and Wynford Eagle par. **4** 117, *v.* Ekwall RN 462–3, Jackson 555–6, CVEP 295. The stream is known alternatively as the Chid *q.v. supra*.

WINSCOMBE WATER (lost), giving name to Winscombe Water Bridge **4** 67; probably a local name for Mangerton River *supra*.

WINTERBORNE, NORTH (affluent of R. Stour), *of* ~, *to* ~, *on Winterburne* 942 (15) *ShaftR* (S 485(1)) (in the Anglo-Saxon bounds of Winterborne Tomson **2** 81), *up anlang streames on Winterburne*, *with Winterburne and þanne andland* [sic] *Winterburne* 943 (15) *ShaftR* (S 490) (in the Anglo-Saxon bounds of Mapperton in Almer par. **2** 55), 'the winter stream', that is one that flows most strongly in winter, from OE **winter-burna.** This river rises near Winterborne Houghton par. **2** 128 and gives name to several other villages on its banks (including Winterborne Clenston, ~ Kingston, ~ Muston, ~ Strickland, ~ Tomson, ~ Whitchurch, and ~ Zelstone) before joining the Stour at Sturminster Marshall. It is possible that an old name for this R. Winterborne may have been *Cerne*, *v.* discussion s.n. Charborough **2** 59.

WINTERBORNE, SOUTH (affluent of R. Frome), identical in origin with the more northerly Winterborne *supra* and first distinguished as *the South Winterborn* in 1774 Hutch[1] **1** 303. This river rises at The Wherry in Winterborne Abbas **4** 72 and gives name to several other villages on its banks, including Winterborne Came, ~ Herringston, ~ Monkton, ~ St Martin and ~ Steepleton before joining the Frome. The alternative early affix for Winterborne Abbas, *Waterless*, recorded from the 13th through to the 18th cent., is telling in indicating the dryness of this stream in some seasons.

WOOD BAR LOOE (a channel in Poole Harbour), *Woodbarloo* 1811 OS, *v.* **1** 166.

WOODWATER (lost, affluent of R. Lodden), *the River or water called Woodwater* 1638 *SxAS*, giving name to Woodwater Fm in Motcombe par. **3** 54.

WOTH (lost), *Woch* (for *Woth*) 12 *CottCh*, *aqua*(*m*) *de Woth'* 1268, 1288 *Ass*, 1349 *Brid*, the earlier name for R. Brit *supra*, either 'the vocal or noisy stream' from OE **wōþ** 'sound, melody', or 'the sweet or pleasant stream' from an OE ***wōþ**, an unmutated variant of *wēþe* 'sweet, mild, pleasant', *v.* Ekwall RN 52, 469. It gives name to Watford Bridge in Bradpole par. **4** 251, Bingham's Fm (earlier *Woth*), Camesworth and Wooth Grange in Netherbury par. **4** 313, 315, 322 and to Watton in Symondsbury **4** 429. The f.ns. *Wothehurn*, *-lond* c. 1300 *Milne* in Moreton par. **1** 138 may contain another instance of this r.n. but are perhaps more likely to be from a surname *de Woth* with reference to one of the places named from the stream.

WRACKLE (affluent of R. Frome), a recent back-formation from Wrackleford in Stratton par. **1** 373.

WRAXALL BROOK (affluent of R. Frome), named from Wraxall par. **4** 73.

WRIGGLE RIVER (affluent of R. Yeo), 1870 Hutch[3] **4** 454 (where it is said to be 'so named from the circumstances of numerous small *vertebræ* (plates of the stems of *encrinites*?) being found at the bottom of the brook. These the inhabitants of the neighbouring villages call Wriggles'). Another suggestion (in a 1981 letter from D. H. Leadbetter) is that it may be named from French *rigole* 'a rill, an irrigation channel'. Perhaps though simply ModE *wriggle* to describe a stream with a winding course. It gives name to Wriggle River Lane and Riggle River Cmn in Leigh par. **4** 207. An earlier name for Wriggle River seems to have been *Humber*, *v. supra* and Heniford in Melbury Bubb par. **4** 211.

WYCH CHANNEL & LAKE, 1811 OS, *Weech Bay & Channel* 1844 *TAMap*, *on wicum*, *anlang wicunstreames* 955 (14) *ShaftR* (S 573), *anland wiken*, *on wicean of wichen*, *on þicean up anlang þicean* [sic] 956 (14) *ib* (S 632) (in the Anglo-Saxon bounds of Corfe Castle par. **1** 5, (*aqua de*) *Wyche* n.d. (15) *ShaftR*, c. 1500 *Eg*, *Wychewater* 1381 (16) *Pitt*, the old name for the Corfe River *supra*. Its origin and meaning are uncertain. It is possibly from OE **wice** 'a wych-elm or other tree with pliant branches' (probably from Germ **wik-* 'bend'), thus perhaps descriptive of a winding stream, cf. the discussion of this name by Kökeritz 124 who notes the assumed existence of an OE **wīc* or **wīce* meaning 'bay or bend', cf. also Zachrisson 19. The river gives name to Wytch Fm in Corfe Castle par. **1** 20, and to the river crossing *Wikenford/Wichenford* mentioned in the Anglo-Saxon bounds of Corfe Castle **1** 18 (cf. Wishford W 230 which is perhaps identical), also to the *TA* f.n. Whitch Meadow in Church Knowle par. **1** 94.

YEO, *Yevel* 852 for 878 (17) S 343 (in the Anglo-Saxon bounds of Long Sutton So), (*on*) *Gifle* 903 for 946–51 (12) *SherC* (S 516) (in the Anglo-Saxon bounds of Thornford par. **3** 376), *oth gifle* 933 (12) *ib* (S 422) (in the Anglo-Saxon bounds of Bradford Abbas par. **3** 294), *Givell* 1243 Ass, *Ivel or Clifton water* 1535–43 Leland, *the Ivil* 1870 Hutch[3], a British r.n. from PrWelsh ***gefl** 'fork', thus 'forking stream' (the Yeo has two main arms, the Yeo itself which rises near Sherborne, and the Cam which rises near Yarlington So), *v.* Ekwall RN 221. The current name Yeo, first noted 1811 OS (unless *Fluvius Yoo* 1325–50 GoughMap belongs here) is probably a back-formation from the p.n. Yeovil So aided by the existence of other stream-names called Yeo (from OE *ēa* 'river') in So and D. The form *Clifton water* 1535–43 refers to Clifton Maybank par. **4** 191, cf. also *Shirburn*(*e*) *water* in the same source with reference to Sherborne par. **3** 355.

ADDENDA and CORRIGENDA
to *The Place-Names of Dorset*, Parts I-IV

THE PLACE-NAMES OF DORSET, PART I
(EPNS Volume LII)

pp. xi–xxxi ABBREVIATIONS AND BIBLIOGRAPHY

AbbotsC	S. D. Keynes, 'The Lost Cartulary of Abbotsbury', *Anglo-Saxon England* 18 (1989), 207–43.
AcctStAug	*Some Manorial Accounts of St Augustine's Abbey, Bristol* (Bristol Record Society), ed. Arthur Sabin, 1960.
Addenda	Addenda and Corrigenda to *The Place-Names of Dorset*, Parts 1–4 (in Part 5).
Arundell	Arundell family papers in Cornwall Record Office (forms very kindly excerpted by Oliver Padel).
AS Mint-Names	J. Carroll and D. N. Parsons, *Anglo-Saxon Mint-Names 1 Axbridge-Hythe*, Nottingham 2007.
Bailey	C. J. Bailey, *The Bride Valley*, Dorchester 1982.
BH	D. Hill & A. R. Rumble (eds.), *The Defence of Wessex: the Burghal Hidage and Anglo-Saxon Fortifications*, Manchester and New York 1997.
CDEPN	V. Watts, *The Cambridge Dictionary of English Place-Names*, Cambridge 2004.
Cochrane	C. Cochrane, *The Lost Roads of Wessex*, Newton Abbot 1969.
Coin, *Coins*	Additional spellings ex inf. Mr William Lean.
CoPNE	*v*. Padel.
Crewe	R. J. Crewe, *Dorset* [rock climbing guide], 1977.
Cunliffe	B. Cunliffe, *Wessex to AD 1000*, London/New York1993.
CVEP	R. Coates and A. Breeze, *Celtic Voices English Places*, Stamford 2000.
Dauzat	A. Dauzat, *Dictionnaire étymologique des noms de famille et prénoms de France*, Paris 1951.
DBPN	A. D. Mills, *A Dictionary of British Place-Names*, Oxford 2003, 2011.
DCoPN	O. J. Padel, *A Popular Dictionary of Cornish Place-Names*, Penzance 1988.
DDuPN	V. Watts, *A Dictionary of County Durham Place-Names*, Nottingham 2002.
DLPN	K. Cameron and J. Insley, *A Dictionary of Lincolnshire Place-Names*, Nottingham 1998.
DLnPN	A. D. Mills, *A Dictionary of London Place-Names*, Oxford 2004, 2010.

Du	County Durham, with page reference, either PN NbDu or PN Du.
Eagles	Bruce Eagles, *From Roman Civitas to Anglo-Saxon Shire: Topographical Studies on the Formation of Wessex*, Oxford 2018.
Edwards	H. Edwards, *The Charters of the Early West Saxon Kingdom*, 1988.
EITN	B. Cox, *English Inn and Tavern Names*, Nottingham 1994.
Everitt	A. Everitt, *Continuity and Colonization: the Evolution of Kentish Settlement*, Leicester 1986.
Fellows-Jensen (1972)	G. Fellows-Jensen, *Scandinavian Settlement Names in Yorkshire*, Copenhagen 1972.
FieldHist	J. Field, *A History of English Field-Names*, London 1993.
Finberg (1964)	H. P. R. Finberg, *Lucerna*, London 1964.
Gazetteer	*A Gazetteer of Dorset Place-Names*, Yeovil 1984.
GazetteerSuppl	*A Gazetteer of Dorset Place-Names – A Supplement*, Dorchester 1998.
Gelling	*v*. LandscapePNs, PNsLandscape, Signposts.
Grinsell	L. V. Grinsell, *Dorset Barrows*, Dorchester 1959, *Supplement* 1982.
Ha	Hampshire, with page reference PN Ha.
Hinton	D. A. Hinton, *Saxons & Vikings*, Dorchester 1998.
Hoskins	W. G. Hoskins, *The Westward Expansion of Wessex*, Leicester 1960.
Insley (1994)	*Scandinavian Personal Names in Norfolk*, Uppsala 1994.
JC-K	Spellings for Hazelbury Bryan f.ns. supplied by John Campbell-Kease.
JEPNS	Journal of the English Place-Name Society.
Jönsjö	J. Jönsjö, *Studies on Middle English Nicknames I. Compounds*, Lund 1979.
Keen	L. Keen, 'The Towns of Dorset', in *Anglo-Saxon Towns in Southern England*, ed. J. Haslam, 1984.
Kelly	S. E. Kelly, *Charters of Shaftesbury Abbey* and *Charters of Glastonbury Abbey* (Anglo-Saxon Charters 5 and 15), Oxford 1996 and 2012.
Kerr	B. Kerr, 'Dorset field-names and the agricultural revolution', DoNHAS **82** (1960) 133–42.
Kristensson	G. Kristensson, *A Survey of Middle English Dialects 1290–1350: the Southern Counties*, Parts I and II, Lund 2001 and 2002.
LandscapePNs	M. Gelling and A. Cole, *The Landscape of Place-Names*, Stamford 2000, new edn. 2014.
L	Lincolnshire, with page reference PN L.
Lei	Leicestershire, with page reference PN Lei.
Meaney	A. Meaney, *A Gazetteer of Early Anglo-Saxon Burial Sites*, 1964.
MillsWt	A. D. Mills, *The Place-Names of the Isle of Wight*, Stamford 1996.

Myres	J. N. L. Myres, *Anglo-Saxon Pottery and the Settlement of England*, 1969.
NDEFN	P. Cavill, *A New Dictionary of English Field-Names*, Nottingham 2018.
Newman–Pevsner	J. Newman and N. Pevsner, *The Buildings of England: Dorset*, Harmondsworth 1972.
Nf	Norfolk, with page reference PN Nf.
NPP	A. R. Rumble and A. D. Mills (eds.), *Names, Places and People: an Onomastic Miscellany in Memory of John McNeal Dodgson*, Stamford 1997.
O'Donovan	M. A. O'Donovan, *Charters of Sherborne* (Anglo-Saxon Charters 3), London 1988.
Padel CoPNE	O. J. Padel, *Cornish Place-Name Elements* (EPNS 56/7), Cambridge 1985.
Penn	K. J. Penn, *Historic Towns in Dorset*, Dorchester 1980.
Perceptions	J. Carroll and D. N. Parsons (eds), *Perceptions of Place*, Nottingham 2013.
pers. comm.	Personal communication.
PN Ch	J. McN. Dodgson, *The Place-Names of Cheshire*, Parts 1–5(I) (EPNS 44–8, 54), Cambridge 1970–81, Part 5(II) (EPNS 74) with A. R. Rumble, Nottingham 1997.
PN Du	V. Watts, *The Place-Names of County Durham*, Part 1 (EPNS 83), Nottingham 2007.
PN Ha	R. Coates, *The Place-Names of Hampshire*, London 1989.
PN L	K. Cameron, *The Place-Names of Lincolnshire*, Parts 1–7 (EPNS 58, 64, 65, 66, 71, 73, 77, 85), Nottingham 1985–2010.
PN Lei	B. Cox, *The Place-Name of Leicestershire*, Parts 1–7 (EPNS 75, 78, 81, 84, 88, 90, 91), Nottingham 1998–2016.
PN Nf	K. I. Sandred (and B. Lindström), *The Place-Names of Norfolk*, Parts 1–3 (EPNS 61, 72, 79), Nottingham 1989–2002.
PN R	B. Cox, *The Place-Names of Rutland* (EPNS 67/68/69), Nottingham 1994.
PN Sa	M. Gelling, *The Place-Names of Shropshire*, Parts 1–7 (EPNS 62/3, 70, 76, 80, 82, 89, 92), Nottingham 1990–2018.
PN St	now also J. P. Oakden, *The Place-Names of Staffordshire*, Part 1 (EPNS 55), Cambridge 1984.
PNsLandscape	M. Gelling, *Place-Names in the Landscape*, London 1984.
QuarrCh	*The Charters of Quarr Abbey*, ed. S. F. Hockey, Newport 1991.
R	Rutland, with page reference PN R.
Reaney–Wilson	*A Dictionary of English Surnames*, rev. 3rd edn. by R. M. Wilson, Oxford 1997.
Roberts	G. Roberts, *The History and Antiquities of the Borough of Lyme Regis and Charmouth*, Lyme Regis 1834.
Sawyer	For 'no. 543', read 'no. 534'.
Signposts	M. Gelling, *Signposts to the Past*, 3rd edn., Chichester 1997.

SR	Add: A. R. Rumble (ed.), *The Dorset Lay Subsidy Roll of 1327* (DoRecSoc 6), Dorchester 1980.
Taylor	*Dorset* now in new revised edition, Wimborne 2004.
ThornDB	C. and F. Thorn, *Domesday Book: Dorset*, Chichester 1983.
VEPN	D. Parsons and T. Styles with C. Hough, *The Vocabulary of English Place-Names* (in progress), Nottingham 1997–2004.

p. xiv for: ChrP read: *ChrP*

p. 5 s.n. CORFE CASTLE. Add: *Coruisgate* 1387 Higden.

p. 6 line 13, the cross reference to Corscombe should be deleted.

p. 15 s.n. KINGSTON. Add: *Kyngstone* Hy (15) *ShaftR*.

p. 22 s.n. HOUNSTOUT CLIFF. Note: many other (local) names for coastal features are given in *Dorset* [rock climbing guide] ed. R.J. Crewe rev edn. 1977.

p. 34 s.n. LANGTON MATRAVERS. Note: *my salt-cotes adjoining my manor of Langeton* (~ *Alfred de Lincoln*) late12 MontC 118.

p. 34 s.n. CHILLMARK. Add: *stone from Chilmark* 1571 WimCW.

p. 56 s.n. WHITECLIFF FM. Add: *Hwitan clife* 962 BCS 1094 (S 701) (~ catch of 1 fishing boat and the royal toll).

p. 67 s.n. EAST & WEST MAN. Cf. article on Mam Tor in JEPNS **41** pp. 41–8.

p. 76 s.n. RODWELL. OE **rōd**[2] unlikely here.

p. 77 s.n. SALTERNS COPSE, add local reference in 15 ShaftR: *Seman Scoria ii plumba ... ii Wikeworkes*.

p. 84 s.n. KIMMERIDGE. Add: *Cemridge* 1657 Hutch[3].

p. 88 s.n. BRADLE. Add: *the garden near Bradle* late12 MontC 119.

p. 96 s.n. CREECH GRANGE. Add: *Creech Grange Tithing* 1664 HTax.

p. 97 s.n. BRIDEWELL. This could perhaps be ModE **bridewell** 'prison'.

pp. 103–4 s.n. WORBARROW. Smith **2** 46 identifies *Wyrebarowe by Pole* 1462 as a spot on the Canford manor coastline, cf. also the form *Werybarowe* 1544 he quotes from early Poole historian Sydenham.

p. 116 s.n. *SOUTHCOMBE*. On the identification of *S., v.* Taylor DoNHAS **88** 214–5.

p. 122 Add f.n. *Rouwode* 1366 Ipm.

p. 124 s.n. *HAMPSTEDE*. The *wood of H.* is identified with Burngate Wood **1** 132 by Taylor DoNHAS **88** 213–4.

p. 128 s.n. f.n. Shag's Mdw. This gives name to hamlet called Shaggs (thus on OS map 1:25000 1975), differently explained in Brewer's *Britain & Ireland* (2005) p. 993 as possibly from OE **sceacga** 'matted hair, wool, etc', later 'tangled vegetation'.

p. 133 Add: RADAR HILL (current), named from the Radar Station on Bindon Hill.

p. 138 s.n. OWERMOIGNE. *v.* now CVEP 100–5 where R. Coates plausibly suggests a meaning 'wind-gap(s) from Brittonic ***oir** 'cold' and ***drust** 'door', pl. ***dröstow** (British ***ogrudrust-**), referring to the gaps in the hills between the village and the sea.

p. 143 s.n. POXWELL. Cf. C. Hough, *Anglo-Saxon England* 30 (2001) 1–14 for the suggestion that this name and Bowridge **3** 11 may contain ***pohha**, ***pocca** with the sense 'fallow deer'.

p. 150 s.n. SPRATLEY WD. John ~, Stephen (*de*) *Spetelegh'* 1327, ~ *Spertele* 1332 SR may belong here.

p. 151 s.n. f.n. Bull Grd. Add: *Bulcroft* 1397 Pole.

p. 151 s.n. WAREHAM. *v.* Dodgson BH 107–8.
p. 159 s.n. KEYSWORTH. Add: *Okeford Kayresworth'* 1494 *Cecil*. This form supports **cærse** and the link with *Acford*.
p. 164 s.n. NUNDICO. Explained locally as 'where nuns do go' (letter from H. Gee 6.7.78).
p. 165 s.n. TANTINOBY FM, thought locally to be from 'Tain't Nobody's' (letter from H. Gee 4.9.78).
p. 165 s.n. TRIGON HILL, cf. surname *Trigon* 1327, 1332 SR.
p. 166 s.n. WOOD BAR LOOE. Acc. to *Dorset Life* Feb 1999 p. 27 a corruption of 'Woodberry Looe' with ref. to sheep being ferried across Poole Harbour to here for the annual sheep fair at Woodbury Hill **1** 277.
p. 174 s.n. WINFRITH NEWBURGH. Add: *Wynfroth* 1338–40 Glast.
p. 176 s.n. EAST & WEST BURTON. Add: *Burton' Hamond'* 1288 *Ass*. Cf. Thomas *Hamond'* 1332 SR (Chaldon Herring).
p. 178 s.n. GATMORE FM. Cf. *Gatemore* (p), *Gattemore* 1218 FF.
p. 185 Add: *parcum apud la hulle* 1390.
p. 186 s.n. WOODSFORD. John Insley, 'The Etymology of the First Element of Woodsford', JEPNS **12** (1979–80) pp. 59–65, makes a strong case for an OE pers.n. **Weard* as first element.
p. 188 s.n. WOOL. One of the DB *Welle* forms belongs to Wellwood in Beaminster **4** 244.
p. 194 s.n. f.n. *Bakerland*, note mention of a bakehouse in *Wolle* 1358 Misc.
p. 198 f.ns. *Ærnelai*, etc. belong under *Earnley* in Corscombe **4** 290.
p. 200 s.n. NOTTINGTON. Gillian Fellows Jensen in NQ **28** 176 suggests an alternative explanation 'farm at **Hnotting* ('the bare place')' for this name.
p. 202 s.n. NOTTINGTON MILL. Add: ~ *Mills* 1824 DCMDeed.
p. 204 s.n. CHICKERELL. *v.* A. Breeze, 'Celtic philology and Chickerell, Dorset', SoDoNQ 35 (2003) 249–9. Cf. also note under lost r.n. *Chece* (*v.* R.Ns *supra*).
p. 205 s.n. PUTTON. Add: *Poddington* 1550 *Midd*.
p. 205 s.n. VICTORIA INN. Local name for field opp. this inn is *Knights in the Bottom* acc. to informant at DRO: perhaps named from Knight's in the Bottom in Hooke **4** 57?
p. 208 s.n. FRYER MAYNE. Add: *hospital'de Meyne* 1280 *Ass*.
p. 217 s.n. PORTLAND. *v.* Gelling *Signposts* 78–9.
p. 220 s.n. VERNE YEATES. Cf. Fern Down **2** 225 for preferred derivation from **fearn** 'ferny place'.
p. 220 s.n. WAKEHAM. On map of 1693 this place is named *Crab* (ex. inf. Sid Payne, who also refers to book by Mark Vine which states that *Crabchurch* was a password – connection with smuggling?)
p. 229 s.n. f.n. *Gowerth*. Add: *Oarth* 1751 Churchwardens' Acct.
p. 232 s.n. LODMOOR. Add: *Drowned Landes alias Lodmore* 1590 *Arundell*.
p. 237 Add: f.ns. *Le Beache*, *Clarkes Hill* 1590 *Arundell*.
p. 248 Add: *a Stowesfete vsq' Westbroke* 1480 Weld[1], *v.* WESTBROOK **1** 248, for *Stowesfete* cf. High Stoy **4** 163.
p. 259 s.n. CULLIFORD TREE. This barrow is also known locally as 'the Music Barrow' (*Dorset Life* June 2011 p. 29).
p. 262 s.n. WINTERBORNE FARRINGDON. *South' Wynterburne* 1236 Fees p. 1468 probably belongs here (Frank Thorn letter).

p. 275 s.n. *Stretford*. Like Brog Street **2** 16 to be associated with the course of a recently discovered Roman road (letter from N. Field).

p. 281 Add: f.ns. *Culeaze*, *Dawes meade*, *Hamm meade*, *South meade* all 1663 *DCMDeed*.

p. 284 s.n. Winterborne Muston. Add: *Wymbourne* (sic) *Turbeuileston'* 1280 Ass. Also *Est Wynterburne iuxta Warham* 1236 Fees p. 1468 probably belongs here (Frank Thorn letter).

p. 288 s.n. BARROW HUNDRED. Lost names in this Hd: *Belle* and *Whynfhide* 1288 *Ass*.

p. 288 s.n. Affpuddle. Add: *Pudell alias Aff(e)pudell* 1544 *DCMDeed*.

p. 290 s.n. Cull-Peppers Dish. It is noted in *Dorset Life* No. 407 (Feb 2013) p. 48 that the hollow, with a tree within it, resembles a mortar and pestle.

p. 301 s.n. Burleston. Add: *Burdaleston* 1317 *MiltC*.

p. 303 s.n. Dewlish. Add: *Develis* 1264 (1288) *For*.

pp. 304–5 Add: *deuelisc made* 859 ?for 870 (15) *ShaftR* (S 334), *deuliscmad* ?870 (15) *ib* (S 342), *v.* **mǣd** (note added to two AS charters describing the bounds of Cheselbourne).

p. 305 s.n. f.n. Louseleys. Probably = *Lowysley*, *Weslowysleysclyff* 1385 (e15) *MiltRoll* in bounds of Milton Abbas.

p. 306 s.n. Milborne St Andrew. Add: *Mulebourne iuxta Middelton' Abbatis* 1358 *HarlCh*.

p. 309 s.n. f.n. The Mead. Cf. *fossati inter pratum vel ingore de Mylborn'* 1385 (e15) *MiltRoll*.

p. 309 s.n. Piddlehinton. Add: *Pidelinton'* 1288 *Ass*. The form *Pidele Edwaker* (*Wm de Hineton ten'*-) 1201 FF may belong here.

p. 311 s.n. Little Puddle Fm. Grid. ref. should read 717966. The lost name *Combe Deverell* is traced by C C Taylor in DoNHAS 88 pp. 211–2.

p. 316 s.n. *Little Cheselbourne*. The Anglo-Saxon charter describing a small estate (3 *virgas*) *æt ceosol burnan* 965 (10) *Ilch* (S 736) belongs here rather than under Cheselbourne par. **3** 202–8, *v.* Keynes, AbbotsC 16.

p. 317 s.n. Frome Mead. Add: *prat' apud Frome* 1317 MiltC. Probably to be identified with *sex made eres* (?sic for *æceras*) *be frome/frume* 859 ?for 870 (15), ?870 (15) *ShaftR* (S 334, 342), 'six meadow acres by R. Frome' (note added to two AS charters describing the bounds of Cheselbourne).

p. 328 Add: f.n. *Ondhull'* 1358 *Quarr Charters* p.121.

p. 331 s.n. Tolpuddle. Add: *Tolpitell'* 1478 *Brid*.

p. 336 s.n. f.n. Knowlhill Bottom. This f.n. should be cross-referred to Knowle Hill **4** 90.

p. 339 s.n. Charminster. Add to Street-names: *the Shambles in the Market Place* James 1 TRMB.

p. 341 s.n. Cowdon. Add: *Cowdon'* (p) 1268 *Ass*.

pp. 345–6 Add: f.ns. *Bruntelhammesmede*, *le Southfeild* 1467 *Arundell*, *Totcombe* hill, *the towne meadowe*, *Yelcombe* all James 1 TRMB.

p. 346 s.n. Dorchester. *v.* A. Breeze, '*Durnovaria*, the Roman name for Dorchester', SoDoNQ **35** (2001) 69–72.

p. 352 s.n. Icen Way. For sound development cf. f.n. Icely Moor **3** 44.

p. 354 s.n. *the twelve Men way*. This p.n. is located in DoNHAS 124 p. 117.

p. 359 s.n. POUNDBURY. Bill Lean (letter) suggests the Cnut coin-name *Eanbyri* may belong here.
p. 367 s.n. LR & HR BOCKHAMPTON. Note: the form *(æt) Buchœmatune* may belong under Buckham **4** 241.
p. 368 s.n. EGLYSHAM. Jo Draper in article for DoNHAS **116** (1994) 139–40 has *Eclisham* 1399, *Aglishamb* c.1650, *Egglisham* 1662.
p. 373 s.n. GRIMSTONE. Add: *Grimstede* 1291 Tax.
p. 374 s.n. f.n. Grimston Bridge. Add: *hedges from Grimston Bridge to Grimston Backside* 1672 *Sheridan*.
p. 375 s.n. WINTERBORNE ST MARTIN. Add: *Wynterborne Marten* 1510 DCM.
p. 376 s.n. BRONKHAM HILL. Add: ?*Brunkescoumbe* (p) 1323 ADi.
p. 377 s.n. MAIDEN CASTLE. Ptolemy's *Dunium* is more likely to be the fort at Hod Hill **2** 117. For a recent study of this name-type, *v.* R. Coates, 'Maiden Castle, Geoffrey of Monmouth and Hārūn al-Rašīd', *Nomina* 29 (2006), pp. 5–60.

THE PLACE-NAMES OF DORSET, PART II (EPNS Volume LIII)

p. 1 s.n. ALDERNEY HEATH. Acc. to *Dorset Life* Dec. 2006 p. 44, this is a transferred name from Alderney, the Channel Island: the Governor of the Island, John le Mesurier, retired to Dorset in 1825, naming his cottage *Alderney Cottage*.
p. 3 s.n. LAKE FM. There was a large Roman settlement here.
p. 4 s.n. BROADSTONE. Possibly named from 'large stones forming a bridge over a local stream' (*Dorset Life* Feb. 2018 p. 21).
p. 5 s.n. CREEKMOOR. Add: *Crickmore Mill(s)* 1725, 1755 (cited by Michael Eyers, *v.* next name).
p. 7 s.n. WATERLOO. For the history of this name, transferred from an ironworks so called at Abbotts Ann Ha, *v.* Michael Eyers, *Waterloo*, privately printed 2016.
p. 12 s.n. *CHARLTON PARVA*. Affix possibly to distinguish this Eton College manor from its manor of Charlton (Magna) W (ex. inf. Patrick Strong).
p. 16 s.n. BROG STREET, on the course of a recently discovered Roman road (letter from N. Field).
p. 29 s.n LYTCHETT MATRAVERS. The cited Litchett Ha is a ghost name.
p. 31 s.n. LUSCOMBE. The form *Loscumb'* 1244 *Ass* may belong here.
p. 33 s.n. LYTCHETT MINSTER. Add: *Lychet munstre* 1288 *Ass*.
p. 39 s.n. CONSTITUTION HILL. So named on 1785 Survey by Mackenzie (*Dorset Life* Dec. 2010 p. 39).
p. 40 s.n. LILLIPUT. Add: *Elipute* (sic) *Summer House* 1785 (*v.* prec. name).
p. 49 s.n. LIMEKILN. Add: *le Lymhest*, *le Lymputt* 1381 Ipm.
p. 51 f.n. Newfoundland. This and other examples of this transferred name may reflect Dorset's chief overseas trade 1650–1830.
pp. 51–4 Add: f.ns. *Aboveweye furlong*, *Bynetheweye furlong*, *le Bereweye*, *Chichilycros*, *Chippelthorn als. Pykes*, *Colslade*, *Gorgeswode*, *Grene*, *Heldeheye*, *le Holebreche*, *Leghe*, *le Marle*, *Mounheswode*, *Shapecroft*, *le Wereheved* all 1381 Ipm.

pp. 57–8 Add f.ns. in MAPPERTON (Almer): *Blowys Close*, *co'mna voc' le Bus(s)ches*, *Fesantes Close*, *Gays Close*, *le Rother Mar(s)che*, *Wargges Howlde* all 1525 AOMB.

p. 71 s.n. COMBS DITCH. For the possibility of 'Coneks ditch', from a Celtic pers.n., *v*. Breeze, JEPNS **35** pp. 42–3.

p. 80 s.n. *WINTERBORNE PHELPSTON*. Frank Thorn (letter) notes one *Philip* holding 1 knight's fee in *Wynterburne* 1236 Fees p. 1469.

p. 80 s.n. OATCLOSE WD. Add: *Oatclose* 1811 OS.

p. 81 s.n. f.n. Clenstone Coppice. Add: *Clenston Copse* 1771 *DROMap*.

p. 83 s.n. WINTERBORNE WHITCHURCH. Add: *Wynterbourne Gurewamb* Hy (14) *AddCh*.

p. 83 s.n. LA LEE FM. Add: *Lawlye Farm* 1659 (1758) *DROMap*.

p. 84 s.n. WHATCOMBE FM & HO. Add: *Whatcombe Ho* 1811 OS.

p. 85 s.n. f.n. North Fd. Add: *broodenorthfeld'* 1385(e15) *MiltRoll*.

p. 86 Add f.n. *the Redwode* 1385 (e15) *MiltRoll*.

p. 87 s.n. BLANDFORD FORUM. A recent article by Paul Cullen, JEPNS **38** 17–24 makes a strong case for *forum* being a mistaken reinterpretation of *forinseca* 'outside' (cf. Beaminster Forum Hundred **4** 237 for an identical development).

p. 92 Add: f.ns. *Braghforlang*, *Byhyndeheys*, *Catforlang*, *Croftforlang*, *Crysindeslonde*, *la Dene*, *Horscroft*, *le Ladylynch*, *Langlench*, *atte Nassh* (p), *le Perces*, *Sixacres*, *Southmede*, *Stokeleslade*, *Toukeresacre*, *Uppethedoune*, *Whetelond* all 1382 Ipm.

p. 92 s.n. DURWESTON. Add: *Durenwineston*, *Dorewyneston* 1258 *For*.

p. 94 s.n. ST NICHOLAS'S CHURCH. Add: *advocac'ones eccl'iarum de Dyrwyneston' et Knyghton'* 1368 *AD*.

p. 95 s.n. LOWER FIFEHEAD. Add: *Fifhyde quentin* 1258 *For*, also *South Fifett* 1583 *SPDom* which probably belongs here.

p. 99 s.n. f.n. Woodclose. Add: cf. *bosco...de Fifhyde* 1258 *For*.

p. 97 s.n. ZOAR FM. Oliver Padel (pers. comm.) cites four instances of Zoar in Cornwall, all with dissenting chapels: this name is therefore biblical, from Hebrew *Zoar* 'little', the name of a small city at SE corner of Dead Sea, to which Lot and his daughters fled after destruction of Sodom.

p. 98 s.n. f.n. Plympthorn Coppice. Add: *la Plumþorn'* 13 Salkeld. The first el. is thus **plūme**, **plȳme** 'plum'.

p. 99 s.n. f.n. *Mayne Place*. Add: *Mayneplas* 1504 *Arundell*. First el. possibly ME **main** 'demesne'.

p. 99. s.n. HAMMOON. Add: *Hampmowne* 1475, *Hammohon* 1611 *DoCo*.

p. 100 s.n. HASELBURY BRYAN. Now officially Hazelbury Bryan (since 1915). Add: *Haselberewe* 1258 *For*.

p. 101 s.n. *CROCKERN STOKE*. Add: *Stoke in hundredo de pimperne* 1327, *Stoke et Crokkernestoke iuxta Hasilbere* 1333 (m14) *Glast*, Stokestrete 1515 *Ilch*, *Crockern Stoke* 1795 Boswell.

p. 102 s.n. CONEY LANE. Add: *Conny Lane* 1626 JC-K. FRIZZEL'S HILL. Add: from local surname *Frizzell*.

p. 102 s.n. WONSTON. Add: John *Stokke de Wolmeston'* 1415 Seymer.

p. 102 s.n. WOODROW. Cf. *bosco de Haselber'* 1258 *For*.

p. 102 add: *Antelope* 1811 OS.

p. 103 s.n. POVERT BRIDGE. Add: *Polford(e)* 1256 Hutch[3], c.1325 *GlastE*, *Poulford'* e14 *ib*. The first element is thus **pōl** 'pool'.
p. 103 s.n. RIDGE. Add: *Rygge* 1385 JC-K.
p. 103 s.n. STIVVICKS BRIDGE. The form *Trubbeforkebrigge* c.1325 GlastE (perhaps for *Stubbeforke-*?), a point in the bounds of Buckland Newton Hd **3** 239, may belong here, thus (?) 'fork of land with tree-stumps', from **stubb**, **forca**.
pp. 104–5 Add: f.ns. (a) *Combe*, *Dolys*, *Le Fryth*, *Medforlong*, (b) *Brodestokfeld*, *Brynsfryth*, *Byryhay*, *Estehede*, *Ferthdelmen*, *Halfyerdlond*, *Northmerssh*, *le Southbreche*, *Suthmersh*, *Westhyde* all 1385 JC-K.
p. 106 Add f.n. *close called Bragges* 1594 Hutch[3].
p. 110 s.n. PIMPERNE. Add: *Pimperne* Hy (15) *ShaftR*. Also *v.* entry for *Pimpern* in RNs. *supra*.
p. 114 Add f.n. *t'ris Rad'i de la Hide...una hida in villa de Pymperne* 1338 (15) *ShaftR*, v. **hīd**.
p. 117 s.n. HOD HILL. Ptolemy's *Dunium* is likely to belong here, cf. Maiden Castle **1** 377.
p. 117 s.n. LAZERTON. A. D. Mills and Richard Coates, 'Lazerton at Ash, Stourpaine, Dorset' JEPNS **47** (2015) pp. 16–18 make a strong case for ME **lacer** 'maker or seller of cord, etc' as first element.
p. 119 Add f.n. *le Wards Tenement* 1615 Hutch[1].
p. 119 s.n. TARRANT HINTON. Add: (*Tarente*) *Henton*, *Hentoun* 15 *ShaftR*.
p. 120 l.12, for 'Hinton St Mary par.' read 'Hinton Parva par.'.
p. 122 Add f.ns. *Buntiford* 15 *ShaftR*, *ten' voc' Corses* 1509 Hutch[3], *t'ris Partrych* c.1500 Eg.
p. 124 s.n. f.n. Keyns Marshe. Add: *la Mersch* 1366 Ipm.
p. 125 Add f.n. *le Wodehouse* 1372 (c.1407) *Shaft*.
p. 127 Add Tarrant Launceston f.ns. dated c.1265 from *Charters of Abbey of Caen* ed. M. Chibnall (1982) p. 27 (kindly supplied by John Field): *brocforlang*, *Cattesbrein*, *dicforlong*, *Hernebergam*, *sortforlong*, *Staing-*, *stanforlong*, and *Woelinham*.
p. 129 s.n. PARK BOTTOM & WOOD. Add: *parci de Howgh'ton' vocat' Gyffardyspark'* 1385 (e15) *MiltRoll*.
p. 130 s.n. f.n. Kit Barrow. Add: *Kytebargh'* 1385 (e15) *MiltRoll*.
p. 131 s.n. WINTERBORNE STICKLAND. One of the steep lanes climbing out of the village is known as Chalky Path (local).
p. 134 s.n. CHALBURY. Line 19, delete 'with reference to the hill-fort here'. There is no hill-fort here, but at Chalbury **1** 231.
p. 135 s.n. CHURCH. Add: *capellam de Chelesberi* 12 *SherC*.
p. 137 s.n. LEIGH. Add: *Legh' iux' Caneford* 1338 (15) *ShaftR*.
p. 140 Note the *TA* f.n. Slop probably gives name to SLOP BOG (a 56-acre nature reserve).
p. 141 s.n. CHETTERWOOD. OE ***rȳd**, ***rēod** 'clearing' should perhaps be considered for the second element. The form *Wodhose Chetrode* c.1500 *Eg* may belong here.
p. 142 s.n. COCKROW FM. Probably **cocc-rodu**, *v.* **2** 96.
p. 142 s.n. STAPLE CROFT. Add: *La Stapele* 1347 *Arundell*.
p. 143 Add f.ns. (a) *La Proute Crofte*; (b) La Almeslonde, *La West Crofte*, *La Wodehuse* all 1347 Arundell.

p. 147 s.n. WOODCUTTS FM. Add: Clement *de la Wodecote* 1255 Cl.
p. 149 s.n. HINTON PARVA. Add: *Parva Hynneton'* (p) p1483 *Sheen*.
p. 151 s.n. MANNINGTON. Is this rather OE *(ge)mǣnan tūne* 'communal estate'?
p. 158 Add f.n. *Casmore Hill* 1558 DoNHAS **88** 198.
p. 160 s.n. ST WOLFRIDA'S CHURCH. Add: 'the church of *Horton*' 1145, 1163 Hutch[3].
p. 160 s.n. WIGBETH. Perhaps to be associated with the OE bdy mark *weggeleawe* **2** 162.
p. 181 Add f.n. *Wytheghes* 1285 *Pitt*.
p. 183 s.n. WIMBORNE MINSTER. Add: *Wymbernemynistr'* 1338 (15) *ShaftR*.
p. 193 s.n. CRANBORNE CHASE, *v.* Desmond Hawkins *Cranborne Chase* London 1980, especially the appendix on the bounds of the Chase pp. 175–180.
p. 201 line 7 from bottom, read: Deverill.
p. 210 s.n. *HYDE*. The form *W(h)ynshide* 1288 *Ass* may belong here.
p. 218 s.n. WESTWORTH FM. Add: *Westworth* 1664 HTax.
p. 221 s.n. FARNHAM. Add: *fernaan* (sic) Wm 1 (15), *Ferna* Hy 1 (15) *ShaftR*, (*Est* ~, *West*) *Fernham* c.1500 *Eg*.
p. 224 Add f.ns. (a) *Manlondesdowne* (b) *Thornedowne* both 1467 *Arundell*.
p. 231 s.n. f.n. *Woodles*. Add: *La Wodelese* 1338 (15) *ShaftR*.
p. 233 s.n. TRICKETT'S CROSS. This post-war estate, which once had a bad reputation, was renamed Heatherlands c.1980.
p. 235 s.n PENTRIDGE. Add: *Pentrich*, *Pentryngton* 1338–40 Glast and *v.* Padel CoPNE xvi, 222.
pp. 235–6 s.n. BOKERLY DITCH. Add: *Blakedounesdich* 1338–40.
p. 242 f.n. (*le*) *Castell* probably refers to Castle Lane **2** 183.
p. 254 s.n. TURNWORTH. Add: *Turnanwrthe* 939–946 (1726) Finberg (S 1729), *Torneworth* 1340 NI, *Turnewurth* 15 *Forde*.
p. 255 s.n. ST MARY'S CHURCH. Add: *eccl'iam de Turneworth'* 1266 (1339) *AddCh*.
pp. 255–6 Add f.ns. *Belakenegoresende*, *Cheldewollesdich'*, *Hurdethorne*, *Lamdich'*, *Lidekeshie*, *Linleghit*, *Longacresend*, *Midlitlecumbe*, *Rame leeshurn*, *Rughedich'*, *Stansuliesende*, *Weghethorn*, *Whettingesthorne*, *la Whitedich* all n.d. (15) *Forde*.
p. 271 s.n. WEST WOODYATES. Add: *Widangete*, *Widingete* 955–9 (1726) Finberg (S 1753).
p. 274 s.n. LONG CRICHEL. The DB form *Chirce* is identified with More Crichel **2** 140 by ThornDB. Tengstrand MN 103 suggests the possibility of an el. ***crȳcel** to explain Crichel.
p. 276 s.n. GUSSAGE. Gelling PNsLandscape 29 suggests OE ***gyse** with noun-forming suffix **-ic**, 'the gusher'.
p. 287 s.n. KNOWLTON CHURCH. Add: 'the chapel of *Cnolton*' 1145, 1163 Hutch[3].
p. 292 s.n. TARRANT MONKTON. Add: *Tarente abbatis de Teokesbir'* 1201 P, *Tarrant Mounckton* 1518 *HarlCh*.

THE PLACE-NAMES OF DORSET, PART III (EPNS VOLUME LIX/LX)

p. 11 s.n. BOWRIDGE HILL. *v.* note s.n. Poxwell **1** 143.
p. 22 s.n. PEACEMARSH. Add: William *Pysmers* 1315, ~ *de Pismersch* 1316 *Brid.*
p. 42 s.n. NYLAND. *v.* M. Ross DoNHAS 115 (1993) p. 162 on the appropriateness of this name to topography.
p. 61 line 7, delete cross-reference to Chilcombe par. (**4** 14)
pp. 84–5 s.n. BUCKHORN WESTON. Add: *Weston Stourton* 1562 Arundell (for the affix cf. Stourton Caundle **3** 276).
p. 86 s.n. GIGG LANE. Add: *Gygge Lane* 1562 *Arundell*; s.n. HOPE FM Add: 'tenement called *Hope*' 1588 *ib.*
p. 100 Add: CLUBMEN'S DOWN (local), named from 'the Clubmen' who once assembled here: they were country folk who opposed both sides in the Civil War, angry at the damage to their crops and livestock, took up arms but were defeated by Cromwell on Hambledon Hill.
p. 113 s.n. SIXPENNY HANDLEY. The boundary point *biken settle* (***bica** or pers.n. *Bica*, **setl**) in the AS charter 956 (14) *ShaftR* (S 630) should be added to f.ns. (*b*) **3** 122.
p. 118, line 3 read: for '**hricg**' read '**hrycg**'.
p. 131 Add: SPREAD EAGLE HILL (local), so named (it is said) because it resembles an eagle with wings spread out when viewed from Abbey Walk, Shaftesbury (*Dorset Life* June 2016).
pp. 139–143 s.n. SHAFTESBURY. *v.* Dodgson BH 106–7. For the pers.n. *Sceaft*, *v.* Feilitzen, 'OE Uncompounded Personal Names and Bynames', *StudNeo* XL No 1 (1968) p. 11.
p. 149 line 9, for 'roundabout' read 'merry-go-round'
p. 159 Heading should read: STURMINSTER NEWTON HUNDRED
p. 160 s.n. YEWSTOCK. Peter Kitson's suggestion that OE **hēafod-stocc** in these names may rather denote '(a certain type of) pollard' should be noted.
p. 173 s.n. CROSS TREE FM. Cf. **cross-tree** 'gallows'.
p. 174 s.n. PILLWELL. Cf. Pill Well **2** 147.
p. 178 s.n. OKEFORD FITZPAINE. Add: *Okeford' fyzppen* 1476 *Brid.*
p. 189 s.n. STURMINSTER NEWTON. Add: *in mercato de Neuwetone* 1288 *Ass*. The boundary point *on hedem'chil* (obscure) in the AS charter 968 (14) *Glast* (S 764) should be added to f.ns. (*b*) **3** 199.
p. 202 s.n. CHESELBOURNE. Note: the charter S 736 cited here belongs under Little Cheselbourne in Puddletown par. **1** 316, *v. supra* and Keynes, AbbotsC 16.
p. 203 s.n. LYSCOMBE. Add: *Luyscombe* 1480 *DCM.*
p. 208 s.n. HILTON. Add: *Heelton* 1349 *Surv.*
p. 209 s.n. ANSTY. Add: *Anestie* c.1220 *Poul* (p), *Anestig'* lHy 3 *CottCh* (p). For further speculation about the meaning of the OE word in this name, *v.* the 2014 edn. of LandscapePNs 66–7.
p. 210 s.n. HATHERLY. Add: *Hath'legh'* 1332 SR (p).
p. 211 s.n. HOOPERS LANE. Marked thus 1811 OS.
p. 240 s.n. ARMSWELL. Add: *Ermynwell* 1308, *Hermyngeswell* 1313, *Harmingeswell* 1314 Drew.

p. 257 s.n. Add f.n. *Le Redemore* 1282, 1307 Drew.
p. 259 s.n. MAPPOWDER. The 1121 form *mapeldream* may be from OE dat.pl. **mapuldrum*.
p. 242 s.n. DUNTISH. Add: 'the park of William (*de Monte Acuto*) at *Dunhyete*' 1249 FF.
p. 244 s.n. MINTERNE PARVA. It is noted in ThornDB (12.16 and note) that the DB holding of *Cerne* (f. 78b) belonging to Milton Abbey is possibly to be identified with Minterne Parva, cf. Hutch[3] **4** 17.
p. 244 s.n. MONKWOOD HILL FM. Add: *Moneke Wodeshulle* 1316, *Monekeneswodehull* 1317 Drew.
p. 253 s.n. f.n. Rugwell. Cf. f.n. Minterne Rigwell **4** 166.
p. 255 s.n. *doggenford*. Add: *Doggeford* 1551 Winch.
p. 256 s.n. AS boundary point *on anne heued*. Possibly referring to Dungeon Hill **4** 164.
p. 257 Add f.n. *Le Redemore* 1282, 1307 Drew.
p. 267 s.n. WOOTTON GLANVILLE. Add: *æt Wudatune ... to Wudetune* 956 *Add* 15350 (BCS 969, S 608). The identification of this estate, and of its bounds described in this charter, once considered doubtful and hence not included in the entries for this parish **3** 267–274, are now thought likely to belong here, *v.* the transcript of the charter with annotation in the section on The Anglo-Saxon Charter Boundaries *infra*.
p. 273, f.n. *Doggevellesheyeshurne*, cf. Dogwell Fm **4** 294.
p. 283 s.n. STALBRIDGE WESTON. Add: *Stalbrigg' Weston* 1386 *AddCh*.
pp. 286–9 Add f.ns. (a) *Bakehouseclose*, *Brodefelde*, *Burycroft*, *Mydelfelde* (b) *Fotelondmede*, *Fotelondysfelde*, *Nytherfelde*, *le Parrok*, *Pyne* all 1450 *Arundell*.
p. 289 s.n. f.n. *Huggescumbe*. Add: *Uggescumbe* 1288 *Ass*.
p. 290 s.n. BEER HACKETT. Add: *Beere* 1323 *Salis*.
p. 291 s.n. TRILL FM & HO. Add: *Trull* 1311 Drew, *Troll* (sic) 1811 OS.
p. 296 s.n. SMITH'S BRIDGE. Add: *Smythes bridge*, *Smethes brudge* 1563 *Digby*.
p. 312 s.n. ATRIM. Add: *Atrambryan* 1313 *Arundell* (p).
p. 319 s.n. PURSE CAUNDLE. Add: *Pourscaundel* 1350 Misc.
p. 323 s.n. THE PARK. Add: *Lytelparke*, *Mochelparke* 1467 *Arundell*.
p. 323 s.n. THORN STUMP. Probably to be associated with *Thornstubbe* 1301 Drew, a boundary point in a description of a 'division between lands of abbot of Cerne in Cerne Abbas and lands of John Plecy in Up Cerne'.
p. 324 Add f.ns. (b) *Fryngemede*, *Lytelbrokemede*, *North-*, *Southorchard*, *Nywemede* all 1467 *Arundell*: s.n. f.n. *Howked Walles*. Cf. f.n. *Bovehokedewale* **4** 182.
p. 324 s.n. f.n. *Huish*. Possibly referring to HUISH FM **4** 176.
p. 336 s.n. CANCER DROVE. Add: *Canshaghe* 1319 *Arundell*.
p. 339 s.n. f.n. Farthing Gate Pce. To be associated with FARTHING GATE **4** 164.
p. 342 s.n. f.n. *Rune hurne*. Now identified with RHYMEHORN **4** 154.
p. 342 s.n. f.n. *Stone Rige*. Now identified with STONERUSH DROVE **4** 155.
p. 344 s.n. BAILEY RIDGE. Add: *Bali* Ed 3 Brad. The forms *spinam de Belya* and *viam de Bulya* (both *sic*) *ib* may also belong here.
p. 346 s.n. f.n. *Hokebrygge*. Add: *Hoka*, *Hoke* Ed 3 *Brad*.
p. 346 s.n. f.n. *Longcroft*. Add: *longa crofta* Ed 3 *Brad*.

p. 346 s.n. f.n. *Shepencroft*. Add: *Scerpecrofta*, *Screpecroftam* (*sic*) Ed 3 Brad, spellings influenced by surname of Richard *Screpekake ib*.
p. 346 Add to Lillington f.ns. (b): (*Northre*)*bacwey*; *la Ber-*, *Burifurlange*; *Binlonde*; *Blakelake*; *Bledelesburgam*; *Brocforlange*; *Brodemede*; *Brodemora*; *Bruernston'*; *Goddon'*; *Gotlega*; *Haggeleslonde*; *Heyfurlange*; *lange-*, *schortehunt'pe*; *Huppeleya*; *Hurnepole*; *langefurlange*; *lega*, *la ley*; *Mercwudeburgam*; *Mildeburgam*; *Mudimore*; *campo del North*; *langessobeham*, *Rindessebeham*; *Rueleishurne*; *campo del Shu* (*sic*, cf. *versus le Suh/le Shu*, ?*australi campo ib*); *Stanchuste*, *Stonchuste*; *hamme sub Stimbe*; *Stonilond*; *Vikleia*; *Vulputeslake*; *campo del West*, all Ed 3 *Brad* (several of the spellings from this source appear to be dyslexic or at least erratic).
p. 346 s.n. LYDLINCH. Add: *Lidelynche Baret* 1426 *HarlCh*.
p. 349 s.n. *RAMSBURY*. Add: *Remmesbury* 1332 SR (p).
pp. 352–4 Add f.ns. (a) *Crokesmede*, *Geales Meade*, *Geles Hylcraft* all 1512, *Mershmede* 1441, *Vitteland* 1512, (b) *Ferthynglond* 1441, *Lytelfrythe*, *Thornelese* 1512, *Underparke* 1571 *Arundell*.
p. 358 s.n. *Lanprobi*. On the significance of this form, *v*. further K. Barker, 'The beginning of Sherborne's history', *Dorset Life* Oct 2004 39–42.
p. 360 Add: Sherborne st.n. Doghouse Lane Gate (Gazetteer 50), cf. DOGHOUSE FM **4** 378.
p. 368 s.n. *Pyddemylle*. Add: *Pudemull*(*e*) 1374, 1385 *AddCh* (p).
p. 370 s.n. HYLE FM. Add: ?*Hugone de la hile* Hy 3 *AddCh* (Corscombe).
p. 372 note the small building still known as the Conduit was originally the *lavatorium* or wash-house for the monks, built *c*.1520.
p. 393 s.n. ADBER. Add: *Attebere* 1268 *Ass* (p).

THE PLACE-NAMES OF DORSET, PART IV (EPNS LXXXVI-VII)

p. 2 s.n. ABBOTSBURY. Add: *æt abbodes byrig* m11 *Ilch* (KCD 942). This document is the foundation charter for the guild at Abbotsbury established and endowed by Orc: *þæ gegyld healle and þone stede æt abbodes byrig*, *v*. also AbbotsC 2–3.
p. 24 Add: TAMARISK FM, originally called Pond Fm, so named from the seashore plant **tamarisk**.
p. 27 line 5 from bottom, read: of Coryates with.
p. 30 s.n. STALL BARN. The reference to Forsberg SNPh **45** No 1 is to an article on the AS bdy mark *on æt stællesbeorh* in which he convincingly argues for reading *æt* as a preposition 'as far as', then translating 'onward as far as the barrow of the stall or stable', from **steall** and **beorg**.
p. 70 Add f.n. *Lurtebury* 1300 Hutch[3] (in bounds of forest of Powerstock).
p. 79 s.n. f.n. Woodville. Add: *Wydewyll* 1407 *Arundell*.
p. 89 s.n. FRAMPTON COURT. Note buildings at priory: *bashalle*, *brewhous*, *deyhous*, *larder*, *yathous* 1380 Misc.
p. 109 Add to f.ns. (b): *certain trees called the Three Sisters* 1675 Ogilby.
p. 135 s.n. DUCK ST. Add: *Duckstreete* 1652 *ParlSurv*.

pp. 138–9 *v.* also M. D. Jones, 'Field Names in Cerne Abbas', DoNHAS 64 (1942) 69–74.
pp. 141–2 Add f.ns. (a) *Blacdon*, *Odesterteslane*, *Portwey* (b) *Buttescombe*, *Courtclose* all 1451 Court Roll (SoDoNQ XXXI 1980 pp. 34–7).
p. 166 line 18 read: **cnæpp**.
p. 186 s.n. BATCOMBE. *v.* G. Kristensson, 'Two English place-name etymologies', SNPh **46** 326–30.
p. 192 line 7 read: DAIRY.
pp. 193–205 s.n. HALSTOCK parish. Add: *æt halȝastoke* 841 (12) *SherC* (S 290). Note: the Anglo-Saxon bounds of Halstock in the charter cited are not all incorporated into the entries for this parish — *v.* the transcript of the charter with annotation in the section on The Anglo-Saxon Charter Boundaries *infra*.
p. 194 s.n. NETHERSTOKE. Add: *æt niþerstoce* 841 (12) SherC (S 290).
p. 211 s.n. WOOLCOMBE. Add: *Owolcom* 1375 Ipm.
p. 229 line 2 from bottom read: Thornford par.
p. 235 Add f.n. (b) *Lyemede* 1467 *Arundell.*
p. 242 line 6 read: the meaning.
p. 242 s.n. HEWSTOCK. See the Addenda note *supra* s.n. YEWSTOCK **3** 160.
p. 254 Add f.n. (b) *Pestellane* 1542 *Arundell.*
p. 263 s.n. GREENHAM. Perhaps the first el. here is rather the stream-name **Grinde* suggested by Ekwall DEPN for the other So Greenham.
pp. 275–6 s.n. TYTHERLEIGH. See now R. Coates '*Tyther-* as an English place-name element', JEPNS **43** (2011) 35–42.
p. 304 s.n. f.n. *on lyc hagen*. Richard Coates suggests *lyc* may represent lWSax **līc** 'corpse'.
p. 305 s.n. f.n. *tryndlea wege*. For **tryndel** read ***trynd**.
p. 308 s.n. MOSTERTON. Note suggestion in CDEPN and DBPN for Mortehoe D that OE ***mort** may have had a topographical sense 'stump, hill'.
p. 316 s.n. FILFORD. Add: *Filleford* 1305 *Arundell* (p).
p. 320 s.n. SALWAY ASH. Add: from **scofl** and **weg**.
p. 322 s.n. YONDOVER FM, line 8, for 'Yetminster' read 'Netherbury'.
p. 326 line 3 read: PATLEY WOOD.
p. 331 s.n. f.n. Worth. Add: (*le*) *Worthe* 371, 1381 *Arundell.*
p. 336 s.n. NORTH POORTON. This name is discussed by A. Breeze, 'Welsh *pybyr* 'staunch' and Poorton, Dorset', SoDoNQ **35** (2004) 302–4.
p. 353 s.n. BRIDPORT. *v.* A. Breeze, 'The names of Bridport, Bredy, and the River Bride', SoDoNQ (2000) 408–9.
p. 364 Add: 'wall called *Pynnon*' 1373 AD.
p. 386 s.n. DALWOOD. Richard Coates points out that **dæl** may have the sense 'dell, pit, quarry' here.
p. 397 s.n. GUN CLIFF. Add: 'four canon were placed here to defend the eastern end of the town during the Civil War' (pers.comm.).
p. 402 s.n. HACKERIDGE FM. Add: *Suthakerugge* 13 *Arundell.*
p. 405 s.n. SHAVE CROSS. Add: 'wood of *La Shahe*' 13 *Arundell.*
p. 406 s.n. BLACKCROFT. Add: *Blackecrofte* 1559 *Arundell.*
p. 406 line 19 read: CROWCROFT.
p. 406 s.n. DODSEYE. Add: *crucem Dodde* 13 *Arundell,* and cf. *Doddehulle* in (b).
p. 406 s.n. GILLIN'S HEAD. Add: *Gydenhedde* 1559 *Arundell.*

p. 406 s.n. LAND COPPICE. Add: *the Lawnde* 1559 *Arundell.*

pp. 408–411 Add f.ns. (a) *Barne Close, Brocombe Meade* 1559, *la Buthele* 13 *Arundell*, (b) *Bynitheth', Gaydun', Haralond*, 'land which *Pulle* sometime held', *Rodfurlang, La Rudinge, Ruggeweye, Sydeham* all 13 *Arundell, Harperyge Oke, Kingescrofte, the Moore, the Platte, Pul(l)howse* (*Mede*), *Pullmede, Roughcrofte, Shepeham, Sherecrofte, Whytearyshe* all 1559 *ib*, *Sottewelle c.*1300 ib, *Stonbowebrygg* 1384 Misc.

p. 416 s.n. RIDGE BARN. Add: *La Rigge* 1357 *Arundell.*

p. 426 s.n. HR ATRIM. Add: *Westatram* 1355 *Arundell.*

p. 427 s.n. EYPE. Add: *Yep* 13 *Arundell*, *Epp* 1675 Ogilby.

p. 431 delete entry for HR & LR DENHAY FM: this name is more fully and correctly treated on p. 427.

p. 431 s.n. HENWOOD. Add: *Hennewode* 1317 *Arundell.*

p. 432 s.n. RYEBERRY HILL. Add: **beorg**.

p. 434 s.n. f.n. Gore. Add: *La Gore* 1261 *Arundell.*

p. 438 Add: f.n. (b) *Bitelethroteshaghe* 1317 *Arundell.*

p. 443 s.n. MORCOMBELAKE. Note as for Mosterton *supra.*

p. 445 s.n. *STUDLEY.* Its location is revealed in a sale document (DRO D/1095/1) dated 1852 in which a close called *Studleys or Hither Fd or Higher Fd formerly Bluntshay Mead* lay W of Stoke Mill and E of Crabbs Bluntshay (letter from Bob Machin), cf. TA f.ns. Hr & Lr Bluntshay and Crabbs Plot **4** 450–1.

p. 446 s.n. BLACKMORE FM. Add: *La Blakemore* 13 *Arundell.*

p. 448 s.n. OSELHAY FM. Add: *Hoswaldeshee* 13 *Arundell.*

pp. 450–4 Add f.ns. (a) *Hedeswill, Stonyfurlange* 1403, *Thornhull* 13 *Arundell*, (b) *Hyndecombesmore* 1403, *Notecrofte* 13, *Roulegh*(*mede*) 1403 *ib.*

p. 466 s.n. MARSHWOOD VALE. Note in document granting liberties in *valle de Merswode*: right to take 'hares, foxes, cats and partridges' 1317 *Arundell.*

pp. 474–6 Add f.ns. (a) *Coluerwill*, (b) *Balrichgore, Bromcrofta, Chestel, Westbreche, Westcumba* all c.1300 *Arundell.*

p. 478 line 20 read: **græf**.

p. 490 line 3 read: from.

p. 495 line 17 read: DAIRY.

p. 496 line 18 read: DAIRY.

p. 504 read: Tarrant Keynston.

INTRODUCTION TO THE PLACE-NAMES OF DORSET

The county of Dorset has a rich variety of different landscapes, many of striking beauty: rich pastureland in broad river valleys, clay lowland vales, high upland with considerable hills, areas of heathland and woodland, widely contrasting coastal vistas, and with relatively little urban development. As will be seen from the different analyses that follow, the place-names of Dorset reflect closely every aspect of this varied landscape, in all its complexity, as seen through the eyes of its earlier inhabitants. The Isle of Purbeck in the south east corner of the county has its own various landscapes – low-lying heathlands in the north of the Isle, the hogback chalk ridge of the Purbeck Hills, a central clay vale, and a limestone plateau to the south of the vale. Most of the centre of the county is occupied by a large area of chalk downland, rolling chalk downs bounded on the north by a high escarpment that crosses from Melbury Hill to Bubb Down, and dissected by the valleys of the Stour, Frome and Piddle as well as by a series of dry valleys. In the east of the county, in the lowlands drained by the lower courses of the Stour, Frome and Piddle, there are large areas of low-lying pasture and farmland, and also areas of heath and woodland. The north of the county is dominated by the low-lying clay pastureland of Blackmoor Vale, drained by the Stour and its tributaries, and dissected by the low limestone ridge of the Corallian beds that stretches from Buckhorn Weston to Haselbury Bryan. In the west of the county, the low-lying clay Vale of Marshwood is central to the area, but rising above it are some of Dorset's highest Greensand capped hills (including Lewesdon Hill and Pilsdon Pen), and just to the east is another area of steep-sided hills and deep valleys.

By far the majority of major settlement names in Dorset, the names of most towns and villages, are of Old English or Anglo-Saxon origin. This is the dominant stratum in the older place-names of the county, as it is of other southern English counties apart from Cornwall, and is the legacy of the Germanic peoples, the Anglo-Saxons, who first came to settle in Britain from about the 5th century. The Saxon settlement of Dorset was relatively late. Although there is some scattered archaeological evidence for a Saxon presence within the Durotrigan territory from the early 6th century in eastern Dorset, and from the second quarter of the 7th century in the north of the county, the relatively sparse documentary historical and archaeological evidence suggests that the somewhat piecemeal Saxon settlement and takeover of Dorset mainly took place from the mid to late 7th century.[1] There was no doubt some fierce resistance at first (often of a bloody kind, with probable encounters at Bokerly Ditch and Combs Ditch), from the native Brittonic-speaking (Romano-)British population here, perhaps followed later by a transfer of political power and eventual acceptance and assimilation of the new settlers. It is certainly important to recognise that the Romano-Britons already possessed, lived in, and worked on, most of the land in the county, apart from the heavy clay vales of Blackmoor and Marshwood. The Saxons therefore moved into a settled and largely tamed and organised countryside, already ancient with something like four thousand years of human activity stamped upon it — the Neolithic long barrows, the numerous Bronze Age round barrows, the many Iron Age hill-forts (some thirty of them), and other prehistoric earthworks in this county, not to mention the Roman roads and villas and the Roman city of Dorchester itself, are witness to that. Thus the general picture regarding the Saxon arrival in Dorset seems to be one of integration, assimilation and continuity. Dorset became part of the kingdom of the West Saxons (Wessex). But whatever political control was exercised by the incoming Saxons on the (relatively larger) Romano-British population, it is likely that the basic arrangement of settlements, estates and land units, developed from prehistoric times but largely of (Romano-)British origin, remained in place. Settlement patterns were it seems not substantially changed by the incoming Saxons, but under their political rule were no doubt modified, refined and eventually fixed. New settlements would still of course be made, as older estates were divided and as new ground was cleared or brought into cultivation, but for the most part the pattern of settlement and land use was well established and would continue to evolve in a gradual way as it had done for centuries.

There are various significant developments in Saxon Dorset that are reflected in its place-names. Although there is some evidence of early (Celtic) Christianity in Dorset, at Sherborne and Wareham, and at the Roman villa sites of Maiden Newton and Hinton St Mary, the more visible effects of Christianity come with Saxon rule. The Church in Wessex already began to be organised during the reign of Ine (688–

[1] For a detailed account and valuable discussion of this evidence, *v.* Eagles chapter 9, 'The Saxon stettlement and takeover of Dorset and the conquest of Somerset', pp. 129–135.

726), the see of Sherborne was founded as early as 705 (Aldhelm, its first bishop, died in 709), and a large 'double monastery' (for men and women) was founded in Wimborne in the same year by queen Cuthburh. Before that, the oldest West Saxon charter surviving in a good text records a grant of land for religious purposes at Fontmell by Cenred, father of king Ine, dated 670–6. Many other such grants followed, and from the 7th century on the influence of the Saxon Church is strongly apparent in the establishment of several large 'minster' churches (such as those at Beaminster and Sturminster Marshall), in the setting up of standing crosses (most eventually replaced by churches), and of course in the creation of monastic foundations (large ones at Abbotsbury, Cerne, Milton, Shaftesbury, Sherborne, and Wimborne Minster, smaller houses at Cranborne, Horton, and Wareham). In addition there is architectural evidence for pre-Conquest churches at Studland, Winterborne Steepleton and (probably) Bere Regis, and Domesday Book records nine churches in existence by 1086, at Bridport, Burton Bradstock, Chaldon Herring, Dorchester, Fleet, Gillingham, Puddletown, Wareham and Winfrith Newburgh. Another important development in the Saxon period was the establishment of four mints in Dorset, at Bridport, Dorchester, Shaftesbury, and Wareham. Set up in the reign of Athelstan (927–939), the strong association of mints and coinage with the growth of trade and the evolution of urban life suggests the importance of all four places as towns and market centres. The abbreviated place-name spellings on the coins minted at these places, dating from the ninth century to beyond the Conquest, provide valuable and interesting material (for these coin forms, *v.* Bridport **4** 353, Dorchester **1** 347, Shaftesbury **3** 139–141, Wareham **1** 152–3). It might also be noted that three of these towns, Bridport, Shaftesbury and Wareham, are listed in the important royal memorandum known as the Burghal Hidage of c.1025, a list of West Saxon *burhs*, fortified towns designed by Alfred to be part of a defensive system, each assigned a number of hides from which can be deduced the length of defended wall.

In addition to all the other rich and abundant source material from all periods, both manuscripts and printed books, available for research into Dorset's place-names, this county is particularly fortunate in having some sixty Latin charters (but with name forms in the vernacular) that have survived from the Anglo-Saxon period, most of them because they were copied into the medieval cartularies of Sherborne, Shaftesbury and Glastonbury abbeys. All of these charters greatly enhance our knowledge of Anglo-Saxon Dorset, but thirty-seven of them are especially valuable and informative because they incorporate full descriptions in Old English of the bounds of the estates granted in the documents. Apart from all the information they contain regarding land use, woodland, tenure, and the vocabulary employed to describe topographical and other features, these charters describe boundaries of estates that are often closely aligned with the boundaries of modern parishes, thus showing that most parishes were probably already defined and in existence by the 9th or 10th century. Because of their importance, these charters are listed, with a full transcription of the Old English bounds of each land grant and with references to the respective parishes where they are treated, in the section The Anglo-Saxon Charter Boundaries of Dorset *infra*.

Although most major names in Dorset, names of towns, villages and other settlements, are first recorded in the Domesday Book of 1086, some hundred or so names are first recorded in pre-Conquest sources, mostly in the Anglo-Saxon charters (though often in medieval copies) referred to, plus a few in the Anglo-Saxon Chronicle. Of these, just three are recorded in (later copies of) 8th-century charters: Charmouth, Lyme Regis, and Pentridge. Another twenty one names are first recorded in 9th-century sources, most of them in (often later copies of) 9th-century charters (Beaminster, Bradford Abbas, Buckland Newton, Cheselbourne, Nether & Over Compton, Fontmell Magna, Iwerne Minster, Plush, Shaftesbury, Sherborne, Sixpenny Handley, Stalbridge, Sturminster Marshall, Tarrant Hinton, Wambrook, West Woodyates, Woolland), some in the late 9th-century MS A of the Anglo-Saxon Chronicle (Portland, Swanage, Wareham, Wimborne Minster). Of the rest, fifty eight major names are first recorded in (often later copies of) 10th-century charters (including names like Abbotsbury, Bloxworth, Cattistock, Cerne Abbas, Corfe Castle, Gillingham, Gussage All Saints, Halstock, Horton, Long Crichel, Milton Abbas, Purbeck, Sydling St Nicholas, Weymouth, Wyke Regis), and just eleven in pre-Domesday 11th-century documents (including Corscombe, Piddlehinton, Portisham and Wool). Some of these names are of Celtic (Brittonic) origin (Cerne, Crichel, Fontmell, Iwerne, Lyme, Pentridge, Tarrant, also in part Charmouth, Corscombe and Sturminster, *v.* next paragraph), but all the rest are of Anglo-Saxon (Old English) origin. Taking into account the fact that all names will always be considerably older than the date of their first documentary record, it is certainly the case that at least some of these Old English major names, whatever their origin and whether habitative or topographical, could have been coined in the early stages of the Saxon settlement of Dorset, in the late 7th or early 8th century. We may also have to reckon with some renaming, especially of the larger settlements, in the early period, so that in some areas the older names may be those of relatively smaller places – a case in point may be some of the names containing the Old English toponymic suffix **-ing**[2] such as Dodding's, Snelling, Uddens, and Fordington ('farm at *Fording*'). These **-ing**[2] names have long been recognised as of an early type (*v.* for instance PNBrk 821–2).

However, among the oldest place-names in Dorset are certainly those of Celtic (Brittonic) origin.[2] Such Brittonic names as have survived are relatively few in number, but they are particularly interesting and significant because they provide strong evidence of continuity and contact between the Brittonic-speaking Britons and the new English-speaking settlers, the West Saxons. Since these names will probably have been passed on during the Saxon settlement of Dorset in the 6th, 7th and 8th centuries, most of them are probably a good deal older than that. Certainly the Brittonic names of Badbury **2** 177 (*Vindocladia*) and Dorchester **1** 347 (*Durnovaria*) are recorded as early as the 4th century (in the Antonine Itinerary). Most of the surviving Brittonic names are those of rivers and streams or of other natural landscape features, especially hills and woods, many of these having been later transferred to settlements. Rivers and streams of Brittonic (or at least pre-English) origin include *Avon*, Axe, Bride (giving name to Little & Long Bredy and Burton Bradstock), ?Cale, Cam, Cerne (as in Cerne Abbas, Nether & Up Cerne, and Charminster), Char (as in Charmouth), ?Corry, ?*Culme*, Devils Brook (as in Dewlish), Divelish, Fontmell Brook (as in Fontmell Magna & Parva), Frome (giving name to Chilfrome, Frampton, Frome St Quinton & Vauchurch), *Humber* (× 6), Iwerne (as in Iwerne Courtney, Iwerne Minster, and Steepleton Iwerne), Lim (as in Lyme Regis), Lodden, ?Luckford Lake, Lydden, ?Parrett (as in South Perrot), ?*Pimperne* (as in Pimperne), ?Stirchel, ?Stour (giving name to East & West Stour, Stourpaine, Stour Provost, Sturminster Marshall & Newton), Tarrant (giving name to eight parishes, Tarrant Crawford, ~ Gunville, ~ Hinton, ~ Keynston, ~ Launceston, ~ Monkton, ~ Rawston, ~ Rushton), *Terrig*, *Toller* (as in Toller Fratrum & Porcorum, Tollerford and Toller Whelme), Trent Brook (as in Trent), Wey (as in Broadwey, Upwey, and Weymouth), ?Winniford, and Yeo. Other original stream-names of Brittonic origin are those giving name to Corscombe **4** 287, Winfrith Newburgh **1** 174 and Wynford Eagle **4** 117, possibly also to Carey and Keysworth **1** 158–

[2] For a useful map and list of these, *v.* Eagles xviii–xxii.

9, Charborough **2** 59, and Loders **4** 481 (on all these names, *v.* RNs. *supra*). The parish-name Long Crichel **2** 274 is from PrWelsh ***crūg** 'mound, hill, barrow' with 'explanatory' **hyll**, and other names that are probably analogous include Creech Hill **2** 268, Church Hill **4** 123, Churchill **2** 211, **4** 277, Crouch Hill **3** 385, and Crook Hill **1** 205, **4** 290. The same word gives name to East Creech **1** 89 (here with reference to Creech Barrow **1** 91) and Crook Furze **4** 4 (associated with an Iron-Age hill-fort). The parish-name Pentridge **2** 235 is from PrWelsh ***penn**[1] 'head, top' (with PrWelsh ***tïrk** 'of the boar'), and a few other names contain the same element, including Pen Hill (earlier *Seaxpenn*) **3** 81, Penn **4** 457, Pen Wood **4** 165, and (probably) Pilsdon Pen **4** 413. The parish-names Lytchett (Matravers & Minster) **2** 29, 33, and (East & West) Orchard **3** 133, together with Chetterwood **2** 141, are from PrWelsh ***cę̄d** 'wood', and the parish-name Chideock **4** 375 is from an adjectival derivative PrWelsh ***cę̄diog** 'wooded'. Two more parish-names that have a Brittonic origin are Broadmayne **1** 337 (from PrWelsh ***main**[2] 'rock, stone') and Owermoigne **1** 138 (probably British ***ogrodrust-** 'wind-gap(s)), and other names that may contain a Brittonic (or at least a pre-English) element include Combs Ditch **2** 70 (giving its name to a Hundred), Cosmore **3** 242, East & West Man **1** 67, Lodmoor **1** 232, Pert Copse **2** 269, and Sharnhill **3** 246. In addition to these, the city of *Dunium* (from British ***dūno-** 'fort'), mentioned by Ptolemy in the 2nd century as the capital of the Durotriges, has been identified with Maiden Castle **1** 377 (although it is perhaps more likely to be the Iron-Age hill-fort at Hod Hill **2** 117). The interesting occurrence of the form *Lanprobi* in Sherborne **3** 358, a Brittonic name from PrWelsh ***lann** 'church enclosure', is evidence for the presence of a British church here, and *Eglysham* **1** 368 is probably from PrWelsh ***eglẹ̄s** 'church'. Also to be noted is the parish-name Chilcombe **4** 14, which may contain a pre-British word ***ciltā-** 'steep hill'.

Only a few names in Dorset, as in most other English counties, are French in origin, in spite of the far-reaching effects of the Norman Conquest on the English language in general. It is clear that by the late 11th century, most settlements and landscape features already had established names, and the new Norman aristocracy only rarely gave French names to parts of their estates: Chantmarle **4** 126 and Organ Ford **2** 34 may well be examples of names actually transferred from France. There are other interesting indications of French influence. The survival of the French definite article in La Lee **2** 83–4 (with no alternation with the English word) is noteworthy, as is the use of the French article in many of the early forms for Nyland **3** 42. The various parts of the Norman castle at Corfe are (perhaps unsurprisingly) solidly French in origin: *Butavaunt*, *Gloriette*, *Malemit*, *Plentey*, *Sauveray* **1** 11. *Beawffrount* **3** 399 and Belle Vue **1** 57 are purely French names of a familiar type (**bel**, **front**, **veeir**). In many names French terms often alternate with English ones in the medieval period, as in Caundle Marsh **3** 316 (**mareis**), Cliff Coppice **3** 36 (**falaise**), Verwood **2** 256 (**be**(**a**)**u**, **bois**). A remarkable instance of this kind of alternation is the middle element in Piddletrenthide **4** 167, where the OFr word **trente** 'thirty', alternating with ME **thrittig** from the 14th century, not only persists in the current form of the name but also becomes an alternative name for R. Piddle! Another interesting case of persistent French influence is the parish-name Corfe Mullen **2** 15, where the affix from OFr **molin** 'mill', referring to the valuable mill here mentioned in Domesday Book, is recorded in spellings consistently from the late 12th century to the present and never alternates with the English word.

However by far the strongest French influence on Dorset place-names is apparent among those of manorial origin, in which estates were named from the Norman-French personal names or surnames of their owners, for instance Druce Fm **1** 316 (from *Drew Bardolf* 1272, one of the family giving name to Bardolfeston **1** 315), Font le Roi **3** 331 (from the *Fauntleroy* family here in the 13th century), and Scoles Fm **1** 19 (from the *Scovill* family here at least as early as the 13th century, from the p.n. Escoville in Normandy). In particular the influence of this powerful French-speaking Norman aristocracy is revealed in the way the names of some of the great feudal families were affixed to the names of the manors in their possession. These manorial additions, whilst often serving to distinguish one manor from another with an identical name, result in a great many hybrid 'double-barrelled' names which contribute considerable variety and colour to the map of Dorset. They also reflect the rich and complex manorial history of some estates. Some of the land-owning Norman families are represented in more than one name, as in Bradford Peverell **1** 334 and Newton Peveril **2** 48 (family of *Peverel* here from 13th century), Charlton Marshall **2** 11 and Sturminster Marshall **2** 46 (from the powerful *Mareschal* family, Earls of Pembroke, recorded here in 1204), Fifehead St Quintin **2** 95 and Frome St Quintin **4** 93 (family *de Sancto Quintino* from p.n. St Quentin in France), Okeford Fitzpaine **3** 178 and Wootton Fitzpaine **4** 455 (family of *Fitz Payn* here from 13th century), Langton Matravers **1** 33, Lytchett Matravers **2** 29 and Worth Matravers **1** 63 (family of *Mautravers* of whom Hugh *Maltravers* named in DB). Some of these manorial additions are now compounded with the original name itself, as in Hammoon **2** 99 (family *de Moion* from p.n. Moyon in Normandy), Owermoigne **1** 139 (family of *Moigne* here from 13th century), and Stourpayne **2** 116 (OFr *Paien*). Other manorial affixes from Norman-French family names include Clifton Maybank **4** 191 (*Maubank*), Hinton Martell **2** 146 (*Martel*), Kingston Russell **4** 18 (*Russel*), Sandford Orcas **3** 389 (*Oriescuilz*), and Turners Puddle **1** 294 (*Tonere*). Many such family names (like some already mentioned) are themselves derived from French place-names and lie behind several more manorial affixes, among them Compton Valence **4** 82 (*de Valencia*, earls of Pembroke, from p.n. Valence), Coombe Keynes **1** 114 (*de Cahaignes* from p.n. Cahaignes), Fifehead Neville **2** 95 (*de Neville* from p.n. Néville or Neuville), Haselbury Bryan **2** 100 (*de Bryene* from p.n. Brienne), Iwerne Courtney **3** 37 (*de Curtenei*, earls of Devon, from p.n. Courtenay), Kington Magna **3** 41 (earlier affix *Plukenet*, from p.n. Plouquenet), Litton Cheney **4** 23 (*Cheyne* from p.n. Quesnay), Tarrant Gunville **2** 242 (*de Gundeville* from p.n. Gondouville), Winfrith Newburgh **1** 174 (*de Neuburgh* from p.n. *le Neubourg* in Normandy), Wootton Glanville **3** 267 (*de Glanvile* from Glanville in Normandy), and Wynford Eagle **4** 117 (*de Aquila* from p.n. L'Aigle also in Normandy). In addition, many names that no longer have manorial affixes had them in medieval times, so that Farnham **2** 221 for instance had the early affixes ~ *Deverel*, ~ *Cary*, ~ *Gounvyle*, and Upwey **1** 245 was variously *Waye Pigace*, ~ *Hamundevill*, ~ *Rabayne*, and ~ *Bayhouse* (this last from the p.n. Bayeux in Normandy). As already noted, these affixes, and the dates at which they are recorded in use, reflect the manorial history of the places named in a significant way.

Mention must also be made of the interesting presence of Latin among the place-names of Dorset. There are no names of purely Latin origin, but it is clear that in the name Portland 'estate attached to *Port* 'the harbour''), the Latin word *portus* lies behind the OE word **port**, just as the Latin word *vicus* lies behind OE **wīc** in names like Week Street Down and Witchampton which refer to Romano-British settlements (*v.* further discussion *infra*). There are however a number of names with distinguishing Latin affixes added in medieval times. This clearly reflects the widespread use of Latin by church and state – it was the principal language of law and administration in medieval times. Presumably so many of these Latin additions survive because they became fixed and accepted in legal and administrative documents, even though one suspects they may never have been used very much in popular speech. Many of them give an indication of ownership, as by the king (Bere Regis, Melcombe Regis, Wyke Regis, all anciently royal demesne), by an abbey (Bradford Abbas, Cerne Abbas, Compton Abbas, Melbury Abbas, Milton Abbas, Winterborne Abbas), or some other religious foundation (Whitchurch Canonicorum, this incidentally producing Dorset's longest name).The affixes Magna 'great' and Parva 'little' occur in several pairs of names to distinguish adjacent manors or villages with a shared name, among them Fontmell Magna and Parva, Minterne Magna and Parva. In Ryme Intrinseca the unusual affix, distinguishing this place from the former manor of Ryme Extrinseca, is administrative, and in Blandford Forum, the affix distinguishing this place from Blandford St Mary is a reference to the early and important market here (though for a contrary view about its origin, *v.* Addenda *supra*). Perhaps the most memorable of the contrasting Latin affixes in a pair of names are those in Toller Fratrum and Porcorum ('of the brethren' and 'of the pigs') — there is perhaps a little mischievous sense of humour at work here.

Among the habitative names of Anglo-Saxon origin in Dorset, by far the most common element is OE **tūn** 'farmstead, manor, estate, village', as indeed it is in many other counties. Well over a quarter of Dorset's parish-names are from this word, but of this number only about a half are recorded in Domesday Book or earlier. These have a variety of first elements. Names with a descriptive adjective, denoting location, age or shape, include Buckhorn Weston (**west**), Hinton St Mary (**hēah**), Horton (**horu**), Langton Herring (**lang**), Maiden Newton and Sturminster Newton (**nīwe**), Milton Abbas (**middel**), and Sutton Waldron (**sūð**). There are several names from **tūn** compounded with words for natural features: Alton Pancras (**ǣwyll**), Clifton Maybank (**clif**), Compton (× 5, **cumb**), Holton (**hol**), Moreton (**mōr**), and Wootton Glanville & Fitzpaine (**wudu**). Five names are from words relating to tenure or possession by an individual or a group, these being Allington (**æðeling**), Charlton Marshall (**ceorl**), Hinton Martell (**hīwan**), West Knighton (**cniht**), and Preston (**prēost**). Three have a river-name as first element, Burton Bradstock (R. Bride), Frampton (R. Frome), and Puddletown (R. Piddle). Only three names have OE personal names in the genitive or possessive case as first elements, Durweston (*Dēorwine*), Kinson (*Cynestān*), and Manston (*Mann*), but to these should be added the five parish-names from **-ingtūn** 'estate called after or associated with (the person named)': Chedington (probably *Cedd* or *Cedda*), Ibberton (*Ēadbeorht*), Lillington (*Lylla*), Osmington (*Ōsmund*), and Poyntington (*Punt*). Among the rest, Stanton relates to stony ground (**stān**), Shipton Gorge must have been a specialised sheep farm (**scēap**), and two names refer to trees,

Mapperton (**mapuldor**) and Silton (**sealh**). The parish-name Bothenhampton (**boðm** 'valley bottom') and Brockhampton in Buckland Newton (**brōc** 'brook') are from the compound term **hām-tūn** 'home farm or settlement'. A particularly interesting name is Witchampton, since this is 'estate of the dwellers at a **wīc-hām'** (the only occurrence of this significant word in Dorset place-names).

As already mentioned, a good number (almost half) of the parish-names derived from the word **tūn** are not recorded in Domesday Book or earlier sources, but only in later documents. It is clear that some of these names, like Castleton (recorded from 1332, named from its 12th-century castle) were coined well after the Conquest, others though perhaps belong to the Saxon period but were not recorded at an early date, like Kingston Russell (**cyning**, from 1212), Litton Cheney (**hlȳde**, from 1204), and Stratton (**strǣt**, from 1212). Of those named from persons, Godmanstone and Leweston contain OE pers.ns., whereas Bryanston and Catherston Leweston are from the names of their Norman-French lords. Several more of these parish-names in **tūn** are first recorded after 1086 because in Domesday Book itself the respective manors to which they refer are simply named from the rivers on which they lie. There was clearly a need in the centuries following the Conquest to distinguish between the different manors lying on the same river, and this was done with reference to manorial ownership or to other features of the estate, thus the parish-names Anderson (simply *Wintreburne* DB, *Andreweston* 1331, from the church dedication) and Athelhampton (*Pidele* DB, *Athelamston* 1285, from the OE pers.n. *Æðelhelm*). Most of the parish-names of this type have retained the river-name in their current forms, notably the series of parishes named from the two Winterborne streams: Winterborne Clenston, ~ Herringston, ~ Houghton, ~ Kingston, ~ Monkton, ~ Muston, ~ Tomson, and ~ Zelstone, all of these late names in **tūn** referring to early manorial ownership. Winterborne Steepleton however is named from its church steeple (*Wintreburne* DB, *Stepelton* 1219, **stīepel**), as is Iwerne Steepleton (*Werne* DB, *Stepleton* 1234). Named from the same river, Iwerne Courtney had a 14th-century alternative form *Schyreuetone* ('sheriff's estate', **scīr-refa**) that has persisted to the present day as Shroton. Piddlehinton is named from a religious community (*Pidele* DB, *Pidel Hineton* 1244, **hīwan**). Apart from these parishes named from rivers, other places too are listed with a simplex or shared name in Domesday Book but are then distinguished with a later **tūn** name, in the case of Hampreston (*Hame* DB, *Hamme Preston* 1244, **prēost**) as a kind of affix, in Shillingstone (*Alford* (for *Acford*)

DB, *Akeford Skelling* 1220, *Shillyngeston* 1444, OFr pers.n. *Schelin*) becoming the name of the parish. Alongside these parish-names, there are numerous examples among the minor names of post-Conquest **tūn** formations, among them names like Barnston (named from the DB tenant), Bhomston (earlier *Frome Bonevileston*), Gatemerston (from a surname *Gatemore*), Herrison (from the surname *Harang*), Lazerton (earlier *Iwernelazerton*), Madjeston (from a ContGerm pers.n. *Malger*), Quarleston (earlier *Wynterburne Quareleston*), Ranston (earlier *Iwerne Randelleston*), and Waterston (earlier *Pydele Waltereston*).

In contrast to the large number of **tūn**-names, names derived from Old English **hām** 'homestead, manor, estate, village', in general shown to be an earlier name-forming element than **tūn** and characteristic of the earliest Anglo-Saxon settlements, are relatively very rare in this county. Even though it is often difficult to distinguish names in **hām** from those containing the topographical element **hamm** discussed *infra*, taking all considerations into account, there is among the parish-names only one certain example of **hām**, Gillingham ('of Gylla's people'), to which can be added a few possible or probable examples: Edmondsham (the OE pers.n. *Ēadmōd* or *Ēadmund*), Farnham (**fearn**), Pulham (**pōl**), and Wareham (**wer**). The significant name Witchampton has already been mentioned: its first element makes reference to a **wīc-hām** 'a village associated with a Romano-British settlement, a *vicus*'. Apart from these, other names ending in *-ham* such as Rampisham and Tyneham are more likely to be from Old English **hamm**.

Of the other habitative elements, there are some twelve names derived from the word **worð** 'enclosure, enclosed farmstead or settlement', four of them parish-names all of which are mentioned in Domesday Book. Bloxworth and Lulworth (East & West) are named from their early owners (OE *Blocc* and *Lulla*), Turnworth is named from its thorn-bushes (**þyrne**), and Worth Matravers was originally a simplex name with the manorial affix added later. The related word, **worðig** with a similar meaning is found in only one parish-name, Hamworthy, a simplex name from **hamm** to which *-worthy* was added only as late as the 15th century. The interesting element **wīc** 'specialised farm', also 'trading settlement', occurs in three parish-names. Shapwick would have been an important sheep farm (the minor name Butterwick ('butter farm') in Folke parish is no doubt analogous). The meaning of Swanage (recorded from the late 9th century) is uncertain, but it could well have been 'swan farm, swannery' (**swan**[1]) rather than 'farm of the swains' (**swān**[2]). The simplex name Wyke Regis (recorded from the late 10th century) refers to a significant trading settlement here, no doubt associated with shipping and fishing. It is also interesting to note that there are no less than six other examples of the simplex name Wyke among the minor names (in the parishes of Abbotsbury, Castleton, Gillingham, Gussage All Saints, Halstock, and Preston), all of them probably having the sense of 'specialised farm or trading place'. A further instance of the element **wīc** occurs in the significant name Week Street Down in Gussage St Michael, where the word may well refer to a Romano-British village (i.e. a *vicus*, this being the origin of the OE word) as in the name Witchampton already noted.

Another important OE habitative term **stoc** 'outlying farmstead or hamlet, secondary or dependent settlement' appears in over a dozen names, eight of them parish-names of which seven are recorded in Domesday Book or earlier sources. Burstock and Cattistock are named from their early owners (*Burgwynn* or *Burgwine* and *Catta*), Halstock is named from its association with the monastery at Sherborne, and Powerstock has an obscure first element. Stockwood has **stoc** as first element (it probably began as a simplex name to which **wudu** was added). There are then three more originally simplex names, all with later distinguishing affixes: East Stoke, Stoke Abbott, and Stoke Wake. The significant word **burh** (dative case **byrig**) 'fortified place' (but with a wide range of applications from Iron Age hill-forts to fortified houses) occurs in a good number of names (often with reference to earthworks or other archaeological sites). There are seven parish-names among them. In the name Abbotsbury (recorded from the mid-10th century) **burh** seems to have the meaning 'fortified house or manor', and this may also be the sense in Netherbury. In Melbury Abbas (recorded 956) and the threesome group of Melbury Bubb, ~ Osmond and ~ Sampford, the sense seems to be 'multicoloured fortified place' (from the OE adjective **mǣle**), but the exact reference is uncertain. The same is true of the parish-name Chalbury (recorded 946, from an OE pers.n. *Cēol*). Shaftesbury (recorded from the mid-9th century, from a pers.n. *Sceaft* or from the word **sceaft** 'shaft' used figuratively of the steep-sided hill on which the town stands) was one of the Anglo-Saxon **burh**s listed in the Burghal Hidage. In the case of Spettisbury however (from **speoht** 'woodpecker' perhaps used as a pers.n.) the reference is to an Iron Age hill-fort. The related habitative term **burh-tūn** 'fortified farmstead, farmstead near a fortified place' is found in five names, two of them parish-names, Bourton and Long Burton. Both these names, and indeed the other three instances of Burton among the minor names, are first recorded in the early 13th century, and only one of them, Burton in Charminster parish, is associated with an archaeological site (in this case Poundbury Camp).

A number of parish-names refer to churches. No less than eight are named from the OE word **mynste**r 'church of a monastery, large or important church', seven of them mentioned in Domesday Book or in earlier sources. Four refer to the rivers on which the respective settlements are situated: Charminster (R. Cerne), Iwerne Minster (R. Iwerne, with **mynster** as affix already present in the Domesday Book form), Sturminster Marshall and Sturminster Newton (both named from R. Stour). Charminster was the best endowed prebend of Salisbury, Sturminster Newton was an early possession of Glastonbury abbey, and Iwerne Minster belonged to Shaftesbury abbey. Also with affixed forms are Lytchett Minster (not on record before 1244) and Wimborne Minster (with reference to the *mynstre* already mentioned c.1000 which was associated with the Anglo-Saxon nunnery here). Of particular interest are the two parish-names Beaminster and Yetminster, both perhaps named from the persons who founded or endowed the churches here (*Bēage* a woman and *Ēata* a man). Several other parish-names are from the word **cirice** itself. Hawkchurch is named from an OE pers.n. **Heafoc*, perhaps again a reference to the founder of the church. Whitchurch Canonicorum is probably 'the white church' (**hwīt**). The same name occurs as an affix in Winterborne Whitchurch (found from the 13th century), and two more parish-names have an affix from **cirice**: Church Knowle (added from the 14th century) and Frome Vauchurch (**fāh** 'multi-coloured', from the 13th century).

There are no instances in Dorset of the so-called 'folk-names' in -**ingas**, and only one example of -**ingahām**, Gillingham already noted ('homestead of the family or followers of a man called *Gylla') recorded from the late 10th century. Only a few other names refer to family or tribal groups, most notable being the parish-name Folke, a unique simplex name, not found elsewhere in England, from the OE word **folc** 'folk, the people of a tribe or family, the common people' here perhaps used elliptically to denote 'estate held by the people, land held in common'. Another OE group word **sǣte** 'settlers, dwellers' occurs of course in the county name Dorset ('the *Dorn* people'). Apart from that, there are also a few interesting names that probably contain the OE word **hǣme** 'the inhabitants of a place', typically combined in the genitive plural case with **tūn** in the form **-hǣma-tūn**, among them Ashington ('farmstead or estate of the ash-tree dwellers'), Bockhampton ('of the beech-tree dwellers'), Brockington ('of the dwellers by the brook'), Shilvinghampton ('of the dwellers on the slope or shelf'), and Stubhampton ('of the dwellers by the tree-stump') as well as the parish-name Witchampton already mentioned ('of the dwellers near a Romano-British village'). Four of this group of six names are mentioned in Domesday Book.

Many names (both major and minor names as well as field-names) refer to routes, roads, tracks and lanes. The important element **strǣt** 'Roman road' gives name to the parish-name Stratton (where the Roman road from Dorchester to Ilchester was joined by a branch from Stinsford). There are also two examples of the name Stratford (**2** 4 and **4** 275), both named from fords where rivers were crossed by Roman roads. Other names too from this word refer to Roman roads, including Abbott Street **2** 164 and The Street (in Charmouth **4** 372). Cole Street **3** 69 may be named from the transport of charcoal (cf. Colway noted *infra*). (Of course in later names, the ME word **strete** came to mean 'straggling village, hamlet' as in Brog Street **2** 16, Hammond Street **3** 261, and *Knightstreet* **3** 169, or simply had the general sense 'urban road' found even in late OE). Of the many places named from the word **weg** 'way' (the general OE term for a road), some refer to ancient routeways and ridgeways, among them Crockway **4** 102, Farway **4** 274, Whiteway **3** 201 (giving its name to a Hundred), Ridgeway **1** 249 (here applied to a Roman road), Ridgeway Hill **1** 92 and Ridgeway Lane **3** 161. Others are named from the characteristics of the road, like High Stoy **4** 163 ('stony'), Holway **4** 128 ('hollow'), *Mudway* **2** 203 ('muddy'), Sandway **3** 3 ('sandy') and Whiteway **1** 90 ('white or chalky'), or from the products carried along it, like Colway **4** 394 ('charcoal'), Fishway Lane **4** 18 ('fish'), and *Higweg* **3** 137 ('hay'). Several names derive from **port-weg** 'road to a market town', such as Partway **4** 497 and Portway **1** 179. Of the other places named from their situation on roads or tracks, two of the most interesting are Hartfoot Lane **3** 209 (from **here-pæð** 'army path, highway') and Ansty **3** 209 (from the much-discussed term **ānstiga** 'path for one, narrow track'). For further instances of minor names and field-names (as well as street-names) derived from the elements cited as well as from other words for roads and paths, including **aley**, **drove**, **lane**, **pæð**, and **stīg**, *v*. Elements *infra*.

The numerous Dorset places that take their names from fords show the early (and continuing) importance of river crossings for communications and trade. Such places where land routes (often ancient ones) crossed river routes, with all their strategic, economic and commercial potential, were clearly often prime sites for early (as well as later) settlement. It is therefore not surprising that besides the many minor names derived from this word there are many major names too, including no less than eleven parish-names, among them three (Bradford Abbas, Okeford Fitzpaine and Thornford) recorded in the early 10th century, whilst all the rest are first recorded in Domesday Book. These are all compounds, with a variety of first elements. Some depict the character of the ford (Bradford Abbas, Bradford Peverell, Sandford Orcas, West Stafford), others refer to trees or other local features (Hanford, Child Okeford & Okeford Fitzpaine, Thornford), others to wildlife (Blandford Forum, ~ St Mary & Langton Long Blandford, Stinsford), and just two contain the personal names of early landowners (Canford Magna and Woodsford). Other significant major names from this element include Filford ('hay'), Huntingford ('of the hunters'), Twyford ('double'), Walford ('difficult to cross'). Particularly interesting is Fordington ('settlement at the ford place'). Also important in referring to river crossings, many other places take their name from bridges. The parish-name Stalbridge ('bridge built on posts or piles') is recorded as early as 998, and names like Stanbridge ('stone built'), Stockbridge ('made of logs') and Woodbridge (probably 'wooden') also refer to the construction of the bridge in question. For the many other minor names and field-names derived from the words **brycg** and **ford**, *v*. Elements *infra*.

As with other counties, the place-names of Dorset relate very closely to the landscape they describe, seen in all its aspects and variations through the eyes of its earlier inhabitants. The importance of river valleys for early settlement — providing fertile soils, a good water supply, and a working environment for various activities — is reflected in the number of Dorset places named from the rivers and streams on which they are situated. Indeed Dorset is notable for the high proportion of its towns and villages named from its rivers. No less than three parishes are named from the river Bride, four from the Cerne, four from the Frome, two from the Gussage Brook, three from the Iwerne, six from the Piddle, six from the Stour, as many as eight from the Tarrant, three from the Wey, seven from the North Winterborne and a further six from the South Winterborne. In Anglo-Saxon sources, as well as in Domesday Book and other early medieval records, many of the various estates that lie behind the village names are often simply named from their situation on the respective river. As a rule, only at a slightly later date are the estates distinguished by an affix, thus Tarrant Hinton is *Terente* 871–877 (15), *Tarente* 1086, *Tarente Hineton* 1280, Turners Puddle is *Pidele* 1086, *Tunerepidel* 1242, and Winterborne Stickland is *Winterborna* 1068–84, *Winterburne* 1086, *Winterburn Stikellane* 1203.

In a few instances of places originally named from one of the rivers, a distinguishing affix or alternative form of medieval date has since taken over and the river-name dropped, thus Anderson is *Wintreburne* 1086, *Wynterborn Fifasse* 1268, *Andreweston* 1331, and Athelhampton is *Pidele* 1086, *Pidele Aleume* 1250, (*Pidele*) *Athelamston* 1285.

Besides the places named from the above mentioned major rivers and streams, there are several others that take their names from a river or stream still bearing a version of its original name. These include Charmouth, Dewlish, Ebblake, Fontmell Magna and Parva, Lyme Regis, South Perrott, Trent, and Wytch. In addition to these are the many Do places named from a river or stream that is now unnamed (or named differently, often with a relatively recent back-formation), among them Camesworth, Cheselbourne, Colesbrook, Corscombe, Cranborne, Furzebrook, Gorwell, Honeybrook, Langbourne, Litton Cheney, Loders, Lowbrook, Milborne St Andrew & Stileham, Northlake, Oborne, Pink Lake Coppice, Seawell, Sherborne, Shortlake, Toller Fratrum and Porcorum, Trill, Wimborne Minster and St Giles, Wynfrith Newburgh, Wynford Eagle, and (possibly) Atram, North Poorton and Powerstock. It should perhaps be noted again here that many of the river-names and stream-names mentioned in this paragraph and the previous one, as having given their names to places, are of Celtic (or at least pre-English) origin (e.g. Cerne, Frome, Iwerne, Tarrant, Toller, Trent, Wey), as indeed are others that lie behind names like Dewlish, Fontmell, Lyme Regis, Winfrith Newburgh and Wynford Eagle (*v.* paragraph on Celtic names *supra*). Further information about these (and other) river-names and stream-names in Dorset, whether lost or still in use, can be found in the entries for the place-names cited in Parts 1–4 as well as in the section River-Names *supra*, *v.* also particularly the elements **brōc** and **burna** in the Elements section *infra*.

Of course besides the rivers, streams and fords, many other watery features contribute to the place-names of Dorset. Among the names of parishes, Charmouth and Weymouth are at the mouths of rivers (*v.* **mūða** in Elements *infra*), Alton Pancras refers to the source of a river (*v.* **ǣwyll**), Fleet is on a notable channel or inlet (*v.* **flēot**), Askerswell, Warmwell and Wool relate to springs (*v.* **wella**), Caundle Marsh, Margaret Marsh and Marshwood refer to marshland (*v.* **mersc**) and Morden to marshy ground (*v.* **mōr**), Almer and Ashmore have their ponds (*v.* **mere**), and Bradpole, Poole and Radipole their larger stretches of water (*v.* **pōl**). The element **hamm** (often difficult to distinguish from the habitative word **hām** 'homestead' discussed *supra*) can also be included here, since it often meant 'land in a river-bend, land hemmed in by water or marsh, river-meadow' in early names. The parish-names Hammoon, Hampreston, and Hamworthy were all originally simplex names with a distinguishing affix added at a later date. In Portisham the word seems to have the sense 'land hemmed in by higher ground'. The parish-names Rampisham and Tyneham may also be from **hamm** in one or other of its senses. Another important element **ēg** 'island, raised or dry ground in marsh' is not found in Dorset parish-names but occurs in its basic sense in Brownsea Island **1** 43 (with an OE pers.n. **Brūnoc*) and Furzey Island **1** 13 (**fyrs** 'furze'), and in its extended sense in a few minor names such as Chaffeymoor **3** 3 (**ceaf** 'chaff'), Eye Mead **2** 166 (originally a simplex name), and Pipsford **4** 291 (earlier *Pippeseia*, from the OE pers.n. *Pipp*). The related compound word **ēg-land** with similar meanings is found in Nyland **3** 42 (a DB manor). Other interesting major names that belong here include Melplash ('multi-coloured pool', *v.* **plæsc**), Moorbath ('bathing place', *v.* **bæð**), Plush ('pool', *v.* **plysc**), Shitterton ('dirty stream', *v.* **scitere**), Slepe ('muddy place', *v.* **slǣp**), Toller Whelme ('river-spring', *v.* **ǣwylm**). There are of course many more major names, minor names and field-names derived from the elements cited, as well as from other words like **bæc/bece**, **celde**, **ēa**, **fenn**, **funta**, **lacu**, **lād**, **læcc/læce**, **mōs**, ***pidele**, **pyll**, **rīð/rīðig**, **sǣ**, **slōh**, **sol(h)**, **wæd**, **wæsse** and **wæter**, listed in the Elements section *infra*. As with other types of landscape, it will be noted that the Anglo-Saxons made use of a wide variety of vocabulary to describe specific wet or watery features in a precise way.

By far the most commonly found word for 'valley' in Dorset place-names is OE **cumb**, with no less than sixteen instances among the parish-names and many more occurring in other major names, minor names and field-names. The word **cumb** was used particularly of 'a short broad valley, usually bowl- or trough-shaped with three fairly steeply rising sides' (Gelling & Cole LandscapePNs 103), and is common in the settlement names of all the SW counties (it is even more frequent in Devon). Among the Dorset parish-names (six of which are first recorded in Anglo-Saxon sources, four of them first in Domesday Book), there is a single simplex name (Coombe Keynes), as many as four instances of Compton (~ Abbas, Over ~ & Nether ~, ~ Valence, West ~, 'valley settlement'), and the rest are also compounds with a variety of qualifying first elements. Some depict the character of the valley (Chilcombe, Thorncombe, Watercombe, Whitcombe), three refer to crops or products (Bincombe, Melcombe Horsey, Melcombe Regis), two contain the personal names of early landowners (Batcombe,

Bettiscombe), Corscombe is 'valley of the *Cori* stream', and Motcombe is 'valley for meetings'. Many other major names contribute to this interesting variety, including Kingcombe ('wall-germander'), Loscombe ('pig-sty'), Nettlecombe ('nettles'), and Renscombe ('raven'). The other main Old English word for 'valley' (typically one that is long and narrow) is **denu**, but this is represented by only one Dorset parish-name, Tincleton 'valley of the small farms'. The word **dæl** 'small valley' also occurs once among the parish-names, as the qualifying element in Dalwood (now Devon). There is also just one instance of the word **boðm** 'valley bottom', as the first element of the parish-name Bothenhampton. The word **corf** 'a cutting, a pass' that gives name to the parishes of Corfe Castle and Corfe Mullen can also perhaps be included here, as can the word **healh** 'a nook or hollow' found in Wraxall ('buzzard'). And two other parish-names refer to valleys or hollows, both of them simplex and both applied figuratively to the topography, Cann (from **canne** 'a can or cup') and Chettle (from **cetel** 'a kettle'). For further instances of minor names and field-names derived from **cumb** and **boðm/botm** as well as from other words describing valleys and hollows like **byden**, **crundel**, **dæl**, **dell**, **gwyle**, **hol**[1], **slæd**, and **trog**, *v.* Elements *infra*.

Dorset place-names derived from words for hilly features and higher ground are also varied and subtly differentiated in meaning. As many as nineteen parish-names denote settlements named from words referring to a raised feature in the landscape. The important word **dūn**, used particularly for 'a low hill with a fairly level and fairly extensive summit' (Gelling & Cole LandscapePNs 164) and often considered a relatively early name-type applied to already established (pre-English) settlement sites, is found in four parish-names, Chaldon Herring ('calves'), Haydon ('hay'), Morden ('marshy ground'), and Pilsdon ('stake' or 'peak'). There are also four parish-names from the word **hyll** itself, Colehill (first element uncertain), Goathill ('goats'), Hillfield (second element **feld**), Marnhull (first element uncertain), and Pamphill (a pers.n. **Pempa*). A fifth name Crichel (shared by the parishes of Long & More Crichel) is particularly significant, for this is from the Brittonic word ***crūg** 'hill' with an 'explanatory' **hyll** added. Other hill-names of Brittonic origin are Pentridge and (in part) the *Seaxpenn* that gives name to Sixpenny Handley, both from PrWelsh ***penn** 'head, top'. Among the other parish-names, four are from **beorg** 'rounded hill', also 'barrow, tumulus' (Chelborough, Seaborough, Symondsbury, and (probably) Todber), three are from **cnoll** 'hill-top, hillock' (Chetnole, Church Knowle, and Puncknowle), and two are from **hlinc** 'bank, ledge' (Lydlinch and Sydling St Nicholas, this last recorded in an early 10th cent. charter). Broadwindsor is from **ōra** 'river-bank' (here with a **windels** or winding device). The simplex name Steeple is from the word **stīepel** in the sense 'steep place'. Other interesting major names that belong here include Charborough (from **beorg** with a r.n), Creech (from **crūg**), and Eggardon (from **dūn** with a pers.n.). For further instances of major names, minor names and field-names derived from the above elements, as well as from other words for hills, hill-slopes, banks and mounds, including **bæc**, **clif**, **cōc**, **copp**, **ecg**, **hēafod**, **helde**, **hlāw**, **hlið**, **hōh**, **hrycg**, **næss**, **ofer**, **ōfer**, **pēac**, **pīc**, **scelf**, **sīde**, and **tōt-hyll**, *v*. Elements *infra*.

Many Dorset settlement names refer to woods and woodland clearings. Although many forested areas had been cleared before the Saxons arrived, with established farms and estates that would in many cases be renamed, some new settlements and clearings were no doubt made and woodland continued to be managed as a valuable resource. Some twenty parish-names denote settlements named from words referring to woodland features in the landscape. Nine names derive from the word **wudu** itself, Dalwood ('small valley'), Gorewood (triangular plot of ground'), Marshwood ('marsh'), Stockwood ('secondary settlement'), Verwood ('fair'), and there are four names where the word is the qualifying element, Woodlands ('estate or arable land in or near wooded area') and Wootton Fitzpaine, Wootton Glanville (in Blackmoor Forest) and North Wootton (all 'woodland estate'). The important word **lēah** 'woodland clearing', often considered a relatively early name-type since it may refer to ancient woodland, is found in Leigh, West Parley ('pear-tree') and Sixpenny Handley ('high'). Among the other parish-names, three are from **bearu** 'small wood' (Haselbury Bryan and (probably) Beer Hackett and Bere Regis), two are from **holt** 'single-species wood' (Alderholt and Holt, the latter giving name to Holt Forest **2** 151), and one is from **hyrst** 'wooded hill' (Holnest). Three names are particularly significant since they derive from Brittonic elements: Lytchett (shared by the parishes of Lytchett Matravers & Minster) and Orchard (East & West) are from PrWelsh **cę̄d** 'wood', and Chideock is from an adjectival derivative **cę̄diog** 'wooded'. Other interesting names that belong here include Adber (from **bearu** with a pers.n.), Bedchester (from **hyrst** with a pers.n.), Bradle (from **lēah**, woodland here is mentioned in Domesday Book), Wilkswood (from **wudu** with a pers.n.), and Woodsend (a reference to the forest of Gillingham). Several large royal hunting forests and chases were established in the early 13th century

(Blackmoor Forest **3** 274, Cranborne Chase **2** 193, Holt Forest **2** 151, and the forests at Blagdon **2** 207, at Gillingham **3** 12, at Marshwood **4** 402, at Powerstock **4** 62 and in Purbeck **1** 3), each no doubt incorporating areas of earlier woodland (*v.* **chase**, **forest**). For further examples of major names, minor names and field-names derived from these elements, as well as from other words denoting woodland and clearings in woodland, among them **fyrð** 'land overgrown with brushwood', **grāf(a)** 'grove, coppiced wood', **hangra** 'wooded slope', **sceaga** 'small wood', **wareine** 'game preserve', and **weald** 'forest', *v.* Elements *infra.*

A good number of settlement-names refer to the different kinds of subsistence farming practised in early times, whether related to arable, meadow land or pasture. The important word **feld** itself, originally used for 'open country, open land cleared of trees and probably used for pasture', and often considered an early name-type applied to already established (pre-English) settlement sites, is found in the parish-name Hillfield (first recorded 934). Four parish-names derive from the word **land** 'estate', also 'cultivated land', Portland ('estate at Port/the harbour'), Studland ('estate with a horse-stud'), Woolland ('cultivated land with meadow'), and Woodlands ('estate or cultivated land in or near wooded area'). The parish-names Wimborne Minster & St Giles have the same first element as Woolland, ***winn**[1], **wynn** 'meadow'. Other interesting names that belong here include Duntish ('hill pasture', from **etisc**), Guppy, Hursey and Marshalsea (all from **(ge)hæg** 'enclosure' with pers.ns., just three of the many minor names from this word particularly frequent in the west of the county), Hethfelton ('estate by the heathy open land', from **feld**), Look (from **lūce** 'enclosure'), and Newland (**3** 268, **land** in Blackmoor Forest newly cleared for cultivation). For further instances of minor names and field-names derived from these elements as well as from other words relating to agriculture and farming practices, such as **bǣr** 'woodland pasture', **bere-ærn** 'barn', **brǣc** 'land broken up for cultivation', **croft** 'enclosure', **gāra** 'triangular plot', **howe** 'enclosure', **lǣs** 'pasture', **mǣd** 'meadow', **pearroc** 'small enclosure, paddock', **penn** 'pen for animals', and **slǣget** 'sheep pasture', *v.* Elements *infra.*

Some names make reference to the speciality of the particular farm, estate or piece of ground, to the crops produced or the animals reared there. The parish-name Bincombe refers to beans (**bēan**), Chaldon Herring to calves (**cealf**), Goathill and Tyneham to goats (**gāt**, **tige**), Haydon (probably) to hay-making (**hēg**), Melcombe Horsey and Regis to dairy farming (**meoluc**), Shapwick and Shipton Gorge to sheep farming (**scēap**), Studland to a horse-stud (**stōd**), and West Parley must have been known for its pear orchards (**peru**). Blandford Forum & St Mary are named from the fish caught there (**blǣge** 'blay, gudgeon'), and Swanage may be named from its swannery (**swan**[1], **wīc**). Among other major names, Orchard in Church Knowle speaks for itself (**orceard**), named as it is from an orchard mentioned in Domesday Book. Names referring to various other fruits, crops and products include Barcombe and Berwick (**bere** 'barley'), Butterwick (**butere** 'butter'), Cashbrook, Cashmoor and Caswell (**cærse** 'watercress'), Filcombe (**fileðe** 'hay'), Honeybrook and Honeycomb (**hunig** 'honey'), Limbury and Linton (**līn** 'flax'), Peacemarsh (**pise** 'pease'), Plumber (**plūme** 'plum'), Ryall and Ryeberry (**ryge** 'rye'), Waddon (× 2, **wād** 'woad'), Watcombe (× 2), Whatley and Whetley (**hwǣte** 'wheat'). Several other names refer to animals bred and reared besides those already mentioned, cows in Cowdon and Cowgrove (**cū**), hogs in Hogchester and Hoggen Down (**hogg**), lambs in Lambrook (**lamb**), cattle in Notton (**nēat**), and swine in Swineham (**swīn**). In addition eels could be caught at Arfleet and Elsford (**ǣl** 'eel'), there was fishing at Fishy Mead (**fisc**, (**ge**)**hæg**), there were vineyards at Wennet Hill (**wīn-geard**) and at other places, doves were kept at Culvers Fm (**culfre-hūs** 'dove-cot) and elsewhere, Loscombe (× 3) and Luscombe refer to pig-styes (**hlōse**), Shearplace Hill and many other names refer to sheep-farming, and swans were bred and kept at Abbotsbury Swannery from at least the late 14th century. Naturally there are many other references to farming and husbandry among the minor names and field-names, derived both from the various words already cited and from others, *v.* Elements *infra*.

Besides farming and agriculture, many other activities, crafts, occupations, trades and industries are referred to in the place-names and field-names. There are references to pot-making in names like Crockern Stoke **2** 101, Crockerton Hill **2** 209, Crokerford **3** 7, and Potterne **2** 257 (*v.* **crocc**, **croccere**, **pott** in Elements *infra*), to brick-making in Brickyard Lane **3** 3 and f.ns like Brickly **3** 25 (*v.* **brick**, **brickkiln**, etc), to tile-making in Tiley **4** 163 (*v.* **tigel**), to leather tanning in Tan Lane **3** 4 and Tanyard Fm **4** 408 (*v.* **tanhouse**, **tanyard**), to cloth bleaching in Buckhouse Fm **4** 406 and (probably) Blashenwell **1** 8 and Buckhorn Weston **3** 84 (*v.* **buck-house**, **bouker**, ***blǣcen**), to fulling and weaving in Tokenhills **4** 171 and Loomcroft **4** 346 (*v.* **tokyng**, **lome**), to rope-making in Rope Walks **4** 358, to brewing in Malthouse Fm **3** 21 and Mash Water **3** 212 (*v.* **malt**, **mash**), to charcoal burning in Cole Street **3** 69, Colway **4** 395 and (probably) Cole Wood **1** 191 and Coll Wood **2**

56 (*v.* **col**[1]), to salt-making in Saltings **1** 24, Salterne **1** 45, Salterns **1** 77, **2** 38, Saltforde **4** 395 (*v.* **salt**, **salt-ærn**), and to soap-making in Soap Ash **4** 111, Soap House Md **4** 425, Soapers Pool **3** 340 (*v.* **soap**, **sāpere**). Merchants are referred to in Mangerton **4** 318 (*v.* **mangere**) and loads were hoisted at Broadwindsor **4** 254 and Wyndlam **3** 18 (*v.* **windels**). There are many references to stone quarrying in the Isle of Purbeck and on Portland (e.g. London Door Quarry **1** 23, Purbeck Stone Quarries **1** 58, Quarr **1** 65, Breston Quarries **1** 221) and to the shipping of the stone from these quarries at various quays and piers such as those at Ower Passage **1** 23 and King's Pier **1** 222 and from the cliffs at Tilly Whim **1** 59 (*v.* **quarr**, **quarr(i)ere**, **stān**, **passage**, **key**, **pier**, **whim**). There are frequent references to other quarries elsewhere in the county, as well as to chalk and gravel pits (as in Wimborne St Giles **2** 268 and Holnest **3** 337), to stone pits (as in Standpitts Lane **3** 23), to mineral extraction (as in *Alum* **1** 51, *Mines heath* **2** 40), and to limekilns (as in Castleton **3** 307) (*v.* **cealc**, **gravele**, **stān**, **pytt**, **līm**, **cyln**, **āst**). References to mills are particularly frequent, some recorded at a very early date. The parish-names Milborne St Giles & Stileham **1** 282, 306 are named from a mill stream (*Muleburne* 934), and a *mylebroce* is recorded in 1024 in Portisham **4** 32. Otherwise many mills are mentioned in Domesday Book, as in Abbotsbury Mill in Abbotsbury **4** 6, and Bere Mill in Shillingstone **2** 239 (*v.* **myln**), and the names Currendon **1** 57, Cornford **3** 384 and Coringdon **4** 289 also refer to mills (*v.* **cweorn**).

Natural history is also well represented among the county's place-names and field-names. At least a dozen different species of tree are evidenced. Particularly frequent are names from **āc**, **æsc** and **þorn**, but for other trees found in the names, *v.* **ellen**, **elm**, **holegn**, **mapuldor**, **sealh**, **wīðig** in Elements *infra*. Most of these tree species were of course valued for their timber or for their various other uses. Plants are also often referred to, as in names from **boðen**, **hramsa**, **netel**. As for wildlife, animals evidenced in the names include wildcats, dogs (also wild?), badgers and wolves, *v.* **catt**, **dogga**, **brocc**, **wulf** in Elements *infra*. Birds referred to include owls, woodpeckers, buzzards, sandpipers or dunlins, and woodcocks, v. **ūle**, **fīna**, ***wrocc**, **stint**, and **cocc**. Other creatures referred to are frogs and toads (**frogga**, **tadde**), and even insects are evidenced (**pēo**, **stūt**[1]).

As might be expected in this county with its large royal hunting forests (in Purbeck and at Blackmoor, Blagdon, Cranborne, Gillingham, Holt, Marshwood and Powerstock as already noted) and its many medieval deer parks and warrens, a good number of names refer to hunting, both within the forests and parks and in other woodland areas.

The various divisions of the forests are often named, as for Cranborne Chase (several parks and walks), Gillingham Forest (*Woods-end-walk*, etc) and Purbeck (*the Est-*, *Westbaillywike*). More importantly, the bounds of most of the forests survive from an early date (Blackmoor **3** 274, Cranborne **2** 193, Gillingham **3** 13, Holt **2** 151, Powerstock **4** 62), these providing a rich source for early minor names and f.ns. Typical significant names within the forests include Hartmoor **3** 265, Hartley **4** 162 ('hart moor and clearing'), Hay Wood **3** 270 ('wood fenced off for hunting') and Lipgate **3** 265 ('leap-gate for deer and other animals'), all these associated with Blackmoor Forest. Hartland and *Harttesknolle* **1** 75 ('hart's land and hill-top') are in Purbeck, and Bokerly Ditch **2** 235 ('buck ditch') and Hawkhill ('hawk hill') **2** 197 are associated with Cranborne Chase. Forest Fm & Side **3** 12 are named from the Forest of Gillingham. *Forstershay* **4** 401 refers to the important office of forester. Particularly interesting are Rocombe **4** 154 ('roe-buck valley') and Huntingford **3** 14 ('hunters' ford') since these provide the starting and finishing points for early perambulations of the Forests of Blackmoor and Gillingham respectively. The possible identification of Huntingford with the Domesday manor held by Edward *venator* the hunter is to be especially noted. Several other names refer to hunters, among them Hunters Bridge (earlier *Huntyngford*) **3** 333, Huntick **2** 30, Huntley **4** 275, and Hawker's Hill **3** 53. There are also other names, besides those already noted, referring to the animals hunted including Delcombe **3** 221 ('deer valley'), Hendover **4** 187 and Hinkhams **4** 442 ('hind ridge and valley'), and Buckham **4** 241 (probably 'buck homestead or enclosure'). Another significant name that may belong here is Wyld **4** 457 (probably 'a trap for animals'). There are also frequent references to warrens, often in association with the forests like the royal warren in Purbeck **1** 3, sometimes independent like the one in Bere Regis **1** 280 owned by the abbess of Tarrant, first recorded in 1546 but known to exist in 1402. Various later references to hunting of different kinds include names like Decoy Pond **4** 7, Doghouse Fm **4** 378, Foxbury **1** 296, Kennel Quarter **2** 247, Larkinglass Fm **3** 53, and Shothole **2** 92.

Some of Dorset's many medieval deer parks (for a detailed list and description of these *v.* Cantor & Wilson **1–4**) are recorded from quite an early date. Typical, and claimed to be the best preserved in the county, is Harbin's Park in Tarrant Gunville **2** 247, first recorded in 1280. Perhaps the earliest on record (in 1162) is that at Sherborne **3** 305. Several of the deer parks were within the forests, like those in the forests at Blagdon **2** 207 and Gillingham (this giving name to Park Fm **3** 16).

The deer park at Milton Abbas **3** 223 is to be associated with the name Delcombe ('deer valley') in that parish, and that in Stockwood seems to have been identical with the 'wood' in the name of the parish. Other deer parks were in Alderholt **2** 196, Marshwood **4** 404 and Woodlands **2** 287 (note all in parishes with 'woodland' names), and there was a particularly important one at Duntish **3** 242. For the many minor names and field-names associated with these deer parks, *v.* **park** in Elements *infra*.

Apart from the parish-names in **mynster** 'large or monastic church' and **cirice** 'church' already referred to, many other major names, as well as minor names and field-names, make reference to the Church and to the various aspects of the Christian religion as it was practised in earlier times. The great landowning and often wealthy monastic and religious foundations in Dorset (at Shaftesbury, Wimborne, Tarrant, Cerne, Cranborne, Forde, Frampton, Sherborne, Milton, Abbotsbury, Horton, Wareham) are particularly important and influential throughout the medieval period, and there are numerous references among the names to these institutions and the lands they possessed, as well as to lands belonging to other abbeys and monasteries outside Dorset. Lands and estates associated with or held by abbots are referred to in names like Abbotsbury **4** 2, Abbotts Wootton **4** 445 (~ Abbotsbury), and Cerne Abbas **4** 134; by abbesses in Abbot's Court **1** 284 (~ Tarrant) and Compton Abbas **3** 99 (~ Shaftesbury); by monks in names like Monkton Wyld **4** 457 (~ Forde Abbey), Monkwood **4** 403 (~ Sherborne) and Tarrant Monkton **2** 292 (~ Cranborne Priory and Tewkesbury Abbey); by priests in names like Hampreston **2** 224 (~ Wimborne) and Preston **1** 230 (~ Salisbury); by bishops as in Bishop's Down **3** 330 and Bishops Caundle **3** 311 (both ~ Salisbury); by nuns as in Minchington **3** 115 (~ Shaftesbury); and by canons in Whitchurch Canonicorum **4** 438 (~ Salisbury). Knights Hospitallers are referred to in names like Fryer Mayne **1** 208, Knighton **3** 290, and Toller Fratrum **4** 109, and other friars in Friary Hill **1** 351. A 13th-century hermitage gives name to the small parish of Hermitage. All early references to churches in the source material have been noted, thus revealing the large number of churches and chapels recorded in towns like Bridport **4** 354–7, Dorchester **1** 348–9, Shaftesbury **3** 143–4, Sherborne **3** 357–9, and Wareham **1** 153–4. There are numerous references to holy wells and springs, as in Holwell **1** 240, and to crosses, as in Holyrood **3** 154 (in Shaftesbury) and Rood Hill **4** 162 (probably alluding to a standing cross). Allusions to the ceremony of beating the bounds at Rogationtide include names like Amen Corner **2** 281 and Pistle Down **2** 219, and the name Curfew Plot

4 341 refers to the ringing of the curfew bell. For further examples of place-names and field-names derived from words related to religious matters, including those found in the names cited here, such as **abbas**, **abbatissa**, **abbesse**, **abbeye**, **abbod**, **biscop**, **cros**, **crūc**, **frere**, **hālig**, **munuc**, **myncen**, **prēost**, **rōd**, *v.* Elements *infra*.

Some of the ways in which the name-forms of places reflect their manorial history (especially in the use of affixes) have already been touched upon. The many links with the appropriate entries in Domesday Book are particularly interesting. Indeed the origins of some Dorset place-names relate in a significant way to the information given in the Domesday survey about the manors in question. In some names with a personal name as first element, the person can be identified with an actual tenant or landholder named in Domesday, among them the parish-names Shillingstone **2** 238 (*Schelin*) and Winterborne Houghton **2** 128 (*Hugh*), as well as Barnston **1** 88 (*Beorn*), Herston **1** 55 (*Her*), and *Odenham* **2** 189 (*Ode*). Other names with a significant association between the place-name and Domesday tenants include the parish-names Chilfrome **4** 80 (first element *cilda* refers to the three thegns who held the manor jointly TRE), West Knighton **1** 207 (first element *cnihta* relates to the two thegns who held the manor TRE), Stour Provost **3** 73 (links with French abbey in early affix), and Sutton Waldron **3** 80 (affix relates to DB tenant-in-chief), other examples being Hyde Fm **4** 89, Kingsdown **3** 191 and King's Mill **3** 169 (relating to land and mill held of the king in DB), Rolls Bridge **3** 22, and Woodyhyde **1** 66 (the 'widow' of the name can be identified). There are a number of other names with a significant relationship between their origin and the DB entry for the manor in question. Salterne **1** 45 and Saltings **1** 24 can be related respectively to the thirty-two saltpans at Studland and the thirteen salt-makers at Corfe mentioned in DB. At Bradle **1** 88 and Wilkswood **1** 36 (both 'wood' names) specific areas of woodland are noted in the DB entries for these manors. At Tincleton **1** 329 (a parish-name, 'valley of the small farms') the assessment of the DB manor at only two hides is significant. The origin of the name Galton **1** 139 ('farmstead subject to tax or rent') is probably related to the specified rent paid by four free men in DB. The early affix of French origin in the parish-name Corfe Mullen **2** 15 clearly refers to the unusually valuable mill here noted in the DB entry for this manor.

Many of the names that may refer to sites of archaeological interest have already been mentioned, among them the five names in **burh-tūn** and the seven parish-names in **burh** (one of which, Spettisbury, certainly refers to an Iron Age hill-fort). The important word **burh** quite often refers to prehistoric earthworks or fortified places of various kinds, so that many other place-names and field-names from this element have archaeological significance. Place-names referring to Iron Age hill-forts include Badbury Rings **2** 177, Banbury Hill **3** 183, Bulbury Camp **2** 33, Chalbury **1** 231, Duds Bury **2** 231, Flower's Barrow **1** 124, Mistleberry Wood **3** 116, Penbury-Knoll **2** 237, Poundbury Camp **1** 359, and Rawlsbury Camp **3** 210. Various other ancient earthworks are referred to in **burh**-names like Buzbury Rings **2** 123 (an Iron Age enclosure), Maumbury **1** 357 (a henge monument of Neolithic date which later served as a Roman amphitheatre), Melbury Hill **3** 130–1 (an undated enclosure), and Woodbury Hill **1** 277 (an unspecified earthwork). Several other key words are used in naming features of archaeological interest. OE **ceaster** gives name to the Roman city of Dorchester and is also found in several field-names including Chesters Well **3** 72 and Chinchester **1** 304 (both sites of Roman settlements), Chestus **1** 41, and Staunchester **4** 203. OE **beorg**, already noted as a word used for a natural hill, very often refers to barrows and tumuli (a sense already found in Old English), as in the Hundred names Loosebarrow **2** 54 and Modbury **4** 120 and other names like Bulbarrow Hill **3** 233, Dunbury **2** 129, Longbury **3** 21, and Seaborough **4** 492 as well as in many other minor names and field-names. The element **dīc** is found in Bokerly Ditch **2** 235 (of Romano-British origin), Combs Ditch **2** 71 (an Iron Age and Romano-British boundary bank and ditch that gives name to a Hundred), Grim's Ditch **2** 236 (another prehistoric earthwork) and in other names of archaeological interest such as Walditch **4** 472. The word **castel** has a variety of applications, sometimes referring to Iron Age hill-forts such as Abbotsbury Castle **4** 6, Lambert's Castle **4** 146 and Maiden Castle **1** 377, sometimes to medieval and later castles such as Castle Close **1** 159, Castle Hill **3** 154, Corfe Castle **1** 10, and Rufus Castle **1** 224. The word **hring** in the names Ringmoor **2** 254 (referring to an ancient enclosure) and Ringstead **1** 212 is of interest and is often used of earthworks in names like The Rings **1** 24 (a 12th-century earthwork). The occurrence of the significant elements **wīc** and **wīc-hām** in the names Week Street Down **2** 144 and Witchampton **2** 260 respectively, in both cases with reference to Romano-British settlements, have already been noted. So too have some of the names referring to Roman roads and ancient routeways like Stratton **1** 373, Stratford **2** 4, Crockway **4** 102 and Ridgeway **1** 249, but there are many other names of this kind derived from words like **strǣt**, **weg** and **here-pæð** to be found in Elements *infra*. Otherwise, features of archaeological interest, often ancient ones, are named using a great variety of terms and phrases, some of them of relatively recent date. Examples include Bestwall **1** 157 (with reference to the 10th-century ramparts of the **burh** of Wareham), Chissels **4** 490 (a Roman site), Dungeon Hill **4** 164 (an Iron Age hill-fort), Great Bournes **3** 36 (a Roman villa site), Grey Mare & Her Colts **4** 54 (a Neolithic chambered long barrow), Hambledon Hill (an Iron Age hill-fort), Hell Stone **4** 32 (a long barrow). There are also names that may invite archaeological curiosity, such as Drakenorth **4** 60 (from **draca**, **hord**) and Gaulter **1** 96, Golter **1** 41 (both from **gold-hord**).

A number of names make reference to a variety of broadly administrative or legal matters. Assembly places are referred to in the parish-name Motcombe **3** 47 (possibly a Hundred meeting-place), Modbury **4** 41, and Modbury Hundred **4** 120 (all from **mōt** 'meeting'). Several names refer to look-outs and beacons. Watch places include Worbarrow Tout **1** 103, Wardstone Barrow **1** 112, and Wor Barrow **2** 269 (all probably from **weard** 'watch). Besides these, look-out hills are referred to in Hambury Tout **1** 131, Nettlecombe Tout **3** 219, Tout **1** 235, Tout Hill **1** 150 (repeated in the well-known Shaftesbury st.n. **3** 147), Tout Quarries **1** 224, The Toyte **4** 449, and Tut Hill **3** 318 (all from **tōt(e)**). Kick Hill **1** 115 may also refer to a look-out hill. Many names refer to the beacon system, a series of beacons in use at the time of the Armada, among them The Beacon **1** 269, Beacon Hill **1** 320 and **4** 219, Beacon Knap **4** 41, Lytchett Beacon **2** 36, and Melbury Beacon **3** 131 (this one perhaps on the site of an Anglo-Saxon *ealde ad* 'an old funeral pyre or beacon'). The parish-name Buckland Newton **3** 239 and Buckland Ripers **1** 239 are from the significant term **bōc-land** 'land granted by charter' (interestingly, the Anglo-Saxon charter dated 941 for Buckland Newton survives, *v.* List of Anglo-Saxon Charters infra). Other specialized administrative terms are found in names like Little Francis **1** 268 (from ME **fraunchise**), *Fryland* **4** 169 (**frēo-land**), and Bailie Gate **2** 47, Bailey Ridge **3** 344 (from ME **baillie**). A few names apparently refer to land disputes, among them Chesters Coppice **4** 3, Cuzenage Coppice **2** 30, Fitzworth **1** 13, Hockerill **1** 240, Sackmore **3** 174, and of course f.ns. like Contraversy **1** 281. Dowerfield **4** 52 refers to a dowry. Many names refer to rents, taxes and other payments, such as Galton **1** 139, Guineagore Lane **3** 326, Penny Farthing Wood **4** 124, Penny Hill **4** 421, Pennymoor **3** 69, Scotland **1** 25, and Smokeham **1** 291.

An interesting group of names refers to places considered haunted or to other superstitious beliefs. There are several names from the word

pūca 'puck, goblin', among them Pucksey Brook (*v.* RNs *supra*), Putlake **1** 39 (said to describe a 'puckish' stream that often floods), Puckstone **1** 49, Pug Pit **1** 192, and Pug's Hole **3** 54. Other names like Robin's Barrow **1** 322 (perhaps also an allusion to Puck), *Pokelmour* **4** 35 (from **pūcel** also 'goblin'), Bugbarrow **1** 278 and Bugley **3** 11 (possibly from **bugge** 'hobgoblin'), *Gobelynshole* **4** 35 (from **gobelin**), *Pexy's hole* **2** 94 (from **pixie**), and *Valgirimede* **4** 12 (from **wælcyrige** 'witch') probably have similar connotations. Old earthwork remains were often given names that evoke their strange (and potentially threatening or evil) presence in the landscape. Examples include Satan's Square (earlier known as *The Devil's Trencher & Spoon*) **3** 81, Giant's Grave & Stones **3** 218, Giant's Grave & Trencher **1** 58, and of course Grim's Ditch **2** 236 (a name found in other counties for prehistoric remains, with Grim being used here as a name for the Devil). Other landscape features are referred to in Devil's Copse **4** 420, Devil's Den **1** 47 (on the coast), Devil's Kitchen **4** 85, Devil's Staircase **1** 98 (a steep lane), *Deoulepole* **3** 317 (devil's pool'), and most notably in the names for the chalk sea-stacks Old Harry & his (now collapsed) Wife and the adjoining Old Nick's Ground 1 48 (named from familiar words for the devil, perhaps because of their danger to ships). It should perhaps be noted that the r.n. Devil's Brook does not really belong here, since this form is simply a result of folk etymology, *v.* RNs *supra.* Gentler superstitions are referred to in Fortuneswell **1** 219 and Elwell **1** 247, but perhaps the most interesting of these names is Drakenorth **4** 60 (probably 'dragon's hoard', an allusion to an ancient Anglo-Saxon belief).

Some names are grim reminders of the violent punishments meted out to wrongdoers in earlier times. The sites of gallows, often on or near parish boundaries, are recalled in names like Gallica Bridge **4** 219, Gallows Corner **3** 215, **3** 223, Gallows Gore **1** 67 (with evidence of hangings here), Gallows Hill **1** 141, **1** 278, and Gallows Plot **3** 307. Other references to execution include Bloody Bank **1** 161, Worgret **1** 74 (recorded in Domesday Book, from **wearg-rōd** 'gallows'), and (probably) the three names Headstock **4** 496, Hewstock **4** 242, and Yewstock **3** 160 (all from the interesting element **hēafod-stocc** thought to mean 'post on which the head of a beheaded criminal was exposed', but cf. Addenda **3** 160 for a different interpretation). Names of an equally gruesome kind are those apparently referring to out-of-churchyard burials or alluding to some tragic occurrence, among them Dead Man **2** 212, Deadman's Coppice **2** 202, and Dead Woman's Stone **1** 90. The coastal name Charnel **1** 104 may be a figurative use of this word to describe a place where bodies lay after a shipwreck. The names Bridewell **3** 51 and Bucky Doo **4** 357 may refer to prisons, and there are also references to criminality. Banbury Hill **3** 183 may have sheltered a murderer, Sugar Hill **2** 77 may allude to highway robbery, and there are probably references to smuggling in coastal names like Brandy Bay **1** 104 (French brandy was the most commonly smuggled cargo in the 18th and early 19th centuries), Buttery Corner **1** 67, Half Moon **1** 58, Mutton Cove **1** 223, and Parson's Barn **1** 48, whilst Smugglers Cave **1** 133 is more explicit.

A good few names allude to fairs and merry-making and to the sports and pastimes practised at different periods. The recurring name Fair Field refers to fairs, often established in medieval times, at among other places, Blandford Forum **2** 90 (recorded 1279), Corscombe **4** 294, Dorchester **1** 359, and Lyme Regis **4** 397 (granted 1268). Names such as Marley **1** 177 and Merryfield **4** 128 and **4** 407, all recorded in the 14th century, probably refer to places where merry-making took place, and there was presumably dancing on Dancing Hill **3** 306, a name recorded from 1563. The names of many early inns and taverns are recorded in the larger towns like Blandford Forum **2** 89 (*le Crowne* 1465), Bridport (*le George* 1464, *the Bulle* and *the Swan* 1543, *the blacke bull* 1563), Dorchester **1** 355–6 (*la Nywehyn* 1398, and four from the 17th century), Gillingham **3** 10 (*Nywe Inne* 1411), Poole **2** 44 (nineteen from the 17th century), Shaftesbury **3** 151 (*Platellesynne* 1407, *le Swanne* 1460, *le Bell* 1480, *La Rame* 1514, *the New Inne* 1538, *the Anteloppe* 1553, *the Lamb* 1557, *lez Sterre* 1574, and six from the 17th century), Sherborne **3** 366–7 (*Juliannys Inne* 1446, *le Georges Inne* 1454, *la Hert* 1476, *the new Yn* 1535–43, *le Bell* 1538, and seven from the 17th century), Weymouth **1** 254–5 (fourteen from the 17th century), and Wimborne Minster **2** 187 (*the George* 1524, *the crystoffer* 1542, and three from the 17th century). Besides those noted already for Sherborne, *taberna de Schyrebourn* 1280, although it has no distinctive name, is the earliest reference anywhere in England to an inn or tavern. Moorbath **4** 428, a Domesday Book manor and identical with Morebath in Devon, may have been so called because there was bathing in the local springs. A spa is referred to in Nottington Spa **1** 202. Mazes are referred to in the names Miz Maze **4** 207, Tray Town **3** 40 and Troy Town Copse **1** 322. Sports are referred to in names like Bowling Green **1** 161 (recorded 1747) and Fives Courts Row **4** 361. Names referring to various other activities and pursuits include Bull Pit **3** 38 (bull-baiting), Ladies Walking Field **2** 28, and Race Down **2** 126 (recalling Blandford races held 1729–1843), whilst Landscombe Lane **3** 244 and the common Love Lane refer to courting spots, and Cuckold's Corner **4** 484 and Cuckold's Pit **4** 278 were perhaps associated with illicit lovemaking.

It is of course the case that many persons and families from different periods of history are recorded and recalled in the place-names and field-names of Dorset as of other counties. Some of the Anglo-Saxon men and women who bore Old English names may have been among the settlers who colonised Dorset from the 7th century onwards, like perhaps the *Gylla* who gave name to Gillingham, but most of them probably occupied their estates at a somewhat later date. About many of them, like the *Cana* who gave name to Canford (and possibly to Canon Hill), nothing more is known, and even some of their personal names are hypothetical. Others, like the *Here* who gave name to Herston and the *Beorn* of Barnston are actually on record as known individuals. Quite a number of these early manorial landowners were women, like the *Ælfrūn* of Afflington, the *Bēaghild* of Belhuish, and the *Tola* of Tolpuddle. Besides these individuals, known and unknown, many families are likewise recalled in the place-names. As already noted, the great landowning families, often of Norman-French descent, have left their distinctive mark on Dorset place-names in the manorial affixes of names like Kingston Lacy, Fifehead Neville and Langton Matravers. Most of these families were influential and well documented. Others, like the families that gave name to places like Bluntshay, Cruxton, Moorcourt and Silkhay, were often rather less notable and only known from local records. For the many Old English personal names, of both men and women, adduced in the place-names and field-names, as well as the personal names of other origins and the surnames of landowning families evidenced, *v.* the section on Personal Names and Surnames *infra*.

Apart from these, many place-names and field-names are more broadly commemorative of personages from different periods of history. They are characterised by their great variety. Among the many examples are Ambrose Hill **3** 305, Bedford's Camp **3** 306, Cull-Peppers Dish **1** 290, Earl's Hill **2** 246, Gaunt's Common **2** 147, Gray's Farm **4** 60, Monmouth's Ash **2** 287, Pennsylvania Castle **1** 223, Prince's Bridge **1** 360, Talbot Village **2** 24, Victoria Buildings **1** 224, and Wyndham's Oak **3** 66. To these should be added the numerous commemorative names alluding to persons, both minor names and field-names, in the parishes of Powerstock **4** 59–70, Corscombe **4** 287–305 and Halstock **4** 193–205 (in the last two parishes the result of systemic renaming by Thomas Hollis to reflect his republican views and his Puritan ideals).

Other names (along with many transferred names, *v. infra*) commemorate various historical events, among them Bloody Bank **1** 161, Clubmen's Down **3** 100 (Addenda), Coronation Plantation **1** 374, Great War & Prisoners of War Plantations **1** 374, Halswell Cliff **1** 67, Jubilee Oak **3** 279, Pageant Gardens **3** 371 and Remedy Gate **2** 269.

The various shapes of natural features in the landscape, as well as of man-made features, invite some imaginative name-giving. The parish-names Cann, Chettle and Swyre for instance, from the OE words **cann** 'can', **ceotol** 'kettle' and **swēora** 'neck', are essentially figurative descriptions of topographical features. There are many more examples of figurative naming, based on shape, appearance or some other fancied resemblance, among the minor names and field-names. Areas of woodland and clumps of trees, often distinctive features in the landscape, have names like Cocked Hat Coppice **4** 90, Double Bow **1** 162, Fiddle Wood **4** 124, Half Moon Clump/Coppice/Plantation **1** 374, **2** 5 (and many other instances), Horseshoe Clump **2** 121, **4** 497, Kidney Clump/Plantation **1** 374, **2** 61, Navvy Shovel **4** 124, Pug's Parlour **3** 131, Serpentine Plantation **4** 171, and Twelve Apostles **4** 265. Figurative descriptions such as these are of course particularly common among the field-names (of the kind Cuckolds Parlor **1** 27, Dripping Pan **1** 79, Dutch Cap **1** 336, Fryingpan **4** 378, and many more). There are also several coastal features that have colourful figurative names, among them Anvil Point **1** 57, Arish Mell **1** 123, Bacon Bluff & Hole **1** 132, The Bear **1** 140 (an offshore rock), The Blind Cow & The Bull **1** 132 (offshore rocks), Bum Point **4** 7, The Calf & The Cow **1** 111 (offshore rocks), Cow Corner **1** 125, Gad Cliff **1** 105, Half Moon **1** 58 (a bay), The Hammer **1** 132 (an offshore rock), The Man o' War **1** 133 (an offshore rock), Pig & Whistle **1** 39, Pinion Rock **1** 133, Pulpit Rock **1** 223, and Sea Horse **1** 133 (an offshore rock). Other names of this figurative descriptive kind are extremely varied in their application. Some refer to ponds or hollows (Bowldish Pond **2** 267, Cull-peppers Dish **1** 290, Punchbowl Coppice **4** 13, Veiny Cheese Pond **2** 276), others to hills (Emmetts Hill **1** 67, Marrowbone **4** 474, Weam Common Hill **4** 138), earthworks (The Grey Mare & Her Colts **4** 54, The Sole **4** 171) or other features (Featherbed Lane **4** 420, Roadstead Farm **4** 379).

Some of these figurative names have a certain humorous quality, and to them can be added a number of names that were probably intended to be jocular, among them Cuckoo Pen **1** 21 and Cuckoo Pound **1** 38, Lollipop Farm **2** 24, Melancholy Lane **1** 76, Nowhere **3** 385, Parson's Barn **1** 48, Ring-Bum Gardens **1** 68, Scratch Arse **1** 39, and Tatten **2** 167. Other names are perhaps simply whimsical, facetious or fanciful, such as Billy Wilkins **4** 99, Bournes **3** 36, Clown Hill **1** 47, Dancing Ledge **1** 38, The Fire Dog **1** 374, Heedless William's Pond **1** 370, Lardenhall **3** 118, Magging Stone **4** 378, Nundico **1** 164, Pineapple Farm **4** 326, Snail Creep **1** 370, and Stapehill **2** 228. There are also many names that are, outwardly at least, complimentary (an element of irony may sometimes have to be reckoned with). Examples include Coking Farm **3** 71 (from ME *cokaygne* 'imaginary land of abundance and bliss'), God's Blessing Farm **2** 154, Golden Bowl **1** 22, Happy Valley **3** 39, Mount Pleasant **1** 39 (and many other instances), Paradise **2** 118 (also frequent), Peace Farm **4** 449, Plenty House **4** 449, and Remedy Cottages **4** 158. Derogatory names (no doubt sometimes referring to poor or unproductive land) are if anything more numerous, among them Addle Path **2** 21, Break Heart Hill **4** 177, Cain's Folly **4** 415, Giddy Green **1** 178, Hill Wood (earlier *Helle*) **2** 30, Hunger Hill **2** 40, **3** 69 and Hungry Down/Hill **2** 108, **4** 64, Labour-in-Vain **4** 39, Long Friday **2** 109, Mount Misery **1** 38, Purgatory **1** 296, The Spittles **4** 398, Starve Acre Clump **4** 13, Starve Acre Lane **4** 226, and Vermin Lane **2** 69. Many more instances of all these types of names, humorous, whimsical, complimentary and derogatory, are of course to be found among the field-names.

In contrast to the decidedly bold and expressive nature of some early naming (as for example in Arish Mell and Scratch Arse *supra*), attitudes to a few names in more recent times may just occasionally seem rather prudish. Several such hints of disapproval of certain names, considered by some wanting in delicacy, have been noted in passing. Foul Hill in Cann **3** 95 has been altered to Foyle Hill on recent maps, a farm formerly called *Maidshole* in Marnhull **3** 174 has apparently been renamed, there was an attempt to have Sodom Lane in the same parish **3** 175 renamed (though this proved unsuccessful), Shitterton in Bere Regis **1** 276 often appears as Sitterton, and the current form of the parish-name Powerstock was officially adopted because the usual earlier form Poorstock was thought by GWR to be a slur on their rolling stock when they came here in 1857. More recently, the roadsign for Piddle Lane in Cerne Abbas **4** 137 is reported to have mysteriously disappeared. This last (though hopefully the roadsign has now been restored) certainly recalls the 1956 controversy regarding the name Puddletown **1** 314, when there was fierce local protest over the move by Dorset County Council to change the *Puddle-* to *Piddle-* in order to better conform with the other village names in the Piddle valley. In this case of course the 'nicer sounding' name (as it was called) won the day and the change was resisted: historically, both spellings are recorded early, and the form *Piddle Towne alias Puddle Towne* from 1665 really says it all.

Folk etymology, and rationalisation, result in some remarkable transformations, some often quite surprising and amusing. Examples might include Caldron **1** 5 (this was *Cauldon* 1557 'cold or cabbage hill'), Devil's Brook (the Celtic r.n. *deuelisc* that gives name to Dewlish **1** 203), Holm & Ivy Farm **3** 93 (*la Holmene Theuele* 1341 'the hollybush'), Necklace **1** 68 (*Nicholesdon* 1306 'hill of St Nicholas'), Suckthumb Quarry **1** 224 (*Sutcombe* 1608 'south valley'), Wavering Lane **3** 23 (*Wyuelynglane* 1313 'lane at the beetle-infested place'), and Wizard Bridge **3** 332 (*Weseford* 1204 'swamp ford'). Most of these rationalisations are relatively recent, but others are old, no doubt often the result of monkish etymologising and invention: Cerne Abbas **4** 134 (interpreted as 'I see God' by early chroniclers), Dorchester **1** 346 (obscure second element of the original British name *Durnovaria* interpreted as OE *-ware* 'dwellers'), Shaftesbury **3** 139 (translated as *oppidum Paladur* 'fort of the shaft' by early chroniclers), and Wimborne Minster **2** 183 (interpreted as *vini fons* 'fountain of wine' in a 9th-century Saint's Life). Some names have been subject to more recent antiquarian etymologising, as in Durweston **2** 92 (*Derwinestone*, the medial *-wines-* taken to refer to the DB vineyard here), Frome Vauchurch **4** 96 (misinterpretation of first element of affix), Gorwell Farm **4** 52 (first element taken to refer to bloody sacrifice), and Poorwood **4** 62 (ancient name *Power* associated with poverty). A few folk etymologies have influenced church dedications, as in Margaret Marsh **3** 164, Melbury Osmond **4** 214, Osmington **1** 212, and Whitchurch Canonicorum **2** 79. Other names where some element of rationalisation or reinterpretation has influenced the modern form include Battleford Farm **4** 273, Beaulieu Wood **3** 241, Bowerswain **2** 279, Buckhorn Weston **3** 84, Cancer Drove **3** 336, Causeway **4** 256, Chetterwood **2** 141, Cock Crowing Corner **4** 430, Coltleigh Farm **4** 306, Cuckoo's Hill **4** 28, Devils Wood Lane **3** 261, Font Le Roi **3** 331, Frying Croft **4** 55, Hartfoot Lane **3** 209, Hawkcombe Lane **3** 99, Hayward Bridge **3** 61, Lamplands Copse **3** 229, Marriage Lane **3** 161, Monmouth's Hill **3** 224, Peaceful Lane **3** 386, Pistol **4** 300, Pop Mallers Coppice **3** 245, Silverlake **3** 305, Sixpenny Handley **3** 113, Terrace Farm **3** 74, Vale Acre **2** 195, Venus Style **4** 249, Wadmill **3** 75, Waverlawns **3** 378, Whistle Barn **3** 323, and Wolfridge **3** 50. It is interesting to note that several such folk etymologies or rationalisations have led to fanciful legends or at least misleading conclusions. Examples include Askerswell **4** 47 ('Ōsgār's spring', held locally to be 'newts' well' from Do dialect *asker*), Blood-Way Coppice **2** 245 (earlier *Fludway*, linked with traditions of battles between keepers and deer-stealers), Coney's Castle **4** 447 (from **coni** 'rabbit', associated with the legend of Conig a Danish chieftain), Gallica Bridge **4** 219 ('gallows acre', recent Latinate rationalisation suggesting Roman origins), the f.n. Hitchens in Leigh par. **4** 208 (curiously associated with a meeting of witches), Hurpston **1** 96 (tradition of harp-like sounds emanating from the Harp Stone here when the wind blows), Kingsettle **3** 50 (from **setl** 'seat, dwelling', linked with a tradition that king Alfred rested here on his way to the battle of Edington), Kingstag **3** 349 (from **staca** 'stake', taken to refer to the legend of the white hart), Pease Leaze **3** 192 (from a surname *Puselegh*, linked with a tradition that Charles I ate green-pease here), Shave Cross **4** 405 (from **sceaga** 'wood', thought to be where monks had their heads shaved), Slaughter Gate **3** 23 (from a surname *Slaunder*, associated with an 11th-century battle), and Yetminster **4** 228 (tradition that first part of name, from OE pers.n. *Ēata*, refers to a gate into Blackmoor Forest).

Folk etymology, together with a certain amount of antiquarian etymologising, results in the phenomenon known as back-formation, this being responsible for the creation of a number of river-names in relatively recent times. Thus Crane came to be the name of the stream at Cranborne **2** 205 because the village name ('crane stream, stream frequented by cranes') came to be understood as 'stream called Crane'. Among the other river-names formed in this way are Allen (from *Aleyn Bridge* 'Ealdwine's bridge' in Wimborne Minster **2** 187), Asker (from Askerswell **4** 47, 'Ōsgār's spring or stream'), Brit (from Bridport **4** 353, 'harbour associated with Bredy'), Chid (from Chideock **4** 375, Brittonic name 'wooded'), Hooke (from Hooke **4** 56, 'river-bend'), Jordan (from Jordan Hill 'river-bend hill' in Preston **1** 231), Simene (from Symondsbury **4** 425, 'Sigemund's hill or barrow'), Trent (from Piddletrenthide **4** 167, 'thirty (OFr *trente*) hide estate on R. Piddle'), Win (from Wimborne Minster **2** 183, 'meadow stream'), Win (from Winfrith Newburgh **1** 174, Brittonic name 'white stream'), and Wrackle (from Wrackleford **1** 373, 'Wrǣcwulf's ford'). For further details and discussion of these names, *v.* section on River-Names *supra*.

The many transferred names in Dorset are of great interest. Some of them indicate the remoteness or small size of a settlement, some are biblical, some historical or commemorative, others perhaps jocular or whimsical. Most are probably relatively recent in origin, but a few are early, such as Alcester **3** 153 (from 1433), Chantmarle **4** 126 (from 1280) and Organ Ford **2** 34 (from 1194). Names commemorating historical events include Alma Cottages & Balaclava Bay **1** 220, Botany Bay **1** 125, Bunkers Hill **1** 148 (and three other instances), Ladysmith **3** 385, Portobello **4** 150, and of course Waterloo **3** 23 (and several other instances). The names of streets and buildings in Bovington Camp (in Wool parish **1** 189) allude to scenes of fighting in the First World War. Names making a biblical allusion include Babylon Hill **3** 295, Gog & Magog **3** 270, Holy City **4** 278, Jericho **4** 397, Jordan **4** 397, Mount Ararat **3** 258, and Zoar **2** 97 (*v.* Addenda). There are several names transferred from places in the British Isles, among them Alderney Heath **2** 1 (*v.* Addenda), Anglesey Cottage **3** 284, Cambridge Wood **1** 47, Lincoln Cliff **1** 47, Londonderry **2** 109, **2** 198, **3** 223, Lonsdale Wood **1** 47, The Mumbles **4** 416, Notting Hill **2** 50, Nutcombe Wood **3** 36, Oxford Wood **1** 48, Pimlico **4** 246, and St Michael's Mount **1** 49. A few of these, such as Charing Cross **4** 189 (and three other instances), Little London **3** 95, **4** 32, Little Lonnen **2** 139, Oxford Street **3** 119, and Piccadilly (Lane) **2** 179, **3** 386 are no doubt jocular or whimsical. Many other transferred names are from various countries and places overseas, some perhaps related to migration, others perhaps referring to remoteness, among them Ballarat **1** 376, California **1** 57, Canada **2** 12, Canberra **1** 234, Carolina **4** 496, Egypt **2** 232, France **2** 117, Hanover **3** 248, Lucerne Lake **4** 99, Malaya **2** 6, Maryland **1** 47, Newfoundland **1** 92 (and five other instances), New Zealand **4** 362, Normandy Farm **2** 132, Pennsylvania **1** 223, Rialto **1** 49, Russia Fds **3** 253, Trinidad Gardens **2** 2, and Venetia Park **1** 50. There are also a few literary allusions among the transferred names, including the parish-name Branksome **2** 1 (Scott's setting for *Lay of the Last Minstrel*), Lilliput **2** 40 (Swift's imaginary country), Mellstock **4** 421 and Melstock **1** 370 (Hardy's fictional name for Stinsford, just one of his many inspired inventions), Moonfleet Hotel **4** 17 (from Falkner's novel), and Tulgey Wood **2** 7 (from Carroll's poem *Jabberwocky*). Such names add an exotic touch to the map of Dorset with all its wonderful, ancient and historical, names.

NOTES ON THE PHONOLOGY OF DORSET PLACE-NAMES

OE *æ* usually becomes ME *a* as in Ash **2** 116, Ashcombe **3** 317, Ashington **2** 2, Ashmore **2** 201 (**æsc**), Haselbury Bryan **2** 100 (**hæsel**), Moorbath **4** 428 (**bæð**), Small Mouth **1** 269 (**smæl**), Watercombe **1** 173 (**wæter**), etc. There are occasional ME *e* spellings for Ash, Ashmore and other names in **æsc**, also for Haselbury Bryan. Names in **æsc** have occasional ME spellings in *Ai-*, *Ay-*, *Ei-*.

OE *ǣ* usually appears as ME *e*, as in Eye Mead **2** 166 (**mǣd**), Hethfelton **1** 146 (**hǣð**), Main Down **2** 224 (**mǣne**), Seacombe **1** 66 (**sǣ**), etc. When shortened in compound names it appears as ME *e* or *a* as in Almer **2** 55 (**ǣl**), Blandford Forum **2** 87 (**blǣge**), Blashenwell **1** 8 (***blǣcen**), Clinger **3** 241 (**clǣg**, the development to *i* is late), Stratton **1** 373 (**strǣt**), Watcombe **4** 123 (**hwǣte**), etc. Hethfelton **1** 146 (**hǣð**) has only *e* spellings, and Alton Pancras **4** 122 (**ǣ-wyll**) only *a*. Names in **hǣme** (Shilvinghampton **4** 28, Stubhampton **2** 245, Witchampton **2** 260) have *e* in DB, then only *a* spellings (no doubt influenced by **hām**). Melbury Abbas **3** 129 and Melplash **4** 318 (**mǣle**) have *e* spellings (the latter has just one *a* spelling and also occasional *u* and *i* from the influence of **mylen**). The simplex name Slape **4** 320 (**slǣp**) has early *e*, then *a* from 1288, but the identical Slepe **2** 35 has first *a*, later *e*.

OE *a* before nasals remains as ME *a* in many names such as Clinger **3** 242 (**hangra**), Cranborne **2** 205 (**cran**), Hambledon **3** 61 (**hamol**), Hammoon **2** 99, Hampreston **2** 224, and Hamworthy **2** 20 (**hamm**), and Lambrook **4** 317 (**lamb**), Mangerton **4** 318 (**mangere**), and Sandford Orcas **3** 389 (**sand**). The simplex name Cann **3** 90 (**canne**) also has only ME *a*. Names with ME *o* spellings from the rounding of *a* include Buckland Newton **3** 239, Portland **1** 217, Studland **1** 43 (**land**), Longfleet **2** 26, Langton Matravers **1** 33 (**lang**), Sandpit **4** 259 (**sand**) and Winterborne Stickland **2** 131 (**lane**). There are occasional ME *au* spellings for some of these names, among them Mangerton, Portland, Sandford Orcas, Sandpit and Studland.

OE *a* in other positions remains as ME *a* in names like Acton **1** 34 (**tacca**), Alderholt **2** 195, Axnoller **4** 241 (**alor**), Catsley **4** 289 (**catt**), Galton **1** 139 (**gafol**), Mappowder **3** 259 (**mapuldor**) and Stalbridge **3**

280 (**stapol**). Axnoller also has ME *o* spellings, and Mappowder has frequent ME *au*.

OE *ā* normally becomes ME *ō* (from the 13th century) as in names like Broadwey **1** 199 (**brād**), Frome Vauchurch **4** 96 (**fāh**), Goathill **3** 382 (**gāt**), Child Okeford **3** 59 and Okeford Fitzpaine **3** 178 (**āc**), Stoborough **1** 73, Stone **2** 167 (**stān**), and Thorngrove **3** 17 (**grāf**). In many compound names *ā* is shortened to *a*, as in Ansty **3** 209 (**ān-stiga**), Bradford Abbas **3** 294, Bradpole **4** 251 (**brād**), Brockhampton **3** 241 (**hām-tūn**), Gillingham **3** 9 (**hām**), Halstock **4** 193 (**hālig**), Hanford **3** 35 (**hān**), Laverstock **4** 339 (**lāwerce**), Stanbridge **2** 149, Stanton St Gabriel **4** 414 (**stān**), and Waddon **4** 30 (**wād**). There are ME *au* spellings for Stanton St Gabriel and Waddon. Bradpole and Hanford have occasional ME *e* and *o* spellings, whilst Breston **1** 221 (**brād**) has *e* from 1582 (for this development, cf. Breadstone Gl **2** 214).

OE *e* remains as ME *e*, as in names like Belchalwell **3** 180 (**belle**), Black Venn **3** 68 (**fenn**), Duntish **3** 242 (**etisc**), Encombe **1** 12 (**henn**), Enmore **3** 48 (**ened**), Kingsettle **3** 49 (**setl**), Nettlecombe **4** 2 (**netel**), and Ringstead **1** 212 (**stede**). Names in **feld** such as Blynfield **3** 90 and Hillfield **4** 157 have ME *e* but also spellings in *ei*, *ey* and *y*. There is late raising of *e* to *i* in Hincknoll **4** 317 (**henn**, *i* spellings from 1605).

OE *ē* remains as ME *ē*, but with shortening in Smedmore **1** 86 (**smēðe**).

ME *er* often becomes *ar*, as in the spellings for Barcombe **4** 123 (**bere**), Barton Hill **3** 153 (**bere-tūn**), Berne **4** 440 (**bere-ærn**), Berwick **4** 41 (**bere-wīc**), Marshwood **4** 400 (**mersc**), West Parley **2** 231 and Parnham **4** 243 (**peru**).

OE *ea* gives both ME *e* and *a* spellings in many names, such as Adber **3** 393 (**bearu**), Burngate **1** 130 (**geat**), Chetnole **4** 189 (OE pers.n. *Ceatta*), Earnley **4** 290 (**earn**), Farrington **3** 38 and Fernbrook **3** 48 (**fearn**), Hurpston **1** 96 (**hearpere**), Shaftesbury **3** 139 (**sceaft**/pers.n *Sceaft*) and Warmwell **1** 170 (**wearm**). Dorchester **1** 347 (**ceaster**) and Sixpenny **3** 104 (**Seaxe**) have only ME *e* spellings, and Buckshaw **3** 384 and Bushey **1** 9 (**sceaga**) only *a*. A few names have the usual ME *e* and *a* but also some *i*/*y* spellings, among them Farnham **2** 221 (**fearn**) and Worgret **1** 74 (**wearg**).

OE *ea* from breaking before *l* + consonant, a feature of the WSax dialect, is evidenced in names containing elements such as **cealc**, **ceald**, **cealf** (all with palatalised initial consonant), **eald**, **healh**, **sealh** and **wealt**. East & West Chelborough **4** 76 (probably from **cealc**) have ME *Chal-*, *Chel*(*e*), *Chau-*. Belchalwell **3** 180, *Chaldecots* **1** 85 and *Chaldwel* **4** 94 (**ceald**) have ME *Chald*(*e*)-, *Chelde-*, and with just a single *Cald*(*e*)- spelling for each name. Chaldon Herring **1** 108 (**cealf**) has mostly *Chauue-*, *Chauve-*, *Chave-*, *Chalve-*, with just three *Cauve-*, *Calve-* spellings. Eldon Hill **4** 489, Yelcombe **4** 138 and Yeldon **4** 433 (**eald**) have *Yel*(*d*)-, *Eld-*. Wraxall **4** 73 (**healh**) and Walton **2** 189 (**wealt**) have ME *a*, with just a single *e* spelling for each name. Silton **3** 64 (**sealh**) has mainly ME *e*, but with many *a*, and then with *i*/*y* spellings from 1327 (raising of *e* to *i*). In simplex names from **healh**, the dative sing. form *heale* when lengthened to *hēale* gives spellings in *i*/*y* alongside *e* in Hile **3** 181 and Hyle **3** 370.

OE *ēa*, when shortened in compound names, gives both ME *a* and *e* spellings in many names such as Redcliff **1** 73 (**rēad**), Shapwick **2** 176 (WSax **scēap**), Sixpenny Handley **3** 113 (**hēah**), and Yetminster **4** 228 (OE pers.n. *Ēata*). Some names have only ME *e*, among them Benville **4** 289 and Binnegar **1** 146 (**bēan**), Easton **1** 219 and Eastington **1** 64 (**ēast**), the last name with a single *i* spelling. Notton **4** 103 (**nēat**) and Rodden **4** 5 (**rēad**) have *a* and *e* but with late ME *o* that persists in the modern form. Simplex names retaining the long vowel include Eype **4** 427 (**gēap**, ME *e* spellings) and Leigh **2** 137 (**lēah**, ME *e* with some *i*/*y*). Bincombe **1** 197 (**bēan**) has ME *e*, *ie*/*ye*, *i*/*y* spellings, Hinton St Mary **3** 159 has ME *e* with *ey* in 1327 and *i*/*y* from 1575, and Redlane **3** 1 has ME *e* and *i*/*y* spellings (all with raising of *e* to *i*).

OE *eo* usually appears as ME *e*. Some names have occasional ME *eo*, *o*, *u* spellings, among them Charlton Marshall **2** 11 (**ceorl**), Hartley **4** 162 (**heorot**), Seaborough **4** 492 (**seofon**), Sturthill **4** 488 (**steortel**), Woodstreet **1** 190 (**steort**) and Yellowham **1** 320 (**geolu**). Other names have occasional ME *i*/*y* spellings, including Chettle **2** 290 (**ceotol**), Melcombe Horsey **3** 216 and Melcombe Regis **1** 251 (**meoluc**), Netherbury **4** 311 and Netherhay **4** 259 (**neoðerra**). Spettisbury **2** 63 (**speoht**) has both *i*/*y* and *ey* spellings. Charlton **1** 339 and Charlton Marshall have ME *a* spellings from c.1300, and Hartley and Spettisbury also have occasional *a*. Names from **beorg** like Charborough **1** 59, Chelborough **4** 77 and Symondsbury **4** 425 have ME *u* spellings from the late 13th century, no doubt partly through influence from **burh/byrig**.

OE *ēo* usually appears as ME *e*, as in Beaulieu **3** 241 (**bēo**), Fleet **4** 16 and Longfleet **2** 26 (**flēot**), Lewcombe **4** 77 and Luccomb **4** 318 (**hlēo(w)**), also in Hampreston **2** 224 and Preston **1** 230, **2** 252, **3** 125 (**prēost**) where early shortening has taken place. Beaulieu and Fleet have occasional *eo*, *oe*, *o*, *i*, *ue*, *u* spellings. Radipole **1** 239 (**hrēod**) has early ME *e* but also *a* from 1237 (probably through influence from **rēad**), Rock Lea **2** 21 has ME *e* and *o*, Rodmore **3** 350 has ME *o*, Red Mead (f.n.) **4** 464 has *u* (also all from **hrēod**). Names from **trēow** like Dunster **4** 401 and Loxtree **4** 85 have ME *-trewe*, *-trowe*, *-tru*(*e*).

OE *i* remains as ME *i*/*y*, as in names like Bindon **1** 130 (**binnan**), Broadwindsor **4** 254 (**windels**), Clifton Maybank **4** 191 (**clif**), Filcombe **4** 376 (**fileðe**), Lily **4** 371 (**hlid**), Minterne Magna **4** 160 (**minte**), Stinsford **1** 366 (**stint**), and Sydling St Nicholas **4** 174 (**hlinc**). Some names have occasional *e* spellings, among them Milton Abbas **3** 219 (**middel**), Purbeck **1** 1 (***bic**), Ringstead **1** 212 (**hring**), Ryme **4** 218 (**rima**), and Shitterton **1** 276 (**scitere**). For Feltham **3** 65 (**fileðe**) *e* is first recorded in 1811. There are two ME *u* spellings for Lyscombe **3** 203 (**lisc**), and the *u* spellings for the places named from R. Piddle (***pidele**), including Affpuddle **1** 288, Piddlehinton **1** 309, Piddletrenthide **4** 167, Puddletown **1** 314, Tolpuddle **1** 331 and Turners Puddle **1** 294, first appear in the 13th century. ME *i* when lengthened in an open syllable in the oblique cases of **clif** gives forms such as Bincleaves **1** 255, Cleevehill **4** 277 and Southcleeves **3** 6.

OE *ī* remains, or is shortened in most compound names, with ME *i*/*y* spellings for both. The long vowel is retained in names such as Hyde **4** 472 (**hīd**), Wyld **4** 457 (**wīl**), and in the first element of Fifehead Magdalen **3** 6 and Fifehead Neville **2** 95 (**fīf**). Names with the shortened vowel include Linton **4** 5 (**līn**), Picket **4** 333 (**pīc**), Swanage **1** 52 (**wīc**), Walditch **4** 472 (**dīc**), Whitchurch Canonicorum **4** 438 (**hwīt**), Whitcombe **1** 259 (**wīd**) and Witchampton **2** 260 (**wīc**). There are occasional ME *e* spellings for Hinton Martell **2** 146 (**hīwan**), Shapwick **2** 176 and Wyke Regis **1** 267 (**wīc**), Sydling St Nicholas **4** 174 (**sīd**) and Withyhook **4** 206 (**wīðig**), also for Sherborne **3** 355 and Sherford **2** 60 (**scīr**), in the last two names persisting in the modern form. Pilsdon **4** 412 (**pīl**) has occasional ME *u*, *yu* and *o* spellings, whilst Fifehead

Magdalen and Neville have some ME *u*, *uy* and *e* spellings for the second element (**hīd**).

OE (WSax) *ie* (from the *i*-mutation of WSax *ea* broken before *l* + consonant) gives ME *i/y* spellings, alongside *e* from non-WSax forms, in names from **wiell(a)**, **will(a)**, **wyll(a)**, **well(e)**, among them Askerswell **4** 47, Holwell **2** 210, Ulwell **1** 56, Warmwell **1** 169 and Woolcombe **4** 211. In some names from this element ME *u* and *o* spellings (from late WSax **wyll(a)**) are also frequent, as in Warmwell **1** 169, Wool **1** 188, Woolbridge **1** 148, and the two instances of Woolcombe **4** 113 and **4** 211. Alton Pancras **4** 122 (**ǣ-wiell**, **ǣ-well**) has late OE *Awul-*, ME *Awol-*, *Awel-*, and Toller Whelme **4** 291 (**ǣ-wielm**, **ǣ-welm**) has late OE *aewylm*, ME *Euulme*, *-ewelme*. Feltham **3** 65, Filcombe **4** 376, Filford **4** 316 (**fileðe**, **feleðe**) all have ME *i/y*. In other names with WSax *ie*, Jordan Hill **1** 231 (**cierr**, **cerr**) has *u*, *e* and *o* spellings, Yard Fm **4** 405 and Yardgrove **3** 172 (**gierd**, **gerd**) have mainly ME *e* but some *i/y* and *u*.

OE (WSax) *īe* gives ME *i/y* spellings, also u, and alongside *e* from non-WSax forms, in names such as Evershot **4** 84 (***scīete**, ***scȳte**), Shipton Gorge **4** 487 (***scīep**,***scȳp**), Steeple **1** 95, Iwerne Steepleton **2** 114 and Winterborne Steepleton **4** 43 (**stīepel**, **stȳpel**, **stēpel**). Hyford **1** 178 (**hīeg**) has only i/y spellings, Nyland **3** 42 (**īeg-land**) and Pymore **4** 469 (**pīe**) have mostly *i/y*, Brownsea **1** 43 and Pipsford **4** 291 (**īeg**) have *ie*, *ye* beside *ei*, *ey*.

OE *o* remains as ME *o*, as in names like Alderholt **2** 195 (**holt**), Cattistock **4** 125 (**stoc**), Dogbury **4** 161 (**dogga**), Frogmore **4** 112 (**frogga**), Holditch **4** 494 (**hol²**), Holnest **3** 336 (**holegn**) and Woodcutts **3** 116 (**cot**). With lengthening, spellings are also ME *o* in names such as Gorewell **4** 52 (**gor**), Church Knowle **1** 87 (**cnoll**), East Stoke **1** 145 (**stoc**) and Woodsford **1** 186 (**ford**). Folke **3** 330 (**folc**) and Throop **1** 290 and **1** 332 (**þrop**) have *o* spellings but also late ME *ou/oo*. Folke also has two ME *a* spellings. ME **croft** often develops to *craft*, *crate* as in Haycraft **1** 64, Moo Crate (f.n.) **4** 25, Whipscrate (f.n.) **4** 301.

OE *ō* remains, or is shortened in most compound names, with ME *o* spellings for both. Names with shortening include Brockington **2** 280 (**brōc**), Colmer **4** 401 (**cōl**), Loscombe **4** 61 (**hlōse**) and Motcombe **3** 47 (**mōt**), but in some names there is shortening to *u*, as in occasional ME spellings for Brimbley **4** 338 (**brōm**), Buckland Newton **3** 239 and

Buckland Ripers **1** 239 (**bōc-land**), Oborne **3** 354 (**wōh**) and Studland **1** 43 (**stōd**), in some of these names persisting in the modern form. Many names have occasional spellings in *ou*, *ow*, *oo* alongside *o* (from c.1300), including Bradpole **4** 251 and Poole **2** 41 (**pōl**), Buckland Newton and Ripers, Hooke **4** 56 and Withyhook **4** 206 (**hōc**), Morden **2** 58 and Moreton **1** 135 (**mōr**), Strode **4** 321 and Stroud **2** 76 (**strōd**).

OE *u* usually appears as ME *o* or *u*, the two spellings (well represented in modern forms) alternating for many names, among them Bokerly **2** 235 (**bucc**), Bowood **4** 314 (**bufan**), Butterwick **3** 331 (**butere**), Compton Abbas **3** 99 (**cumb**), Duncliff **3** 74 (**dunn**), Huntingford **3** 14 (**hunta**), Romford **2** 218 (**hrung**) and Woodlands **2** 284 (**wudu**). Nutford **2** 112 (**hnutu**) usually has ME *u*, *o* but there are two 13th-century spellings in *e*, *eu*. There is late fronting of *u* to *i* in Biddlesgate **2** 207 (OE pers.n. **Butel*).

OE *ū* develops to the sound usually spelt ME, ModE *ou*, *ow* as in names like Bishops Down **3** 331 (**dūn**), Cowgrove **2** 166 (**cū**), Small Mouth **1** 269 (**mūða**) and Southbrook **1** 277 (**sūðan**). When shortened in the first element of compound names, ME spellings are usually *u*, but also with occasional *o* and *ou*, as in Burngate **1** 130 (**brūn**), Duncliffe **3** 74 and Duntish **3** 242 (**dūn**), Plumber **3** 349 (**plūme**), Purbeck **1** 1 (**pūr**) and Ulwell **1** 56 (**ūle**). With shortening in the final element of compounds, the ME spelling is *o*, as in names from **dūn** and the many names from **tūn**, such as Pilsdon **4** 412, Waddon **1** 30, Moreton **1** 135 and Osmington **1** 211. Puddletown **1** 314 usually has ME *o* but also *ou* from 1270. The development of the usual *-ton* to *-stone* (as if from **stān**) in names like Rempstone **1** 16, Shillingstone **2** 238 and Winterborne Zelstone **2** 68 is only recent. Some names from **dūn** like Bindon **1** 130, Chaldon Herring **1** 108 and Eggardon **4** 47 have mainly ME *o* but occasional spellings in *u* and *ou*. Other **dūn** names like Bowden **3** 42, Farrington **3** 38, Linton **4** 5, Morden **2** 58, Norden **1** 16 and Sutton **2** 267 all have ME *-don*, but have been influenced in later spellings (and sometimes early ones) by **denu** and **tūn**. There is fronting of *u* to *i* in Dimstone Hill **4** 307 (**dūn**).

OE *y* and *ȳ* give ME spellings in *u*, *i*/*y*, and occasionally *e*, in many names such as Boveridge **2** 208 (**hrycg**), Gussage St Michael **2** 143 and Gussage All Saints **2** 276 (**gyse**), Kimmeridge **1** 84 (**cȳme**), Litton Cheney **4** 23 (***hlȳde**), Marley **1** 177 (**myrge**), Stalbridge **3** 280 (**brycg**) and Stubhampton **2** 242 (**stybb**). Some names have ME *u* and *i*/*y* but with no *e* spellings, among them Colehill **2** 137 and Pamphill **2** 163 (**hyll**), Pitt **4** 443 and Sandpit **4** 259 (**pytt**), Plush **3** 245 (**plysc**) and Rushton **1** 147 (**rysc**). Hurst **1** 135 (**hyrst**), Marnhull **3** 166 (**hyll**) and Turnworth **2** 254 (**þyrne**) have mostly ME *u* (persisting in the modern forms), and other names have only ME *u*, among them Furze Hill **2** 138 (**fyrs**), Hillfield **4** 157 (**hyll**, with *i*/*y* from 1535) and Perry **2** 195 (**pyrige**). Names with only, or mainly, ME *i*/*y* spellings include Bowridge **3** 11 and Ridge **4** 377 (**hrycg**), Frith **3** 281 (**fyrhð**), Kingston **1** 15 and Kingsettle **3** 49 (**cyning**), Kington Magna **3** 41 (**cyne-**), Purcombe **4** 405 (**pyrige**), Ryall **4** 443 (**ryge**), Tincleton **1** 329 (***tȳnincel**) and Trill **3** 291 (**tyrl**). Bexington **4** 36 (**byxen**) has mostly ME *e* spellings (persisting in the modern form), whilst other names with many *e* spellings, beside *i*/*y* and/or *u*, include Beaminster **4** 239 and Yetminster **4** 228 (**mynster**), Holnest **3** 336 (**hyrst**), Milborne St Andrew **1** 306 and Milborne Stileham 1 282 (**myln**).

OE *b* becomes *p* in Cripton **1** 261 (**cribb**, *Cryp-* 1530) and in the ME spellings of Purbeck **1** 1 (***bica**, *-pyk* 1356–1360) and Stubhampton **2** 245 (**stybb**, *-p-* 1280–1580). ME *p* becomes *b* in Bowridge Hill **3** 11 (**pohha**) and Briar's Wood **4** 225 (**prior**).

OE *c* becomes *g* medially in names such as Bagber **3** 189 (OE pers.n. *Bacca*), Blagdon **2** 207 (**blæc**), Bugden **3** 114 (**bōc**[1]), Cogdean Elms **2** 16 (**cocc**[2]), Doghill **4** 413 (**docce**) and Hogboro **4** 428 (**hōc**).

OE *d* becomes *t* medially in Attisham **4** 255 (OE pers.n. *Æddi*), Farrington **3** 38 and Linton **4** 5 (**dūn**), Litton Cheney **4** 23 (***hlȳde**), Tincleton **1** 329 (**denu**) and Whitcombe **1** 259 (**wīd**), and finally in Povert **2** 103 (**ford**). Epenthetic *d* appears finally in Wyld **4** 457 (**wīl**) and in spellings for Ball Coppice **2** 202 (**ball**).

OE initial *f* is voiced to ME *v* in the spellings of a number of names, among them Black Venn **3** 68, Vale Acre **2** 195, Vembury **4** 45, Venn **4** 429 (**fenn**), Farnham **2** 221 and *Ferngore* **3** 49 (**fearn**), *Fifash* **2** 71, Fifehead Magdalen **3** 6, ~ Neville **2** 95 & St Quintin **2** 96 (**fīf**), Filcombe **4** 376 (**fileðe**), Fox Allers Coppice **3** 228 and Fox Barrow **4** 21 (**fox**), and Frome Vauchurch **4** 96 (**fāh**). This voicing to *v* occurs later in some names, for instance in the ModE spellings for Benville **4** 289 (**feld**), Fitzworth **1** 13 (**fitt**), Forty Pond **3** 3 (**forð-īeg**), Venn Hill **4** 265 (**fenn**), Verne Yeates **1** 220 (**f(i)ergen**), Verwood **2** 256 (**fæger**), Vulscombs **4** 283 (**fugol**), and Waverlawns **3** 378 (**furlang**). In some names there is a further development of *V-* to *W-* in ME and ModE spellings, as in *Ferngore* **3** 49 (**fearn**, *W-* 1332), Fifehead Neville **2** 95 (**fīf**, *W-* 1280), Frome Vauchurch **4** 96 (**fāh**, *W-* 1399), Vale Acre **2** 195 (**fenn**, *W-* 1382), Verwood **2** 256 (**fæger**, *W-* 1575, 1621), and Writh **3** 132, **3** 333 (**fyrhð**, *W-* 1839). The dialectal development of **croft** to *crate*, as in names like Crate Wood **3** 234 and Pettycrate **4** 379, is again to be noted.

OE initial *g* is lost in Lydford **3** 94 (**glida**). Medial *-gc-* becomes *k* in Lankham **4** 71 (**lang**, **cumb**). The dialectal development of **sceaga** (ME *shawe*) to *shave*, as in Shave Cross **4** 405, Shave Hill **3** 87, Shave Lane **3** 83, is to be noted.

ME medial *l* is vocalised to *u*/*w* in Lowbrook **3** 181 (OE pers.n. *Lolla*/*Lulla*) and Mappowder **3** 259 (**mapuldor**), also in the ME spellings for Chaldon Herring **1** 108 (**ceald**).

ME medial *n* sometimes becomes *nd*, as in Handfast **1** 44 (**hān**), Standpitts Lane **3** 23 and Standridge **2** 213 (**stān**). ME medial and final *n* sometimes becomes *m*, as in Dimstone **4** 307 (**dūn**), East Holme **1** 82 (**holegn**, cf. dial. *holm*), Hambury **1** 131 (*hēan* < **hēah**), Limbury **4** 317 and Lymburgh's Fm **3** 169 (**līn** or **lind**), Wimborne Minster **2** 183 and Wimborne St Giles **2** 2 263 (**winn**[1]), and in the ME forms for Stanbridge **2** 149 (**stān**). The reverse of this, a change of ME medial *m* to *n*, is seen in Bronkham **1** 376 (**brōm**), Heniford **4** 211 (**Humber**), and in the ME spellings for Cranborne **2** 205 (**cran**), Henbury **2** 47 (*hēan* < **hēah**) and Linton **4** 5 (**līn**). ME *n* is lost before a labial consonant in High Stoy **4** 163, West Stafford **1** 243, Stallen **3** 325, Stoborough **1** 73, Stoford **4** 192 and Storridge **4** 279 (**stān**). Epenthetic *n* appears medially in Parnham **4** 243 (**peru**), evidenced from 1431. A svarabhakti vowel develops between *r* and *n* in Coringdon **4** 289, Currendon **1** 57 and Farrington **3** 38.

OE initial *Sc-* (ME *Sh-*) occasionally develops to *Sk-* (a feature in Devon p.ns., *v.* PNDe xxxv), as in Scully (f.n.) **3** 29, Skitterlinch (f.n.) **2** 181, *Skuttbargh* (f.n.) **3** 227.

OE *t* becomes *d* medially in Piddles Wood **3** 192 (OE pers.n. *Pyttel*). Initial *t* is lost in Acton **1** 34 (**tacca**), and medial *t* is lost in several late **tūn**-names, among them Anderson **2** 71, Basan **1** 315, Herrison **1** 340 and Knitson **1** 33. Epenthetic *d*/*t* appears medially in Hartgrove **2** 134 (**hara**/**hār**[2]).

OE *þ/ð* becomes *d* initially in Droop **2** 101 (**þrop**) and Durlston **1** 57 (**þȳrel**), and medially in Smedmore **1** 86 (**smēðe**[1]) and Soddon **1** 150 (**sūð**). OE *þ* becomes *t* initially in Turnworth **2** 254 (**þyrne**), and medially in Mosterton **4** 308 (**þorn**). In names from **mūða**, among them Charmouth **4** 370, Small Mouth **1** 269 and Weymouth **1** 250, ME spellings in *-mue*, *-mewe* show AN influence. The development of **hȳð** to *hive* in Langton Hive Point **4** 21 and Rodden Hive **4** 8 is to be noted.

Prosthetic *w-* has been noted in Oakley **2** 266 (**āc**, *W-* 1811), *Brodewoke* (f.n.) **2** 173 (**āc**), *Wodowne* (f.n.) **3** 301 and *Wool Park* (f.n.) **2** 31 (both **eald**), and *Wotecrofte* (f.n.) **1** 329 (**āte**). Inversion of this with loss of ME *w* occurs in Oakers **1** 291 (ME *Wolgaris-*) and Oborne **3** 354 (**wōh**, ME *Wo-*).

Prosthetic *y-* has been noted in the ME or later spellings for Eldon Hill **4** 489, Yelcombe **4** 138, Yeldon **4** 433 (**eald**), Yarlands **1** 205 (**ere-**[3]), Yetminster **4** 228 (pers.n. *Ēata*), Yewstock **3** 160 (**hēafod**), and Yon Barrow **1** 280 (**earn**). An inverted loss of *Y-* occurs in Eype **4** 427 (**gēap**, ME *Yepe*).

Various cases of assimilation (other than those already noted) are found, among them Afflington **1** 14 (*-lfr-* to *-fl-*), Battleford **4** 273 (*-kl-* to *-tl-*), Shatcombe **4** 118 (*-pc-* to *-tc-*), Vale Acre **2** 195 (*-nl-* to *-ll-*), and Wadmill **3** 75 (*-bn-* to *-dm-*).

Metanalysis occurs in a number of names. Most instances arise from the wrong division of ME **atten** and **atter** 'at the' with names that have initial vowels. Examples from **atten** include Nallers **4** 49 (**alor**), Nash Court **3** 170 and Nash Fm **4** 404 (**æsc**), Noake **3** 249 (**āc**), Norchard **4** 416 (**orceard**) and Nyland **3** 42 (**īeg-land**). Examples from **atter** include Rull **4** 421 (**hyll**) and Rye Water **3** 338, **4** 295 (**īeg**). Other than these, interesting cases of metanalysis are Acton **1** 34 (ME *Tacton* with loss of initial *t* due to preceding preposition *at*), Attewell **2** 21 (ME *atte* treated as part of the name), Hickford **2** 27 (initial *Th-* mistaken for def. art. and then dropped), and Evershot **4** 84 (earlier *Teuerset* due to presence of preposition *at*). The affixed forms of the names White Lackington **4** 169 and White Lovington **1** 275 result from metanalysis of the OE pers.ns. evidenced in the early spellings of the names.

Metathesis occurs in the spellings of several names, among them Alderholt **2** 195 (**alor**), Burngate **1** 130 (**brūn**[1]), Corfe Castle **1** 5 and

Corton **4** 26 (**corf**, ME *Crof*(-) spellings), *Ferngore* **3** 49 (**fearn**), Frizzel's Hill **2** 102 (**fyrs**), Girt Fm **4** 85 (***grytte**), Sterte **2** 27 and Woodstreet **1** 190 (**steort**), and Writh **3** 132, **3** 333 (**fyrhð**).

OE grammatical features are (at least in part) preserved in some names. The OE wk. gen. sg. inflexion *-an* appears as medial *n* in names like Cudnell **2** 23 (OE pers.n. *Cuda*), Drakenorth **4** 60 (**draca**) and Totnell **4** 205 (***tote**/OE pers.n. *Tota*). The OE wk. obl. adj. inflexion *-an* survives as medial *n* in names such as Blackney **4** 338 and Blacknor **1** 220 (*blacan* < **blæc**), Broadenham **4** 315 (**brād**), Darknoll **3** 181 (**deorc**), Henbury **2** 47, Hinton St Mary **3** 159 and Sixpenny Handley **3** 113 (*hēan* < **hēah**), Midney **3** 4 (*middan* < **midd**) and Newnham **3** 282, **4** 259 (*nīwan* < **nīwe**). The OE wk. gen. pl. inflexion *-ena* survives as medial *n* in names like Blandford Forum **2** 87 (*blǣgna* < **blǣge**), Huntingford **3** 14 (*huntena* < **hunta**, with development to *-ing*), Penmore **3** 390 (*pēona* < **pēo**) and forms like *All Hallon parishe* **1** 348 (*hālgena* < **hālga**). Particularly common in Dorset from the 14th century are instances of the analogical ME wk. gen. pl. *-ene*, as in the forms for Buckhorn Weston **3** 84 (*boukerene-* < **bouker**), *Crockern Stoke* **2** 101 (possibly *croccerene-* < **croccere**), Fryer Maine **1** 208 (*frerene-* < **frere**), *Fysshernebrigge* **1** 237 (*fisherene-* < **fiscere**), Hoggen Down **3** 223 (*hoggene-* < **hogg**), *Kotterynpoul* **1** 184 (*cotterene-* < **cottere**), Lardenhall **3** 118 (*lordene-* < **hlāford**), Onyx **1** 193 (*hundene-* < **hund**), *Schipenedone* **2** 123 (*shipene-* < **scīep**), Soapers Pool **3** 340 (*soporne-* < **sāpere**), *Thevenwell* **4** 250 (*thevene-* < **þēof**), *Threacrendiche* **1** 173 (*acrene-* < **æcer**), Toller Porcorum **4** 110 (*swinene-* < **swīn**), Winterborne Monkton **1** 266 and Wyld **4** 457 (*monkene-* < **munuc**), and *Wyvenwalleslake* **4** 22 (*wivene-* < **wīf**). Some names have the analogical ME wk. pl. forms in *-en*, as in Goren **4** 420 (*goren* < **gāra**), Hayne **4** 418 and Popehayne **4** 421 (*heghen* < **hæg**), *Millhays Treen* **4** 421 (*trewen* < **trēow**), Threbruggin **3** 200 (*briggen* < **brycg**), *Twohegyn* **4** 174 (*heggen* < **hecg**) and Wellen **4** 422 (*wellen* < **wella**).

Elliptical place-names (and field-names) are numerous. Examples in which the preposition is partly retained include *Bestedon* **4** 126, Bestwall **1** 157 (**bī ēastan**), Bincleaves **1** 255, Bindon **1** 130, 189 (**binnan**), Bowden **3** 42, Bowditch **4** 273, Bowood **4** 314 (**bufan**), Tweenways **4** 422, Twinwood **3** 161 (**betwēonan**), Underdown **3** 296 (**under**), Yondover **4** 322, **4** 484 (**begeondan**). Examples without trace of preposition in the modern forms include Northbrook **1** 322,

Northport **1** 160 (**bī norðan**), Westbrook **1** 248, Westport **1** 160 (**bī westan**).

There is shift of stress in Hammoon **2** 99.

Shortening of vowels and weakening of stress in the final part of compounds results in confusion between different elements, or in substitution of one element for another, in some names, among them Ashmore **2** 201, Enmore Green **3** 48 and Latchmore Pond **3** 53 (all **mere**), Bedchester **3** 103 and Hogchester **4** 456 (both **hyrst**), Catsley **4** 289 and Rock Lea **2** 21 (both **clif**), Charborough **2** 59, Lymburgh's Fm **3** 169 and Modbury **3** 120 (all **beorg**), Court House **3** 377 (**cot**), Dibberford **4** 257 and Piddles Wood **3** 192 (both **worð**), Farrington **3** 38, Morden **2** 58, Norden **1** 16 and Norton **4** 259 (all **dūn**), Goathorn **1** 44 (**ōra**[1]), Haselbury Bryan **2** 100 and Ramsbury **3** 349 (**bearu**), Hayward Bridge **3** 61 and Wizard Bridge **3** 332 (**ford**), Mosterton **4** 308 (**þorn**), Sandley **3** 17 (**hyll**), Tincleton **1** 329 (**denu**), Whitefield **2** 60 and Whitfield **1** 335 (**wella**).

THE ELEMENTS IN DORSET PLACE-NAMES AND FIELD-NAMES (APART FROM PERSONAL NAMES)

The names quoted under each entry are arranged in alphabetical order, with no distinction between compounds and simplex names. Names which survive on modern maps and also lost major names are listed first, followed by representative examples of the use of the element in field-names. Although a concise translation of each element is provided, for a fuller discussion of its significance and use, reference should be made to EPN (*English Place-Name Elements*, amended in JEPNS **1**) and VEPN (*The Vocabulary of English Place-Names*). The elements are often given in an OE, ME or OFr form, but it should be remembered that many of these elements continued as common words in the English language (and therefore in place-name formation), so that many Dorset place-names and field-names are of more recent origin than the periods represented by the head-forms used. Elements marked * are not independently recorded in the form cited, or are hypothetical reconstructions from place-name evidence. Names marked † refer to ancient sites or monuments. Forms from the OE bounds of AS charters are usually included in the entries for the appropriate elements even when they are identified with modern names, e.g. *doggeneberwe* as well as Dogbury are listed under **dogga** and **beorg**.

abbas Lat 'abbot'. Abbott's Wootton Fm, Bradford Abbas, Buckland Newton, Cerne Abbas, Compton Abbas, Corscombe, *Lyme Abbatis* **4** 392, Milton Abbas, Monkton Up Wimborne, Stoke Abbott, Toller Whelme, West Compton, Winterborne Abbas

abbatissa Lat 'abbess'. Compton Abbas, Kingston **1** 15, Melbury Abbas

abbesse ME 'abbess'. *Abbess of Tarents*, Abbot's Court Fm

abbeye ME 'abbey'. Abbey Buildings & Ho, Abbey Church, Abbeycroft Coppice & Down, Abbey Farm, Abbey Ford Bridge, Abbey Grange (× 3), Abbey Ho & Ponds **3** 151, *the Abbey Mill* **3** 367, The Abbey Park, Abbey Rd **3** 359, Abbey Wood, Bindon Abbey, Cerne Abbey, Forde Abbey, Grey Abbey, Milton Abbey, St Mary & All Saints' Abbey; f.n. *Abbeyweye* **2** 250

abbod OE, **abbat** ME 'abbot'. *Abbatston* **1** 307, Abbotsbury, ?Abbots Court, *Abbot's Fee*, Abbot's Hill (Fm), Abbott Street, Abbott's Wootton Fm, Buckland Newton, Stoke Abbott

abele eModE 'the white poplar tree', Alleybones **4** 367 (*v.* NDEFN 1)

above, *v.* **bufan**

***aßon** PrW 'river', in AS charters, *auenes broc* **1** 30, *v.* RNs

āc OE, **oke** ME 'oak'. Ash Oaks Coppice, Big Oak Strip, Boundary Oak, Broad Oak (× 3), ~ Fm, Child Okeford, Coppleridge, *Cowles Oke* **4** 231, Crooked Oak Hill, France Oaks Coppice, *Gildenock*, Gold Oak, Hollow Oak, Jubilee Oak (× 2), Melbury Oak, ?Mudoak Wood, Noake Fm, The Oak, Oak Bed, ~ Coppice (× 2), ~ Ctg(s), Oakey Plot, Oakford Fm, Oak Grove, *Oak Hill* **3** 119, Oak Hill, Oakhills Coppice, Oak Ho, Oaklands, Oak Lane, Oakley (Fm), Oakley Down, Oakley Lane, Oakley Wood, The Oaks, Ochill Wood, Okeford Fitzpaine, Pig Oak, Queen Oak, ?Roke Barn & Fm, Roke Hill, Sedge Oak Coppice, Shillingstone, Shire Rack, Stockbridge Oak, Three Oaks, Waddock Copse, Wyndham's Oak. Freq in f.ns., e.g. *Acford* **2** 172, *Brode Oke Coppice* **2** 205, *Brodewoke* **2** 173, *Burnedoke* **2** 158, Oak Breach **4** 424, ~ Fd **2** 51, *The oke acre* **4** 43. In AS charters, *achirste* **2** 162, *ac hylle* **3** 119, *greatan ac* **3** 289, *leaxan oc* **2** 82, *ligcgendum æc* **4** 454, *litlen ac lee* **2** 266, *oc* **3** 237, *smiceran æc* **4** 454

***ācen** OE 'growing with oaks, oaken', f.n. *Hockenbench* **3** 148

***ācet** OE 'oak wood', f.n. *Ackettes* **2** 172

***āciht** OE 'growing with oaks', f.n. ?Artley **4** 266

ād OE 'burning place, beacon', f.ns. Noad Close **2** 205, Odd Md **3** 109. In AS charters, *ealde ad* **3** 102, 131

adela OE 'a filthy place'. *Addle Path*; f.n. Addle Md **2** 18

admiralty ModE 'admiralty'. *Admiralty Quarries*

æcer OE 'plot or strip of cultivated land'. The Acres, Beer Acre, ?Benecke Wood, Bone Acre Copse, Compton Acre(s), Eight Acre Coppice/Copse (× 5), ~ ~ Plant., Eightacres, Eleven Acre Coppice, Forty Acre Bottom, ~ ~ Plant., Four Acre Coppice, Gallica Bridge, Hold Acre Coppice, Innsacre (Fm), Lease Acre, Long Acre Quarries, Nine Acre Coppice, ~ ~ Copse, ~ ~ Pit, ~ ~ Withy Bed, Nineteen Acre Hanging, Seven Acre Coppice, ~ ~ Plant, ~ ~ Withy Bed, Seventeen Acre Point, Six Acre Coppice (× 2), ~ ~ Copse (× 2), ~ ~ Plant., ~ ~ Wood, Slide Acre Coppice, Starve Acre Clump, Starveacre Lane, Tan Acre Plant., Ten Acre Copse/Coppice (× 3), ~ ~ Gate, ~ ~ Plant., ~ ~ Row, Twelve Acre Coppice, Twenty Acre Coppice (× 2), Twenty Acres Altmt Gdns. Freq in f.ns., e.g. *Acrehey*

2 13, *Bromaker* **1** 260, *Harryesacre* **1** 264, *Wowacre* **3** 35. In AS charters, *akere heueden* **3** 163, *forerð akere* **3** 206, *mæd æcras*, ~ *æcres* **2** 136, *seofan æceras* **4** 374, *sixe acres* **3** 200, *stanhecheres ande* **2** 58, *þry akeres* **3** 139, *twegen ærcas* **2** 136

æcer-mann OE 'ploughman, farmer'. Acreman St. **3** 359, **4** 135; f.n. *Acremanlonde* **2** 172

ǣl OE 'eel'. Almer, ?Arfleet Mill, Elsford Bridge; f.ns. ?Amour Md **3** 55, Eel Fishes **4** 490, Eel Pits **3** 235, Elver Oaks Bottom **1** 271. In AS charters, *elmere* **2** 55

æmette OE, **emmet** ME 'ant'. Emmetts Hill, f.ns. *Emethill* **4** 209, ?Lt Emmet Plant. **4** 308

æppel OE 'apple'. Sweet Apple Copse

æppel-trēow OE 'apple-tree', in AS charters, *apeltren* **3** 205

æpse OE 'aspen tree', f.ns. *Apeshull* **2** 51, Asp Plain **3** 55, ?Culls Apes Coppice **1** 27

ærn, **renn** OE 'a house or building'. ?Arne, ?*Crockern Stoke*, ?Dullar Fm, ?Lintern Lane, Minterne, ?Pimperne, Potterne Fm; f.ns. *Crokkerne fyldys* **2** 192, *Lusarne* **1** 121, Potten Hill **2** 214

æsc OE 'ash-tree'. Ash (× 2), Ash Bed (× 3), Ash Bed Coppice, Ashbed Plant., Ashcombe (Copse, Fm & Wood), Ash Coppice/Copse (× 4), Ashcroft, ?Ashes Fm, Ashey Copse, Ashey Knapp, Ash Fm, Ashgrove, Ash Hills Coppice, Ashington, Ashlands Coppice, Ashleigh, Ashley, Ashley Barn, ~ Ctg, ~ Fm, ~ Wood, Ashmore, Ash Pit, Ash Plant., Ashridge, Ashton Ctgs & Fm, Ashton Fm & Wood, Ashy Coppice, Ashy Drove, ?Axen Fm, Court Ash Ctg, Dole's Ash Fm, *Fifash* **2** 72, Four Ashes, Little Ashes (Coppice), Long Ash Fm, Long Ash Lane, Monmouth's Ash, Naish Fm, Nash Court, Nash Fm, Pill Ash, Salway Ash, Seven Ash Common & Fm, Templeman's Ash, Three Ash Cross, *Weeke ashe* **3** 305, *Windmill Ashes* **3** 222. Freq in f.ns., e.g. Ascombe **3** 212, *Aysdown* **4** 34, *Croketaysshe* **1** 172, *(H)aschemede* **2** 51. In AS charters, *æsc leage* **3** 288, *aescleȝe*, *esc* (*v*. Addenda **4** 193)

æscen OE 'growing with ash-trees'. ?Ashey Knapp; f.ns. Ashen Style **2** 220, Ashen Woods **4** 408, *Ashin Rewes* **4** 70, *Asshenlynches* **1** 297

æspe OE 'aspen-tree', f.ns. *Apeshull* **2** 51, *Apse* **2** 250, Apsey Coppice **2** 204, Asp Plain **3** 55

æt OE 'at', f.n. At Shoot **4** 9 (*v*. **atte**, **atten**, **atter**)

æt-steall OE 'encampment, fortification' or ***ǣt-steall** OE 'feeding stall (for cattle)'. In AS charter, ?*on æt steallesbeorh* **4** 29, *v*. Addenda

æðeling OE 'prince'. Agglestone, ?Allington

ǣwiell, ǣwyll OE 'spring, source of a stream'. Alton Mead Lane, Alton Pancras, ?Devils Wood Lane

ǣwielm, ǣwylm OE 'river spring'. Toller Whelme; f.n. *Ewelme* **3** 256. In AS charters, *tollor æwylman* **4** 291

afore ME 'in front of'. *Auorbrandon* **1** 213; f.n. *Aforehosse Lande* **2** 173

alcove ModE 'a nook', f.n. Alcove Gdn **4** 233

aley ME 'an avenue'. Bat Alley; f.ns. *Les Aleys* **4** 284, *Ally* **4** 302, Pepper Alley **1** 286

allotment ModE. Allotment Fm, Poor Allotment, Turbary Allotment, Twenty Acres Allotment Gdns, Hr & Lr Allotments Plant.

almes eModE 'alms'. *le Almes garden* **1** 157

almshouse ModE. Almshouses (freq), Almshouse Fm, ~ Wood, Gillingham's Almshouses, St Margaret's Almshouses

alor OE 'alder'. Alder Bed (× 3), Alderbed Copse, Alder Coppice/Copse, Alder Fm, Alder Grove (× 2), Alderholt, Alder Mead, Alder Moor (× 6), Aldermore (× 5), Alders Coppice, Alder Wood, Gt & Lt Alderwood Coppice, Aller (Fm), Aller-Moor, Alleymoor Wood, Axnoller Fm, Broad Alders Coppice, Elder Moor, Fox Allers Copse, Nallers Dairy Ho & Lane, Pop Mallers Coppice, Twisting Alders Coppice. Also in f.ns, e.g. Alderhams' Coppice **1** 296, Wisealler Md **4** 425. In AS charters, *aelres* **3** 106, *alor riðe* **3** 233, *alr* **3** 138, *wonalre* **3** 137

***alren** OE 'growing with alders'. ?Alderney Heath; f.n. ?Alderney **1** 59

***alret** OE 'alder copse', f.n. *le Alleret* **3** 175

alum ModE 'alum', f.n. *Alum* **1** 51

amen ModE, f.n. Amen Style **2** 25

amore OE 'a bird, perhaps the yellowhammer', f.n. ?Amery **3** 378

ān OE 'one'. One Elm

anchor ME hermit', f.n. Anchor Mead **2** 191

anchoret ME 'a recluse, a hermit'. Anchoret's Chapel; f.n. Anchor Md **2** 191

andhēafod OE 'headland', in AS charters, *an heued* **3** 200

***andfurlang** OE 'furlong', in AS charters, ?*an furlang* **3** 206 (now **1** 316, *v.* Addenda)

angle ME 'corner', f.ns. Angles **4** 159, Stakeford Angle **3** 293

ānli(e)pig OE 'solitary', in AS charters, *an lipien þorn* **3** 122

ānstīg(a) OE 'narrow path', **anstig** OE 'steep path'. ?Ansty Lane, Hr & Lr Ansty; f.n. Anstice **4** 450. In AS charters, *nearn þan anstigan* **2** 82

anvil ModE. Anvil Point

apuldor OE 'apple-tree'. *Horeappledore* **3** 13. In ΛS charters, *apildore* **3** 254, *haran apuldran* **2** 136, ~ *apoldran* **2** 161, *here apeldren* **3** 128, *ræw apoldran* **4** 454, *suran apœldran*, ~ *apeldran* **4** 375

***ar** PrWelsh 'beside, in front of, facing'. East & West Orchard **3** 133

arable eModE 'land for ploughing, arable', f.ns. Arable Fd **1** 27, Arable Sheephouse **1** 323

arch ModE. Bere Arch, Grange Arch

arewe ME 'arrow', f.n. *Arewe* **4** 204

arrish ModE dial. 'stubble field, aftermath', *v.* **ersc**

assa OE 'ass'. Ass Hill

assart ME 'clearing in woodland'. Sort; f.ns. *The Bushy serte* **4** 132, Sart **4** 391, Sort Md **4** 464

āst OE 'kiln'. *Limost* **3** 307, *Lymeste* **4** 132; f.n. *Hostham* **4** 284

āte OE 'oat'. Oatclose Wood, Oatcroft Lane, Oatlands Fm, Oatridge, *Le Ottlofte* **3** 150. Also in f.ns, e.g. *le Oteberne* **2** 289, *Otebreche* **4** 22, *Ote close* **4** 43, *Otecroft* **3** 33, *Wotecrofte* **1** 329

athwart eMod 'across', f.n. Athwart Legg **2** 105

atte ME 'at the'. ?Adbury, *Attewell* **2** 21, Berne Fm, Chalk Pit **4** 212, Coombe Gate, Cross Hill **4** 178, Ford **1** 201, Ford Fm **4** 324, Gerrard's Fm, Gore's Hill **4** 324, Great Coombe, Ham **4** 418, Hayne Fm **4** 418, Hole House, Hyde Fm **4** 89, (The) Hyde (× 2), Little Ashes Coppice, Lynch Fm, Park Fm **4** 404, Perry Fm (× 2), Pitt Fm, Stockwood Mill, Temple Brook Fm, Wood's Coppice, Yard Fm **4** 405

atten ME 'at the'. Four Ashes, Naish Fm, Nallers Dairy Ho & Lane, Nash Court, Nash Fm, Noake Fm, Norchard Fm, Hr & Lr Nyland; f.ns. *atte Naissh* **4** 167, *Nelbowe* **1** 327, Noad Cl **2** 205

atter ME 'at the'. ?Roke Barn & Fm, Roke Hill, Rull Fm, ?Ryeclose, Rye Water Coppice & Lane, Rye Water Fm

australis Lat 'south'. Alton Pancras, Austral Fm

avenue ModE. The Avenue, Avenue Lodge, ~ Plant. Freq in modern street names as ~ Avenue, e.g Clayford Avenue, Laurel Avenue

āwel OE 'fork of a river', f.n. ?*Auliscombe* **3** 254

baby ModE 'small'. Baby Hyde Crook

back eModE '(land) lying behind'. Back Brook (Engine Stream or ~ ~) (*v.* RNs), Back Drove (Copse), Back Lane (freq), Back Lane Fm, Back Street, Backwater, Back Water Channel; f.n. *le/the backe* **1** 264

backside eModE dial. 'land at the back'. Backside Coppice; f.ns. *le Back*(*e*)*sid*(*e*) **1** 264, **3** 79, **4** 34

backdoor ModE, f.n. Back Door Grd **2** 233, **3** 345
backwater eModE 'side-channel', f.ns. *Le/The Backwater* **2** 90, **3** 310
bacon ModE. Bacon Bluff & Hole
badde ME 'bad', f.n. Bad Md **2** 149
badger eModE. Badgers Bank, ~ Copse, ~ Covert, Badger's Coppice; f.n. Badgers Hole **1** 142
bæc OE 'hill, ridge'. ?*Felback*; f.n. *Backebergh* **4** 270
bæce, bece[1] OE 'stream, stream-valley', f.ns. *Bache* **1** 344, ?Furzey Beach **4** 423, ?*Shulbukesbache* **1** 238
***bæc-hūs** OE 'bakehouse', f.ns. *backehowse close* **3** 286, *le Bakhous* **4** 359, *le Common Backehouse* **3** 103
bær[1] OE 'exposed'. Bare Cross; f.ns. ?Bare Barton **3** 25, *Barefurlong* **3** 380, *Barleg* **1** 119, ?Bear Hams **1** 193
bǣr[2] OE '(woodland) pasture'. ?Barewood Copse, ?Barford Fm/Old Barford, ?Bear Wood, ?Bearhall Fm, ?Beer Acre, ?Beer Hackett, ?Beerhall Fm, ?Beer Lane, ?Beerwood Ctg, *Bere*, ?Bere Fm, ?Berehays Fm, Bere Knap, Bere Lane, Bere Marsh Fm, ?Bere Regis, ?Hogboro' Coppice & Ctgs, ?Ilsington Wood, ?Middlebere Fm, ?Todber; f.n. Beers Md **4** 380
bærnet OE 'land cleared by burning', f.ns. Barnet Pce **1** 106, *La Bernette* **3** 198, ?Burnetts **2** 149
bæst OE 'bast, bark of lime-tree', f.n. *Bastmede* **4** 344
bæð OE 'bathing place, pool'. Moorbath, Penns Bath; f.ns. Bad Barn Cl **3** 108, Cold Bath Gdn **3** 156, *Est-*, *Westbath* **2** 283, Long Bathe **4** 334
***bagga** OE, **bagge** ME 'bag', possibly 'badger' or 'bag-like feature in the landscape, e.g. hillock'. *Baggeridge*, Baglake Barn & Fm, Bagman's Fm, Bagmore Fm, ?Bagshot, Bagwood Close; f.ns. Bags Cl **4** 280, *Cleybagge* **3** 346, Croam Bags **1** 236
baie ME 'berry', f.n. ?*the Baycrofte* **3** 110
bailiwick eModE 'jurisdiction of a bailiff'. *Est-*, *Westbaillywike* 1541 (Isle of Purbeck **1** 3)
baille ME 'palisade', *La Baillie* **4** 235, *la/le Baley* **3** 365
baillie ME 'bailiff', or 'district or jurisdiction of a bailiff'. Bailey Ridge Fm, Bailey's Drove, Bailie Gate & Ho, *la Baillifeschaumbre* **3** 365, *Battyllysbayll* **2** 193, *La Westbaylly* 1402 (Isle of Purbeck **1** 3); f.ns. Bailey Croft **2** 7, *the Baleys Walke* **3** 31
bait ModE dial. 'food (for men and horses)' also 'pasture', f.ns. Bait **2** 259, Bait Leaze **3** 175
balca OE 'ridge, bank', f.n. Baulk **2** 199
***balg** OE 'smooth, rounded'. ?Ballard Cliff, Down & Point

***ball** OE 'hillock, rounded hill'. Ball Coppice (× 3), Ball Hill, ?Balls Lake, Ball Stone, Little Ball, Tolpuddle Ball, Whitham Ball Pit. Also in f.ns., e.g. Gt & Lt Ball **4** 408, Blue Ball **4** 422, Culver Ball **3** 12, Sandy Balls **2** 259

ballast ModE. Ballast Ctg & Knap, Broom Ballast Siding

bān OE 'bone'. ?Bone Acre Copse, ?Great Bournes; f.n. Bonnell **4** 490

bana OE 'murderer' also 'poison'. ?†Banbury Hill; f.ns. ?Bannell Mdw **2** 66

banke ME 'bank, slope'. Badgers Bank, Bank Barrow(s), Bank Plant., †Battery Bank, Bloody Bank, Carrot Bank, Chesil Beach, Gudgins's Banks, Hanging Banks Coppice, Holly Bank Fm, Lily Bank Firs, Sandbanks, Scar Bank Ho, Stour Bank. Also in f.ns., e.g. Brewers Bank **2** 214

bār[2] OE 'a boar (wild or domestic)'. Boar Hill Lodge, ?Bowerswain Fm; f.ns. ?*Boreswill* **3** 31

barelegged ME 'with bare branches', f.n. *Bareleggudthorn* **3** 226

bargain ModE, f.n. *Sorrow bargaine* **4** 11

barge ModE, f.n. Hr & Lr Barge **3** 25

bark-house ModE 'tannery'. ?Barkus Coppice; f.n. ?*Barkeshulle* **1** 202

barli ME 'barley', f.ns. Barley Bread **1** 292, *Barley field* **4** 9, *Barleygore* **4** 476, Barley **1** 136, *Barlyerche* **4** 173

barracks ModE. The Barracks (× 2), Barrack St. **3** 357, Bincleaves Barracks

barre ME 'bar, barrier, gate' also 'bar of sand'. Barbridge, Bar Lane, *Gore Barrs* **4** 21, Halswell Bars, Thickthorn Bar, Wood Bar Looe. Also in f.ns., e.g. Barlands **1** 225, *Barlane* **2** 173, *Barros* **4** 358, *Ferne Barases* **4** 266, Hobbys Bars **2** 69, the Outer Bar **1** 257, *Shute Barrs* **4** 26

barred ModE 'with bars', f.n. the Barred style **3** 323

barren eModE 'infertile', f.n. Barren Close **3** 87

barres ME 'barrier', f.n. *Ferne Barases* **4** 266

basil ModE 'sheepskin tanned in bark', f.n. Bazill Pit Flg **1** 361

bastard eModE, f.n. Bastard Close **4** 151

bat ModE 'bat'. Bat Alley, ?Bat's Head

bāt OE 'boat'. *Boat Cove*, *Boat Fm* **2** 24, *Boat Haul*, Boat Ho (Clump), ?*Boteridge* **1** 85–6

***bata** OE '?rich pasture', '?wooden club'. ?Batcombe

bataille OFr, ME 'battle'. Battle Plain; f.n. Battle Mead **1** 53

batch ModE dial., **bache** ME 'hillock'. The Batch (× 3), *Chydcombebache* **3** 234, Combe Close Batch, Elsdon's Batch, Flintland Batch, Great Batch, Greenland Batch, Hole Batch, Mill

Batch, Newlands Batch, *Piercy Batch* **4** 150, Sandbury Batch, ?Windsbatch. Also in f.ns., e.g. *the Bache* **4** 43, *huggeclyff bache* **4** 173, *Southbach* **3** 227, West Batch **3** 31

bathing ModE. Bathing Ho **4** 3, Bathing Pond

batte ME 'stick, lump', f.ns. Batlands **2** 282, Batt Md **3** 132, the Batts **4** 33, Broad Bat **4** 433, Short Batts **4** 330

battery ModE 'platform or fortified work on or within which artillery is mounted'. †Battery Bank, †*Cromwells Battery* 1844 (Corfe C.)

bay[1] ME 'indentation of the sea'. Arne Bay, Baiter, Balaclava Bay, *le Bay* **4** 384, Bramble Bush Bay, Brand's Bay, Brandy Bay, Chippel Bay, Durlston Bay, Freshwater Bay, Hobarrow Bay, Holes Bay, Kimmeridge Bay, Lytchett Bay, Ower Bay, Parkstone Bay, Poole Bay, Ringstead Bay, St Oswald's Bay, Shell Bay, Studland Bay, Swanage Bay, West Bay (× 3), Weymouth Bay, Worbarrow Bay

bay[2] ModE 'an embankment to form a dam'. Bay (Bridge & Fm), Bay Cottage; f.ns. ?Bay Flg **3** 175, Cloths Bay **4** 462, *mill bayes* **3** 367

beach Do dial. 'topsoil dug out in opening new quarries'. Fancy Beach

beach eModE 'seashore'. Chesil Beach, Cobb Gate Beach, Dungy Beach, Front Beach, Little Beach, Overcombe Beach

(ge)bēacon OE 'a signal, a beacon'. The Beacon, Beacon Corner, ~ Hill (× 4), ~ Knap, Bullpitt Beacon, Coombe Beacon, *Herbekne* **4** 371, Limekiln Beacon, Lytchett Beacon, Melbury Beacon, Salterns Beacon, *Shipton Beacon* **4** 490, Thorncombe Beacon. Also in f.ns., e.g. Beacon Barrow **4** 33, *Beckyn* 2 192, Becken Md **1** 215, *the bekon* **2** 10

bēag OE 'ring', in AS charters, *withi begh* **1** 33

beam OE 'beam', f.ns. *Beme close* **4** 411, *la Ysenebeme* **4** 93. In OE charters, *beam broc* **1** 31 (*v*. RNs)

bēan OE 'bean'. Bainly Bottom, ?Benecke Wood, Benville Fm & Manor, Bincombe, ?Binegar, ?Bingley Fm, ?Binnegar. Also in f.ns., e.g. Beanehill field **2** 180, *Bencroft* **4** 411, Benleigh **3** 161, Bynehill **1** 237

***bearde** OE 'battle-axe' (or **beard** OE 'beard' in some extended sense), f.n. ?*atte Berd* **4** 399

bearu OE 'wood, grove, copse'. Adber, Bagber (× 2), ?Barewood Copse, ?Barford Fm/Old Barford, ?Bearhall Fm, ?Bear Wood, ?Beer Acre, ?Beer Hackett, ?Beerhall Fm, ?Beer Lane, ?Beerwood Ctg, ?Bere Fm, ?Berehays Fm, ?Bere Regis, Colber Crib Ho, Elwibber, Haselbury Bryan, ?Hogboro' Coppice & Ctgs, ?Ilsington Wood, Kittle Barrow, Langbeer Fm, ?Middlebere Fm, Plumber Fm, *Ramsbury*, ?Todber, *Willow beare* **4** 327, *Willow Beer* **4** 422,

Withybere **4** 13; f.n. *le Whitheber* **1** 237. In AS charters, *bridian beara lace* **4** 453, *ciddesbeara* **2** 161, *suðbeara* **2** 162

beat, *v.* ***bete**[2]

bēaw OE 'gadfly', f.n. ?*Netherbawfurlonge* **3** 175

bece[1], *v.* **bæce**

bēce[2] OE 'beech-tree'. (Gt) Beach Coppice, Beech Clump (× 2), Beech Coppice & Lane, The Beeches, Beech Fm, Beech Grove, Beech Ho, Beech Lane, Beech Plant., Beechwood, Brockington Beeches; f.ns. *Beches* **1** 271, North Beach **4** 327

beck ModE dial. 'brook'. Horton Beck (*v.* RNs)

bedd OE 'plot of ground where plants are grown' also 'layer or stratum'. Alder Bed (× 2), Alderbed Copse, Ash Bed (× 4), Ashbed Plant., Causeway Withy Beds, Drinkwater Osier Beds, East Morden Withy Bed, ?Hibbert Bottom, Highbridge Willow Bed, Langford Withy Bed, Lutton Withy Bed, Nettlebed, New Withy Bed, Oak Bed, Rickett's Bed, Warren Withy Bed, Willow Bed, Withy Bed (freq). Also in f.ns., e.g. *Cutterbeds* **3** 26, *lez lekebeddys* **1** 128, *Selbeddes* **4** 223. In AS charters, *holenbedde* **1** 31, *wiþibedde* **1** 33

bed(e)rede ME 'confined to bed'. *la/le Bedredehous* **4** 359

bedlam eModE 'hospital'. Bedlam; f.n. Bedlam **4** 39

beehive ModE 'beehive'. Beehive Fm

begeondan OE 'beyond'. Yondover (Fm); f.n. Yeanhay Plot **4** 383

beggar ModE. Beggar's Lane

bēl[1] OE (**bǣl** WSax) 'fire, pyre'. ?Bellows Cross; f.n. ?Belbury **1** 323

bel[2], **beau** OFr, ME 'fair, beautiful'. Beaucroft, Belle Vue (× 2), Belle Vue Barrow, Bellevue Plant., Belmont Ho, *Beubo(y)s* **2** 256; f.ns. Beauvoir **4** 138, *Beawffrount* **4** 399, Belle Vue **4** 347

belle OE 'bell, bell-shaped hill or other feature'. Belchalwell, Bell Cliff, Bell Hill, ?Bellows Cross, Belstone Covert, Bull Bridge. Also in f.ns., e.g. Belfland **3** 216, *the Bell Acre* **2** 10, Bell Drong **1** 193, Belle **3** 340

bell-flower ModE 'campanula'. Bell Flower Plot

bell-hūs OE 'belfry', *le Belhous* **3** 150

belly ModE, f.ns. Pot Belly and Hollow Belly **1** 94

below ModE, f.n. bilow Glebe 2 Acres **1** 333

belt ModE 'screen of trees, narrow wood'. The Belt (freq), Belt Plant., Bishop's Belt, Church Ground Belt, Hanging Piece Belt, Higher Belt, Hyde Crook Belt, Park Belt, Penfold Belt, Thorncombe North Belt & South Bolt (sic), Wimborne Belt

benc OE 'shelf, ledge'. Bench, Bench Coppice, ?Benchy Hill, Broad Bench; f.ns. *le Benche* **3** 31, High Bench **3** 186, *Hockenbench* **3** 148

beneoðan, **beniðan** OE 'beneath'. *Bynetheclyft* **3** 143, *eynythe Alre* **2** 239; f.n. Neath Way **3** 28

benison ModE 'blessing', f.n. Benison **3** 43

bēo OE 'bee'. Beaulieu Wood, Bee Garden, Bewley Down & Fm; f.n. *le Byorchard* **2** 13

beofor OE 'a beaver'. Befferlands Fm, *Beuerore* **2** 258; f.ns. Beverlands **2** 78, *Beveryscombe* **4** 132

beonet OE 'bent-grass'. Bentover Coppice; f.ns. Bennety Coppice **2** 204

beorc OE 'a birch-tree'. ?Bartleys Row & Wood; f.ns. Bark Hill **3** 185, ?*Burclinche* **1** 313

beorg OE 'hill, mound, barrow'. Adbury, *æt steallesbeorh* **4** 29, *Albretesberge Hundred*, Ball Barrow, Bank Barrow(s), †Barrow Coppice, Barrow Field, Barrow Firs, Barrow Hill (Fm) (× 6), Barrow (or Hundredsbarrow) Hundred, Barrowland Fm, Barrow Plant., ?Barrows Lane, Belle Vue Barrow, Bind Barrow, Black Barrow, †Bolton's Barrow, Bradford Barrow, †Broom Barrow, *Bugbarrow*, †Bul Barrow/Bulbarrow Hill, †Bull Barrow, Burberry Lane, Bush Barrow, *Celberge* Hundred, Charborough Ho, Charlton Barrow, Chebbard Fm, E & W Chelborough, †*chetoles beorge* **2** 236, †Chettle Long Barrow, Chilbury Plant., †Church Barrow, Clandon Barrow, Cold Barrow, †Conquer Barrow, Coombe Beacon, †Creech Barrow, Deverel Barrow, Dogbury, †Double Barrow, †Drinking Barrow, †Dunbury, †End Barrow, Fern Barrow, †Ferny Barrows, Figbury (Barrow), †Fishing Barrow, †Five Barrow Hill, †Five Barrows, Folly Barrow, †Four Barrow Hill, Gainsborough Hill, †Goggin's Barrow, Great Barrow, ?Hadber Coppice, †Hambury Ho & Tout, Hawkesbury (Barrow), Henbury Barrow, †Herringston Barrow, ?Hillberry, Hobarrow Bay, †Howdes Barrow Plant., †Huck Barrow, †Hundred Barrow, *Hunesberge Hundred*, †Icen Barrow, †King Barrow (× 3), King Barrow Quarries, †(The) King's Barrow, Knap Barrow, *Langeberge Hundred*, *Langton Barowe* **2** 40, *Lawrence Barrow*, Limbury Fm, Lobbett's Barrow, †Long Barrow (× 3), ~ ~ Hill, ~ ~ Lane, †Longbury, Loosebarrow Hundred, †Lord's Barrow, Lush Barrows, Lymburgh's Fm, Mag's Barrow, *Modbury*, Modbury (Fm), Modbury Hundred, Mowlam's Barrow, Muddox Barrow Coppice, †the Music Barrow (*v.* Addenda **1** 259), †Nine Barrow Down, Parley Barrow, †Pimperne Long Barrow, †Povington Barrow, †Rainbarrows, Ralph's Barrow, Rawles Barrow, †Rimbrow Coppice, Robin's Barrow, †Round Barrow, *Row Barrow*, Rowbarrow (× 2), Rowbarrow Barn,

Rowbarrow Hundred, Rowbarrow Lane, Ryeberry Hill, Sadborow, Sandbury Batch & Moor, †Sandy Barrow (× 2), †Scrubbity Barrows, Seaborough, †Seven Barrow Plant., †Seven Barrows, †Stable Barrow, Stand Barrow, Stoborough, Stonebarrow (Hill), †Straw Barrow (× 2), ?Swanderbury, Symondsbury, Tadnoll Barrow, †Thorn Barrow, †Thorny Barrow, †Three Barrow Clump, †Three Barrows, †Three Lords' Barrow, Tinker's Barrow, ?Todber, Trent Barrow, †The Two Barrows, †Wardstone Barrow, †Water Barrows, †Windmill Barrow, Woodbury Copse, †Woolsbarrow, †Wor Barrow, Worbarrow, †Yon Barrow. Freq in f.ns., e.g. Barrow **3** 25, Bugbarrow **4** 33, *Childebergh* **4** 10, *Wyreberge* **2** 11, *yelde Barrowe* **4** 43. In AS charters, *bacging berghe* **2** 122, *beorgas* **2** 161, *beorge* **3** 205, *beorhleage* **3** 281 ?*beorhlem* **2** 122, *berendes beorh* **3** 122, *berges* **2** 82, *bergh* **3** 205, *blacmanne bergh*, *bleomannes berge* **3** 205, *braden bergh* **3** 205, ?*burg* **3** 233, *ceatwanberge*, *shete bergh* **1** 303, *chelesbergh'* **2** 122, *chellenberghe* **4** 173, *cylberge* **3** 237, ?*delesburg* **3** 222, 237, *doggeneberwe* **4** 161, *dollen beorge*, *-berch* **3** 102, *graetem beorge* **2** 122, *heandene beorg*, *hendune beorch* **3** 218, *hiþisc bergh* **3** 206–7, *hornes beorȝ*, *-beorh* **4** 303, *motbeorh* **4** 35, *pegan beorh* **3** 122, *Ruanberghe* **3** 257, *rugen bergh* **3** 207, *ruwan beorȝ* **4** 304, *Scherdenberwe* **3** 258, *schilleburghe* **3** 258, *sebergh* **3** 207, *tatanbeorge* **2** 122, *toten berg'* **3** 123, *windberghes* **3** 259, *windee bergh* **2** 58, *worres berg* **2** 122

***bēos** OE 'bent grass', f.n. ?Bestbury **1** 292

bere OE 'barley'. Barcombe; f.ns. *Beere Way* **2** 51, *Berlond* **3** 31

bere-ærn OE 'barn'. Barn Bottom, ~ Close (× 2), ~ Coppice (× 2), ~ Corner, ~ Field, ~ Plant., Red Barn Fm. Freq in modern minor names, e.g. Afflington Barn, Ashley Barn, Bayard Barn, Bincombe Barn, Black Barn, Blagdon Barn, Bottom Heath Barn, Brook Barn, Coombe Barn, Dennet's Bottom Barn, Down Barn, Druce Higher Barn, East Barn, Eweleaze Barn (× 3), Fair Barn, Field Barn, Forston Barn, Gascoyne New Barn, Gate Barn, Grant's Barn, Grove Hill Barn, Higher Barn (freq), Hill Barn (freq), Hog Hill Barn, Holcombe Barn, Home Barn, Hovels Barn, Huish Barn, Lawn Barn, Muckleford Higher Barn, Muston Field Barn, New Barn (freq), North Barn, North Down Barn, Northdown Barn, Owermoigne Down Barn, Painter's Barn, Parsonage Barn (× 2), Parson's Barn, Pen Barn, Pigeon Ho Barn, Pount's Barn, Pulston Barn, Red Barn, Ringstead Barn, Rough Close Barn, Ryeland's Barn, Seager's Barn, South Barn, South Holworth Barn, Spring Garden Barn, Underbarn Walk, Vyse Barn, Warren Barn, Water Barn (× 2), West Down Barn,

West End Barn, Whitcombe Barn, Wolfeton Barn, Wood Barn, Yellowham Barn, Yewleaze Barn. Freq in f.ns., e.g. Barn Cl **2** 36, Barn Fd **2** 199, Barn Grd **4** 18, *Bernhey* **4** 80, *le Oteberne* **2** 289

bere-tūn OE 'barley farm', later 'farm-yard'. (The) Barton, Barton Fm (× 2), Barton Hill, Barton Hill Fm, Coombe Barton, Crocker's Barton, Gough's Barton, Greenhill Barton, Hazel Barton, Hr Barton (Ctg), Hutchins Barton, Kings Barton, Lee Barton, Lower Barton, Milking Barton Coppice, New Barton, Rhode Barton, Yarn Barton. Freq in f.ns., e.g. Bare Barton **3** 25, Barton Plot **4** 433

bere-wīc OE literally 'barley farm', more usually 'outlying part of an estate retained for the lord's use'. Berwick; f.n. Berrick Plot **1** 280

berned, **bærned** OE, **brend** ME 'burnt'. *Hr & Lr Barn Down*, Burnt Firs, Burnt House Bottom; f.ns. *Burne*(*d*)*oke* **2** 158, Burnt Oak **3** 135

berry ModE 'berry'. Berry Coppice

***berse** ME 'enclosed part of a forest', f.n. *Burse Hunte Cross* **2** 173

bersel ME 'an archery butt', *les Barselles* **4** 9

best(**e**) ME 'animal', f.ns. Beast Lease **3** 96, Beast Leaze horn **3** 250

bēte[1] OE 'beet', f.n. *Betacres* **2** 72

***bete**[2] ME, **beat** ModE dial. 'rough sod, turf', f.ns. Beat Close **4** 408, Beat Grd **4** 79, Beatham **2** 104, Beat Leaze **1** 323 [*cf.* **burn-beat**]

betwēonan OE 'between'. Tweenways, Twinwood Coppice & Lane; f.ns. *Bytwene Tweylynches* **1** 133, Twinholds **4** 383, Twinways **3** 389, *Twyne Feylde* **2** 217

bī OE 'by'. *Bestedon*, Bestwall Fm & Ho, East Brook, Northbrook, Northport, Norwood Coppice & Park, Southbrook, Warmwell Down Barn, Westport. Also in f.ns., e.g. *Biwestewode* **2** 54, *By Kinges Barrowe* **3** 263, *bynorthstrete* **4** 22, West Brook **3** 98

***bic**, ***bica** OE 'a bill or beak, a beak-like projection'. ?Bickham Fm, ?Bickham Wood, Isle of Purbeck; f.ns. *Bikenhurste* **3** 199, *Bykenhulle* **4** 58, *Underbikelee* **4** 181. In OE charters, ?*biken settle* (in bounds of Sixpenny Handley **3** 113, *v.* Addenda)

***bīc** OE, **bike** ME 'bees' nest'. ?Bickham Fm; f.ns. Bicks Cl **1** 286, Bincklinch **2** 13, *Longbyklynche* **2** 13

bicce OE 'bitch', f.n. *Byccheneburgh* **3** 255

bīcere OE 'bee-keeper', f.n. ?*Bekercomb* **4** 487

***bīegel** OE 'bend', f.n. Boyle **4** 180

big ME 'large'. Big Burles, Big Copse, Big Covert, Big Leaze Hole, Big Oak Strip, Big Wood (× 2), Poxwell Big Wood; f.n. ?*Biggputtlane* **3** 340. *Cf.* also **bigger** in *Bigger Palmers Herne* **4** 235

biker ME 'dispute', f.n. ?*Bekercomb* **4** 487

bile OE 'bill, beak'. Bill of Portland; f.n. *Bylecumbe* **4** 181

***billing** OE 'hill, ridge', f.n. ?*billingedownecrofte* **4** 348
binde OE 'climbing plant', f.n. Bindleaze **1** 206
***bing** OE '?hollow'. ?Bingledon Wood
binnan OE 'within'. Bincleaves, Bindon Abbey, Bindon Hill, Little Bindon; f.ns. Benalders **3** 101, Benditch **3** 196, Bimport **3** 145
birce OE 'birch-tree'. The Birches, Birch Coppice (× 2), ~ Copse, ?Birches Copse, Birch Hill (× 3), Birch Lane, Birch Plant., Birch Wood (× 2), Burchill, ?Burgess's Fm. Also in f.ns., e.g. *Birche Meade* **3** 6. In OE charters, *biricgaran* **2** 161, *burch linken* **3** 110
biscop OE 'a bishop'. Bishop's Caundle, Bishop's Down, ~ Fm, ~ Road. In AS charters, *bissopes imare* **3** 136
biscuit ModE, f.n. Biscuit Cl **1** 343
bis(s)e ME 'fine linen' or **bis(se)** ME 'brownish or dark grey', f.n. Golden Biss **3** 292
bit ModE 'small piece of land', f.n. Bit **4** 200
bitula OE 'beetle', f.ns. *Bitelegh* **2** 250, Biddle Grove **2** 179
blacksmith ModE 'blacksmith'. Blacksmith's Lane
blæc OE, **blake** ME 'black, dark-coloured'. *Blachford* **4** 406, Black Barn (× 3), Black Barrow, *Black Barrow*, Black Bottom, Black Bush, Black Castle, ~ Ctg, ~ Covert (× 2), Blackcroft Coppice, Black Down (× 5), Blackdown Hill, Blackers Hole, Blackfern Plant., Black Head, Black Heath (× 2), Black Hill (× 5), ?Blackhills Plant., Black Ho, Black Hole, Blackland Plant., Blacklands, Blackland's Close, Blacklands Lane, Black Lane, Black Lawns Coppice, Blackmarsh Fm, Black Moor, Blackmoor Forest & Vale, Blackmore Ctgs, ~ Fm, Blackney Fm, Blacknoll, Blacknor, Black North Plant., Black Rock(s) (× 4), *Black Rock*, Blackrow Cmn & Fm, *Blackstone* **3** 52, Blackven Fm, Black Venn (× 2), *Blackwater*, Blackwater, Blackwater Bridge, Blackwater Fm, Blackwater Ford, Blackwater River (*v*. RNs), Black Wood (Bottom), Blagdon Barn, ~ Copse, ~ Fm & Hill (× 2), Blake Dene, ~ Hill, *Blakelond*, *Blakenotford* **2** 112, Bracklands, Brake Hill. Freq in f.ns., e.g. Blacklands **4** 43, *La Blakehalle* **2** 192, *Blakehey* **3** 31. In AS charters, *blacan dune* **2** 161, *blake ston* **1** 271
***blǣcen** OE 'bleaching'. ?Blashenwell; f.n. ?*Blachenlond* **4** 22
***blæc-þorn** OE 'black-thorn, sloe', f.ns. Blackthorn Fd **2** 62, *Blakethorndone* **4** 181, *la blakethurne* **2** 52. In AS charters, *blaken þornes* **3** 205
blǣge OE 'the blay, the gudgeon'. ?Blandford Forum/St Mary, ?Langton Long Blandford
(ge)bland OE 'storm', f.n. ?Bland **1** 78

blatch eModE 'blacking, black stuff or soot'. ?Blatchford's Fm
bleach eModE, f.n. Bleach Yard **3** 5
bleak eModE 'exposed'. Bleak Ho; f.n. Bleak Hill **2** 32
bleat SCy dial. 'cold, bleak'. Bleet Fm
blessing ModE, f.ns. Blessing Fd, Blessings Md **2** 154
blight eModE 'plant disease', f.n. Blight Acre **4** 380
***blinc** OE r.n. ?'glittering'. Blynfield Fm. In AS charters, *blinches broc* **3** 91–2 (*v*. RNs), *blinchesfelde* **3** 90–1
blind OE 'blind, dark, hidden'. The Blind Cow, Blind Lane (× 7), Blind Walk Plant., *Blindway* **2** 245, *la Blyndbowe* **3** 365, *la Blyndlane* **3** 364. Also in f.ns, e.g. Blind Pitt **3** 175, Blindwell **3** 25, *le Blyndelane* **2** 70
blindhous ME 'asylum for the blind', *Blindhowse* **1** 354, *le Blyndehouse* **3** 150
blockhouse ModE. *the Blockhouse* **1** 254
bloody ModE. Bloody Bank
blow ModE. Blow Hole
blue ModE. Blue Pool; f.ns. Blue Ball **4** 422, Blue Buttons **4** 270, Blue Gate **4** 390
bluff eModE 'headland'. Bacon Bluff
boathouse ModE. Boat House (Plant.)
bōc[1] OE 'beech-tree'. ?Hr & Lr Bockhampton, Bugden Bottom
bocardo eModE 'prison, dungeon'. Bucky Doo Square **4** 357
***bōcen** OE 'growing with beech-trees'. Bugden Bottom; f.n. *Bokenacre* **4** 181
bōc-land OE 'land granted by charter'. Buckland Newton, Buckland Ripers
bog ME 'bog, marsh'. Bloxworth Bog, The Bog (× 2), Bog Lane, Mill Bog; f.n. Bog **2** 171
boga OE 'a bow, a bend, arch'. *la Blyndbowe* **3** 365, ?Boveridge, The Bow, Bow Brook (*v*. RNs), Bowcroft Hill, Bowden Hill, *Bowell Lane* **3** 148, ?Bowshot Fm, *Boywellane* **3** 148, Double Bow; f.ns. Bow **3** 43, *la/le Bowe* **3** 365, Bowden **4** 422, *Boweputtes* **3** 31, Bowhays **4** 380. In AS charters, *bogen wylle* **3** 148
boggy ModE, f.n. Boggy Pasture **2** 62
***boi(a)** OE 'boy, servant'. ?Boywood Fm
boiling ModE. Boiling Rock
bois OFr 'wood'. *Beubo(y)s* **2** 256, Boys Wood
***bola** OE, **bole** ME 'tree-trunk, log', f.n. *Bolecroys* **3** 226, Bowbridge Md **4** 233

bolla OE 'bowl'. ?*Bollane* **3** 364, ?Bowldish Pond, Tibble's Copse & Lane

bon ME 'favour, gift', f.n. *le Boone* **4** 10

bonfire ModE. Bonfire Hill; f.n. ?Bon(i)fore Hill **2** 18

***bor** OE 'hill'. ?*Dimor Boor* **3** 306; f.ns. Boar Flg **4** 131, *atte Bore* **3** 199, Nettle Boor **3** 309

bord OE 'board, plank', also 'border, rim'. Chebbard Fm; f.ns. Board Wood **4** 422, *Bordhegge* **4** 181

borden ME 'made of boards or planks'. Boarden Bridge, *Boarden Bridge* **4** 220; f.ns. Boarden Bridge Mdw **4** 433

border ModE. The Border (× 2), Free Down Border

bord-land ME 'land held by a bordar', f.ns. *Board Land* **3** 374, *Bordelond*(*e*) **1** 120, **1** 172, **4** 43, *Bourdeland*(*es*) **1** 264, **3** 133, *Burdelonde* **4** 43

borealis Lat 'north'. Alton Pancras

***bors** OE 'spiky plant', f.n. *Borsehay dech* **2** 19

bōt OE 'help, remedy, privilege'. ?*Boteridge* **1** 85–6; f.n. *Botmede* **3** 374

bote ME 'boat'. Boat Close; f.n. Boat Cl **1** 50

botele ME 'bottle'. ?Bottlebush **2** 267

boðen OE 'rosemary, darnel; thyme'. Bothenwood

boðl OE 'dwelling, house'. ?Buddel, Buddlewall; f.ns. Botherly **4** 380, Bothhill **3** 161, Bottle Ash **4** 408

botm, ***boðm** OE 'valley bottom'. Barn Bottom, Bothenhampton, Bottom Barn, ~ Coombe Quarries, ~ Coppice, ~ Copse (× 2), ~ Fm, Bottom Ground, Bottom Heath Barn, Bottom Plant., Bottom Wood, *the buttome* **3** 302, Combe Bottom Coppice, Knight's in the Bottom. Freq in modern minor names as ~ Bottom, e.g. Barnes' Bottom, Black Bottom, Black Wood Bottom, Brimstone Bottom, Co(o)mbe Bottom (freq), Creech Bottom, Dennet's Bottom Barn, East Bottom, Fordington Bottom, Forston Field Bottom, Giant's Grave Bottom, Gould(')s Bottom, Grove Hill Bottom, Happy Bottom, Headbury Bottom, Hill Bottom, Holcombe Bottom (× 2), Kit Hill Bottom, Laurel Bottom, Little Puddle Bottom, Long Bottom (× 3), Middle Bottom, Pier Bottom, Poxwell Bottom Dairy, Ridge Bottom, Rushy Bottom, Scratchy Bottom, Seacombe Bottom, Smokeham Bottom, Spring Bottom (× 3), Spring Head Hotel, Squibbs Bottom, Strap Bottom, Stratton Bottom, Watcombe Bottom, West Bottom, Winspit Bottom, Yearlings' Bottom. Also freq in f.ns., e.g. Bottom Md **3** 25, *Broade Bottome* **4** 43, *Cunygere Bottome* **2** 215, *Hebbes Bottom* **3** 219

bouker ME 'a buck-washer, a bleacher'. Buckhorn Weston

bound eModE. 'boundary'. Bounds Lane; f.ns. Bound Breach **4** 422, Bound lane **4** 235

boundary eModE. Boundary Copse, ~ Fm, ~ Oak, ~ Tree

boundstone eModE. Bound Stone Plant.; f.n. *the Bonde stones* **1** 264

bounty ModE, f.n. Bounty **4** 347

boutavant OFr 'that which sticks out'. *Butavaunt* (a tower) 1280 (at Corfe C. **1** 11)

bowed ME 'arched', f.n. Bowl Bridge Md **3** 292

bowe-schote ME 'distance an arrow can be shot'. ?Bowshot Fm

bowl ModE. Golden Bowl

bowling ModE. Bowling Alley Walk; f.ns. Bowling Ally **3** 55, Bowling Md **1** 286

bowling green ModE. Bowling Green(s), Bowling Green Fm, ~ ~ Lane, ~ ~ Wood; f.ns. Bowling Green **2** 44, **4** 248, 328, 462, *The bowling greene* **4** 384, Lt Bowling Green **2** 90

box OE 'box-tree'. ?Boxbury Row, Box Ctg, Boxen Hedge; f.n. Great Box **4** 281

boxen ME 'planted with box-trees'. Boxen Hedge

brād OE 'broad'. ?Bladeley Hill, Bradford Abbas, Bradford Fm, Bradford Lane, Bradford Peverell, Bradle Barn & Fm, Bradpole, Breston Quarries, Broad Aldermore, Broad Alders Coppice, Broad Bench, Broad Breach, Broad Bridge (× 2), Broadclose (Fm), Broad Coppice (× 2), Broad Croft, Broadcroft Quarries, Broadenham Fm, Broadfield (Coppice), Broad Garden, Broad Lane (× 2), Broad Lawn, Broadleaze, Broad Ledge, Broadley Wood (× 2), Broad Lines Plant., Broadmayne, Broadmead (Ctgs, ~ Lane), Broadmoor Coppice, ~ Fm, Broad Oak (Fm) (× 4), Broad Orchard, Broadridge Common, Broad Road Fm, Broadrock, The Broad Stone, Broadstone, Broadwey, Broadwindsor, Broad Witcham, Friar Waddon, Sydling St Nicholas. Freq in f.ns., e.g. *Brademore* **4** 10, *Broadcraft* **2** 19, Broad Flg **3** 25, Broad Md **4** 9. In AS charters, *bradeleighe* **3** 255, *braden bergh* **3** 205 [*cf.* **broad**]

bræc[1], **bracu** OE, **brake** ME 'brake, bracken, brushwood'. ?Bracham Lane, Brake Ctg, Brake Plant., Buckshaw Brake, Caundle Brake, Farthings Brake Coppice, Fernbrook Brake, Foxbrake, Furze Brake (Plant.), Holly Brake, Rough Brake, Shroton Brake, Yards Brake. Also in f.ns., e.g. Brack **3** 135, Brake Grd **4** 39, Brake Wood **2** 25

brǣdu OE 'broad stretch of land', f.n. Bread **3** 299

brægen OE 'brain, hill', f.n. Brayen Cl **3** 196

brandire(n) ME 'a gridiron', f.n. Brandy **3** 272

brandy ModE. Brandy Bay

bray(e) OFr 'mud', f.n. ?*Braylondes* **3** 381

break ModE, f.ns. Break Heart **2** 25, Brickneck **1** 27

breakwater ModE. Portland Breakwater

brēc OE (**brǣc**[2] WSax), **breche** ME 'breaking, land broken up for cultivation'. Boards Breach Copse, Brach Copse, Bratch Coppice & Lane, Bratch Copse, Breach Common, Breach Copse, Breach Ctg, Breach Fm, Breaches Lane, Breach Lane, Breach Plant., Breach Wood, Breeches Coppice, Broad Breach, *Foys Breach* **4** 189, Furze Brach Lane, *Parsons Breach* **4** 190, Shore Breach, Snail's Bridge. Freq in f.ns., e.g. Bratch/Breach **2** 7, Mitchells Breach **3** 5, *Otebreche* **4** 22, *la Westbreche* **4** 19

breden, **briden** OE 'made of planks', f.n. *Bredenewelles mores hafede* **4** 302. In AS charters, *brydenan briccge* **2** 158

***brēme** OE 'broom, place where broom grows'. Broomhill Fm **1** 175; f.n. ?*Bremehull* **1** 113

brēme OE, **breme** ME 'rugged', f.n. ?*Bremehull* **1** 113

brēmel, **brǣmel**, **brembel** OE 'bramble'. Bramble Bottom, ~Bush Bay, ~ Close, ~ Combe Copse, Bramblecombe, Bramble Coppice, Bramble Hew Coppice, *Breme(l)clift* **4** 90, Brimpit Fm. Also in f.ns, e.g. Bramble Croft **3** 224, Brambly **3** 25, Brimble Down **3** 101, Hay brimble **1** 41

***brēmelig** OE 'growing with brambles', f.ns. *Bremelifurlang* **2** 173, *Bremly Closs* **1** 128, *Bremlyforlong* **1** 182

brend, **brent** ME 'burnt', *v.* **berned**

brēosa OE 'gadfly', f.ns. ?Breastlands **2** 104, *Brese Milles* **4** 453

brēr OE 'briars'. Briar Thicket, Brierhill Wood

brerd OE 'border, hill-side', f.n. ?*Brerdesfold* **3** 374

bretherhede ME 'brotherhood, guild', *the Bretherhed Howse* **4** 487

brewery ModE. Brewery Ho

brew(ing)-house eModE 'brewery'. *Brew(e)hous* **4** 487, *the Brewinge howse* **1** 134; f.n. Brewhouse Cl **3** 40

brick ModE, **brike** ME. Brick Hill (Plant.), Brick Ho (× 3), *Brickhouse* **3** 62, Brick Kiln Plant., Brickplace Copse, Dodding's Brick Yard & Fm; f.ns. Brick Coppice **2** 288, Brickly **3** 25

brickkiln ModE. Brick Kiln (× 2), Brickkiln Coppice, Brick Kiln Drove (× 2), Brickkiln Ho, ~ Lane, Brick Kiln Plant.; f.n. Brick Kiln Bottom **2** 199

brickworks ModE. Blackhill Brick Works, Jubilee Brick Works, Lytchett Brick Works, Manning's Brick Works, Puddletown Brick Works

brickyard ModE. Brickyard Copse, ~ Ctgs, ~ Lane, Morden Brickyard
bridale ME 'a wedding feast', f.n. *langebrideall* **4** 222
briden, *v.* **breden**
bridewell eModE 'a prison'. Bridewell (× 2), Bridewell Lane, ?Bridewell (Plantations) (*v.* Addenda **1** 97); f.ns. *Brid(e)well* **1** 254, **3** 365, Bridewell **4** 114
brigdels OE 'bridle', f.n. Bridle Marsh **1** 78
brimstone ModE. Brimstone Bottom
broad ModE 'broad expanse of water'. The Broad [*cf.* **brād**]
brōc OE 'brook'. Back Brook (Engine Stream or ~ ~) (*v.* RNs), Bibbern Brook (*v.* RNs), *blinches broc* **3** 92 (*v.* RNs), Bow Brook (*v.* RNs), Brockfield (Bottom), ?Brockham, ?Brockham Lane, Brockhampton Fm, Brockington Fm, ?Brog Street, Broke Lane, Brook Barn, Brook Cottage, Brookfield, Brook Ho (× 3), Brook Knowle, Brooklands (Pit), Brook Lane, ?Brooks Copse, Brookside (× 3), Broxhill, *Byle Brook* (*v.* RNs), Cashbrook Ctgs, Caundle Brook (*v.* RNs), Chivrick's Brook (*v.* RNs), *Colebrok'* (*v.* RNs), Colesbrook, Collyer Brook (*v.* RNs), *Combesbrok* **3** 191 (*v.* RNs), Cookwell Brook (*v.* RNs), Corry Brook (*v.* RNs), Crichel Brook (*v.* RNs), Darknoll Brook (*v.* RNs), Deansbrook (*v.* RNs), Devil's Brook (*v.* RNs), East Brook, Eastbrook Fm, ~ Ho, *estbroke* **4** 395, Fern Brook (*v.* RNs), Fernbrook Fm, Filley Brook (*v.* RNs), Fontmell Brook (*v.* RNs), Furzebrook, *Garbroke* (*v.* RNs), Gussage Brook (*v.* RNs), Holbrook, Holebrook Fm & Water (*v.* RNs), Honeybrook Fm, Key Brook (*v.* RNs), Kingsbrook, Knighton Brook (*v.* RNs), Lambrook, Lowbrook Fm, Lowbrook Fm, Manston Brook (*v.* RNs), Middlebrook Cottage, Millbrook Ho, Northbrook (× 3), *Northbrook or Norbrook Fm*, *Palbroke*, *Plusshebroke* (*v.* RNs), Pucksey Brook (ib), *Selwaie Brook* (ib), Sleep Brook (ib), Slod Brook (ib), Southbrook, Stanbrook, *Swerebrok* **3** 188, Tadnoll Brook (*v.* RNs), Temple Brook (*v.* RNs), Temple Brook Fm, Trent Brook (*v.* RNs), Twyford Brook (ib), *Vrtebrok* (ib), Wadbrook, Wambrook, West Brook (*v.* RNs), Westbrook Dairy/Fm/Ho., Wraxall Brook (*v.* RNs). Freq in f.ns., e.g. *Brokelond* **3** 226, *Brokfurlang* **1** 260, *le Brooke* **4** 10, *Cowbroke* **1** 264, *Watebrouke* **4** 454. In AS charters, *auenes-*, *aueres broc* **1** 30, *beam broc* **1** 31, *blinches broc* **3** 91–2, *broc* **1** 32, *broces* **1** 14, *broke* **2** 82, *byrran broc* **4** 302, *cagbroc* (*v.* RNs.), *cealdan broc* **2** 161, *Conibrok'* **3** 191, *hlorbroc* (*v.* Addenda **4** 193), *Holambrok'* **3** 348, *holan broc* **2** 161, **4** 193 (*v.* Addenda), **4** 374, *holebroke* **3** 109, *iþern broc* (*v.* RNs.), *liscbroc*, *lisebroke* **3** 203 (*v.*

RNs), *mearc broces heafde* **2** 136, *melenbroc* **3** 236, *myle broc* **4** 32, *secg broc* **2** 158, *smale broke* **3** 109, *stanbroc* **3** 258, 289

brocc OE 'badger'. ?Brockham, ?Brockham Lane, Brock Hill, ?Brog Street; f.n. *brockfurlongefeild* **1** 345

brocc-hol OE 'a badger hole, a sett'. Brockhill, Brockhole Lane, Brockwell Coppice; f.ns. *Broc(c)hole* **2** 125, *Brockhold* **3** 199

brocen OE 'broken', f.ns. Broken Cross **1** 343, *Broken Wall* **1** 229, *Brokyndiche* **1** 326

brōm OE 'broom'. Bramdon Lane, Bran Ledge & Point, Brimbley Fm, Brimley, Bronkham Hill, Broom, Broom Barrow, Broom Bridge, Broom Cliff, Broom Close (× 3), Broom Hill (Lane), Broomhill, Broomhill Coppice, ~ Copse, ~ Drove, ~ Fm (× 2), Broomhills, Broom's Fm, ~ Plant., Brumham Coppice, ?Burnt Bottom. Freq in f.ns., e.g. *Bromdun* **4** 10, *Bromeclose* **2** 38, *Bromhull* **4** 34, Broom Croft **4** 433. In AS charters, *bromhulle* **3** 255, *brumdun* **4** 302

***brōmig** OE 'growing with broom'. Broomhill Coppice; f.n. *Bromyknapp* **4** 10

brū OE 'brow of a hill, declivity, projecting edge'. Allan's Brow, Axen Brow, Bowdens, Brow Copse; f.ns. Brow **2** 32, **4** 39, Brow Coppice **4** 337

bruiere OFr 'heath', f.n. Brewen **3** 314

brūn OE 'brown'. Brown Down Plant., Brownsall Hundred/Brunsell's Fm & Knap, Burngate (Fm); f.ns. *browne grove* **4** 437, *Brownhill* 2 173

brushet Do dial. 'brushwood', f.ns. Brushets **1** 93, Brushett(s) **4** 131, 433, East Hill Brushet **4** 166

brycg OE 'bridge'. Abbey Ford Bridge, Alderholt Bridge, Alexander Bridge, Atrim Bridge, Axe Bridge, Bagber Bridge, Bar Bridge, Hr & Lr Barbridge, *Barnaby Bridge*, Bay Bridge, Benville Bridge, Bibbern Bridge, Blackmore Ford Bridge, Blackwater Bridge, Blandford Bridge, Bluntsmoor Bridge, Boarden Bridge, Bockhampton Bridge, Boddel's Bridge, Bourton Bridge, Bridge, Bridge Close, Bridge Ctgs, Bridge Fm (× 6), Bridge Lane, *The Bridge* 1586 (at Corfe C. **1** 12), Bridge Wood, Broad Bridge (× 2), Brockhampton Bridge, Broom Bridge, Broomhill Bridge, Buddle Bridge, Bugley Bridge, Bull Bridge (× 2), Cale Bridge, Canford Bridge (× 2), Causeway Bridge, Cecily Bridge, Chamberlayne's Bridge, Charmouth Bridge, Chilbridge, Church Bridge, Clapton Bridge, Cornford Bridge, Cotton Bridge, Cow Bridge, Crawford Bridge, Damas Lawn Bridge, Dark's Bridge, Devil's Bridge, Dixon's Bridge, Dummett's Bridge, Dunn's Bridge, Durweston

Bridge, East Bridge, Ebblake Bridge, Elsford Bridge, Emlands Bridge, Eye Bridge, Farrington Bridge, Ferry Bridge, Fiddleford Bridge, Fitche's Bridge, Five Bridges, *Fold Bridge* **4** 195, Forde Bridge, Frome Bridge, Fryer's Bridge, Gallica Bridge, Gascoyne Bridge, Gillingham Bridge, Green Pit Bridge, Grey's Bridge, Halscombe Bridge, Harbour Bridge, Hart's Bridge, Hay Bridge, Hayne Bridge, Hayward Bridge, High Bridge, Hixon Bridge, Holme Bridge, Hooke Bridge, Horn Bridge, Hornshayes Bridge, Hosey Bridge, Hummer Bridge, Hunters Bridge, Hunting Bridge, Hurst Bridges, Julian's Bridge, Kettle Bridge, *Kingesbrigge* **3** 11, King's Bridge, King's Mill Bridge, Kingstag Bridge, Kit Bridge, Kitford Bridge, Knapp Bridge, Lacey's Bridge, Landshire Bridge, Lawn Bridge, Lecher Bridge, Legg's Bridge, Lodden Bridge, Long Bridge (× 2), Longford Bridge, Longham Bridge, Lytchet Bridge, Madjeston Bridge, Magg's Bridge, Manston Bridge, Marsh Bridge (× 2), Merriot Bridge, *Milbridge* **4** 277, Gt & Lt Mohun's Bridges, Narrow Bridge Coppice, Netherhay Bridge, Netherstoke Bridge, New Bridge, *New Bridge*, Newlands Bridge, North Bridge, Orchard Water Bridge, Organ Ford Bridge, Oxbridge, Packers Bridge, *Pain's Bridge*, Parson's Bridge, Peggs Mill Bridge, Pin Bridge, Pinnocks Moor Bridge, Piper's Mill Bridge, Poll Bridge, Pond Head Bridge, Poole Bridge, Povert Bridge, Prince's Bridge, *Pymersmyllebrugge* **4** 469, Red Bridge (× 2), Rock Lea Bridge, Rolls Bridge Fm, Rolls Mill Bridge, Roman Bridge, Romford Bridge, Rowden Bridge, St Edward's Bridge, St Helen's Bridge, St Leonard's Bridge, St Mary's Bridge, Salkeld Bridge, Sandford Bridge, Seaborough Bridge, Sharford Bridge, Shedrick Bridge, Sherford Bridge, Shitterton Bridge, Silverbridge Fm, Smallbridge Fm, Smith's Bridge, Snatford Bridge, South Bridge (× 2), Spickett's Bridge, Stalbridge, Stanbridge, Stanbridge Mill, Stepping Stone Bridge, Stivvicks Bridge, Hr & Lr Stockbridge Fm, Stockham Bridge, Stockley Bridge, Stoford Bridge, Stour Bridge, Stroud Bridge, Stubb's Bridge, Tanpits Bridge, Three Guinea Bridge, *Titherley Bridge* **4** 279, Ton Bridge, *Toppole Brygge* **3** 291, Town Bridge (× 2), Trent Bridge, Trill Bridge, Twofords Bridge, Walford Bridge, Wareham Bridge, Warr Bridge, Watford Bridge, Westbridge Fm, Westover Bridge, Whetley Bridge, Whitchurch Bridge, White Bridge (× 2), White Mill Bridge, Winscombe Water Bridge, Wizard Bridge, Woodbridge (× 2), Wool Bridge/Woolbridge, Wools Bridge/ Woolsbridge, Yeovil Bridge. Freq in f.ns., e.g. *Brygeheys* **2** 45, *Keynesbrugge* **3** 33, *Maydenbrigge* **4** 11, *Strowbrege* **4** 35,

Threbruggin **3** 200. In AS charters, *brydenan bricgge* **2** 158, *Cumtun bricgge* **3** 325, *oxene bricge* **3** 163, *stanene bregge* **1** 32, *wde bricge, -brigthe* **3** 105

bryd OE 'bride'. ?Bridewell (Plantations)

***brȳd** OE 'surging', ***brȳde** OE surging stream'. ?Bridewell (Plantations) [*v.* Addenda **1** 97]; f.n. *Brudcombe* **4** 58

***Brydi** OWelsh 'gushing or surging steam'. R. Bride (*v.* RNs, giving name to Little and Long Bredy and Burton Bradstock). In AS charters, *bridian beara lace* **4** 453

bryne OE 'burning', f.ns. Brinstead **2** 220, *Burne Downe* **1** 94. In AS charters, ?*birninge*, *Byrnynge* **3** 254

bubble ModE. Bubble Spring

bucc(a) OE 'a buck, a male deer' also 'he-goat'. Bokerley Ditch, *Bucke mead* **2** 257, ?Hr & Lr Buckham, Buckhill Stall, Buckland Down & Hill, Buck's Head; f.ns. ?Book Park **3** 67, *Bucheclive* **3** 199, *Buckebere* **4** 204, Buck Mdw **2** 199, Bucknam **3** 314, *Bukhay* **4** 384

buck-house eModE 'house for buck-washing or bleaching cloth'. Buckhouse Fm

bucking ModE 'the process of steeping cloth'. *Bucking House* **4** 359

bufan OE 'above'. *bovenalre* **2** 239, ?Boveridge, Bowden, Bowditch or Burridge, N & S Bowood; f.ns. *a boue cliue* **4** 10, *Above Coombe* 1 221, *Bovebury* **1** 260, *Boveditch* **1** 345, Bowditch **1** 270

bugge ME 'a boggart, a hobgoblin'. ?Buckshaw Ho, *Bugbarrow*, ?Bugley; f.ns. *Bugdowne* **4** 227, ?*Bugputtlane* **3** 340, Bugbarrow **4** 33, 221

building ModE. The Buildings (× 3), Building Garden Coppice, ?Buildings Copse. Freq in modern minor names as ~ Buildings, e.g., Down Bldgs, Higher Bldgs (× 2), Middle Bldgs, New Bldgs (freq), Sleight Bldgs, Sprake's Bldgs, Victoria Bldgs, Vine's Down Bldgs, Winfrith Down Bldgs

bula OE 'bull'. Bolehays Copse, ?Bowldish Pond, Bowleaze, Bul Barrow/Bulbarrow Hill, Bulhams, The Bull, ?Bull Barrow, Bullhill Copse & Fm, Bull Pit, Bullpitt Beacon, Bullpits, *Bull Stake* **1** 168, 353, Bully Wood. Freq in f.ns., e.g. *Boleacre* **4** 411, *Bolehegh* **3** 213, *Buldown* **3** 226, Bullpits **2** 191, *Bulpoole* **4** 476

bulluc OE 'bullock', f.n. Bullocks Well **3** 212

bulwerk ME 'rampart, fortification'. The Bulwarks, *Clanbulwarke* **1** 355; f.ns. *le Bolewarke* **4** 376, *the Bulwarkes* **1** 254

bum ModE, **bom** ME 'buttocks'. Bum Point, Ring-Bum Gdns; f.ns. Bomb Hill **4** 297, ?Bombs Lane **1** 226, ?Bumshill **1** 128

bungalow ModE. The Bungalow, Hilltop Bungalows, Woodrow Bungalow

bunting hous ME 'room for bolting meal'. ?*le Bontynghous* **4** 7

buoy ModE, f.n. Groves Buoy **1** 227

bur eModE 'butterbur', f.ns. Burfurlong **3** 135, Burr Mdw **3** 121

būr[1] OE 'cottage, dwelling, bower'. Bower Hill Plant., ?Bower Point, Burberry Lane, ?Burcombe Fm, ?Burwood, ?Bushey; f.n. *Harvest Bower* **1** 95. Also in f.ns., e.g. *la Boure* **2** 215

(**ge**)**būr** OE 'a peasant'. ?Bushey

***burg** OE, **burgh**, **borowe** ME 'burrow'. ?Burrow, Fox Barrow Fm, Foxbury, ~ Coppice, ~ Lower Hedge, Musberry Lane; f.ns. Burrow Cl **4** 423, 347, ?Cattlebury **4** 450, *Holeweysburwgh* **4** 181

burgage OFr, ME 'burgage', f.n. *la/le Burgage* **2** 184

burh (dat.sg. **byrig**) OE 'fortified place, earthwork, fortified house or manor'. Abbotsbury, †Badbury Rings, †Banbury Hill, Barcombe, Berie Well, Berry Fm, ?*Boltbury* **3** 154, The Borough, †Bulbury Camp, ?Burcombe Fm, ?Burl Fm & Moor, ?Burrow, ?Burwood, †Buzbury Rings, Chalbury (× 2, †**1** 231), †Duds Bury, East Borough **2** 184, Ensbury, †Flower's Barrow, Hembury Fm & Mill, Henbury Manor, †Maumbury, †*mealeburg* **3** 116, †Melbury Abbas, Melbury Bubb, Melbury Osmond, Melbury Sampford, †Mistleberry Wood, Netherbury, *Netherbury*, Newberry Gdns, †Penbury-Knoll, †Poundbury, †Rawlsbury Camp, Shaftesbury, †Spettisbury, ?Swanderbury, Upbury Fm, †Weatherby Castle, West Borough **2** 185, Woodbury Coppice, †Woodbury Hill. Also in f.ns., e.g. Berry Fd **2** 7, *biriacre* **1** 69, *Buryrydyng* **1** 271, *Bynorthebury* **2** 85, *Luuenesburg* **2** 216, *Northebury* **3** 33. In AS charters, *brugh doune* **3** 255, *cheoles burge eastgeat* **2** 134, *ealde berig* **3** 206 [*cf.* **bury**]

burh-tūn OE 'fortified farmstead', or 'farmstead near a fortified place'. Burton (× 2), Burton Street, East & West Burton, Little Burton, Long Burton; f.n. Bourton **3** 121

(**ge**)**būr-land** OE 'land occupied by peasants', f.n. ?Burland **2** 51

burna OE 'stream'. Bibbern Brook (*v.* RNs), R. Bourne (*v.* RNs), Bourne Fm, Cheselbourne, *Little Cheselbourne*, Cranborne, *Craneburn* (*v.* RNs), Forest Bourne Fm, Langbourne, Lodbourn, Lodburn, *Lorteborn* (*v.* RNs), Ludbourne Rd **3** 361, Milborne St Andrew, Milborne Stileham, Milburn Lane, Oborne, Sherborne, *Wimborne* (*winburnan*) (*v.* RNs), Wimborne Minster and St Giles, R. North Winterborne (*v.* RNs, giving name to Winterborne Clenston, ~ Kingston, ~ Muston, ~ Stickland, ~ Tomson, ~ Whitchurch, ~ Zelstone), R. South Winterborne (*v.* RNs, giving name to Winterborne Abbas, ~ Came, ~ Herringston, ~ Monkton, ~ St Martin, ~ Steepleton). Also in f.ns., e.g. *Bornethornes* **2** 143, *Bourn lake* **2** 200, Burncombe **4** 433, *Burneheye* **3** 31, *Craneburn* **3** 32. In AS charters, ?*birninge*, *Byrnynge* **3** 254, ?*bournandign'* **3** 255, *bydeburnan* (*v*. RNs), *cyselburnan*, *Ceosol burnan*, *chiselburne* **3** 202 (*v*. RNs.), *mylebroce* **4** 32, *mylenburnan* **3** 298, *mylenburnna* **3** 380, *neodbournen* (*v*. RNs)

burn-bake ModE 'land cleared by paring and burning'. Burnbake; f.ns. Burnbake, Burn Bake **2** 85, 109, **3** 25, Burnbeck **2** 214

burn-beat ModE dial. 'turf cut and burned in preparing land for cultivation', f.ns. Bourn Beat **1** 323, Burnbeat **4** 280, Burnt Beat Cl **1** 264, Burnt Pea **4** 200

burning ModE. Burning Cliff

burn-stōw OE 'place where people assembled at a stream, bathing place', or 'watering place for cattle'. Bourne Fm. In AS charters, *burnestowe* **2** 122, *burnstowe*, *burestowe* **3** 205, **4** 173, *burstowe* **3** 205

bury ME 'manor house, centre of an estate, urban area outside the main part of the town'. Eastbury, Eastbury Ho, Newbury Ho, *Northbury* **4** 123, *Suthbury* **4** 123, Vembury Dairy, Westbury **3** 363, Westbury Fm; f.ns. Berry Fd **2** 7, 171, **3** 121, *Burylane* **3** 31, *Burywode* **3** 79 [from OE **byrig** dat.sg. of **burh** *q.v. supra*]

burying ModE. Burying Ground **1** 27

busc OE 'bush'. Black Bush, Bottlebush Clump, *Bush*, Bush Barrow, The Bushes, ?Bushes Barn & Pit, *Bushfield Fm* **3** 46, Bush Road Ctgs, Common Bushes, Deanend Bushes, Grosvenor Bushes, Hinton Bushes, Hog Hill Bushes, Holmbush, The Holm Bushes/Holm Bushes Fm, *Holm Bush Knap* **4** 397, Holmbush Plant., Kine Bush Lane, Mill Bush, Moncton Bushes, Roundbush Wood, Shady Bushes, Sparrowbush Fm, *Stawelbusshe* **3** 325, Stourpaine Bushes, Wagbush Coppice. Also in f.ns, e.g. *Buschebryche* **3** 108, Bushy Cl **1** 286, **2** 51, *Gospell Bushe* **2** 174, *Smythes Bushe* **2** 216

bushi ME 'bushy'. Bushy Eweleaze, ~ Huscombe, ~ Park, ~ Row; f.ns. *Busshyclose* **3** 340, *Bussylese* **3** 319

butere OE 'butter'. (*The*) *Buttercrosse* **3** 148, Butter Rock, Butterwick Fm; f.ns. Butter Print **3** 55, 97

***butta** OE 'butt, stump'. ?Butt Lane Hollow; f.n. ?*la Butweye* **2** 99

but(te) ME 'archery butt'. Butt Fm & Lane, ?Butt Lane Hollow, ?The Butts, ?Butts Lane, ?Hillbutts, Old Butts (× 2); f.ns. Butts **1** 304, **4** 248

butte ME 'short strip at right angles to others'. ?Butt Lane Hollow, ?The Butts, (?)Butts Lane (× 2), ?Butts Quarries, ?Hillbutts. Freq in f.ns., e.g. Butland **2** 204, Butt Close **4** 433, Lt Butts **4** 42, ?*la Butweye* **2** 99

buttery eModE 'store-room, pantry'. Buttery Corner, Buttery Street; f.n. Best Buttery **2** 241

buttuc OE 'buttock', f.ns. *le Bottokes* **1** 326, Buddick **4** 297, Buddock **4** 58, *Buttockespytt* **4** 222

butur, **bittor** ME 'bittern', f.n. Bitters Gall **1** 78

bydel OE 'beadle', f.ns. Beetles Md **1** 78, Bidlond **3** 254, *Bydelhill* **3** 374

byden OE 'a vessel, a hollow'. ?Bedmill, Bibbern Brook (*v.* RNs); f.ns. Bedwell (Cl) **2** 18, **4** 208, *Bydecumbe* **4** 14, ?*Bydemylne* **2** 192. In AS charters, *budencumbe hracan* **3** 128, *bydeburnan* (*v.* RNs)

byge[1] OE 'a corner, an angle, a bend'. Bye Fm; f.ns. *La Bie* **1** 113, Buy Acre **4** 297, *le Bye* **1** 168, Bylands **4** 434. In AS charters, *bige* **1** 31

byht OE 'bend'. Egmont Bight

***byrde** OE 'border, bank'. ?Stivvicks Bridge (but now *v.* Addenda **2** 103)

byrgels OE 'burial', in AS charters, *æþenam byrigelsas* (*v.* Addenda **3** 267), *haþene berielese* **4** 173

byrgen OE 'grave, burial place', in AS charters, *burghen* **3** 255

(**ge**)**byrst** OE 'landslip, chasm', f.ns. ?Bristland **3** 378, *La Burste* **4** 399

*(**ge**)**bysce** OE 'copse of bushes', f.n. ?Biscombe **4** 380

bytme, **bytne** OE 'valley bottom'. Home Bottom Coppice; f.ns. *la Bitene* **3** 58, Bittom **2** 106, Bottom Hill **4** 159, *Byttemne* **2** 173 [*cf.* **botm**]

byxen OE 'growing with box-trees'. E & W Bexington Fm

cabbage ModE (**caboche** ME). Cabbage Hill, Wooden Cabbage Fm

cabin ModE. The Cabin

cacche ME 'to catch', f.n. Catcham **1** 361

cǣg OE 'a key' or ***cǣg** OE 'stone'. ?Key Brook (*v.* RNs); f.n. Key Hay **3** 83. In AS charter *cagbroc* (*v.* RNs.)

cæppe OE 'cap'. ?Tyneham Cap

cærse, **cerse** OE 'cress'. Cashbrook Ctgs, Cashmoor, Caswell Fm & Wood, ?Keysworth Fm (*v.* Addenda **1** 159); f.ns. Carsals **4** 266, *Carswellesyate* **3** 255, *La Cresseye* **1** 249, Crislain **3** 293

cage ME 'cage', (*le*) *Cadge* **4** 241, *le/the Cage* **3** 310, **4** 359

caggy ModE dial. 'decaying, tainted, stagnant'. Caggypole Fm

cāl, **cawel** OE 'rape, cole, cabbage'. ?Caldron Barn Fm, ?Cole Combe; f.n. ?*Colehey* **2** 70

calenge, **chalenge** OFr, ME 'challenge, dispute', f.ns. *Calenge* **1** 120, *Calengleyn* **1** 345, Challenge **4** 462
***camm** PrWelsh 'crooked'. ?R. Cam (*v*. RNs)
cammoc OE 'cammock', f.ns. Cammick Hay **3** 43, Cammic Hill **3** 87
camp[1] OE 'open land, field', f.ns. *Campeden* **1** 364, ?Compse **2** 199, ?Quomp **2** 26, *Westercomp* **1** 70
camp[2] eModE 'military encampment, army camp', also 'ancient earthwork'. †Bedford's Camp, Bovington Camp, †Bulbury Camp, ?Camp Ctg, †Camp Down, †Chalbury Camp, †Duds Bury Camp, †Rawlsbury Camp, Wool Camp
***canc** OE 'a steep, rounded hill'. Cank Fm/High Cank. In AS charters, *canuc*, *canu rih* (*v*. Addenda **3** 267)
Candlemas eModE, f.n. Candlemas Md **4** 450
canel ME 'canal, channel', f.ns. Canal withey bed **1** 78, *Canellake* **3** 156
canne OE 'a can, a cup', 'a hollow, a deep valley'. ?Cancer Drove, Cann
can(n)e ME 'cane, reed'. ?Kine Bush Lane; f.ns. ?*Caneryhule* **1** 120, Canvilles **4** 68
canonicus Lat 'a canon'. Whitchurch Canonicorum
canoun ME 'canon', f.n. *Canonbern* **1** 128
cap ModE 'cap-like covering, top stratum'. ?Golden Cap
***cape** OE 'a look-out place'. ?Caphays Dairy & Drove
capel ONFr, ME 'chapel', ?*Les copell* **3** 249
capon ME 'castrated cock', f.ns. Capon Hays **1** 378, *Capon Hill* **3** 263
capped eModE 'with a cap-like top'. Cobbett Hill
capstan ModE 'capstan', f.n. Captains Fd **1** 40
caraway eModE. Carraway Lane; f.n. Carraway **3** 70
***carn** PrWelsh 'cairn, rock, heap of stones'. R. Cerne (*v*. RNs, giving name to Cerne Abbas, Nether Cerne, Up Cerne and Charminster), ?*Cherne* (r.n. **2** 59), R. Char (*v*. RNs, giving name to Charmouth), Sharnhill Green
carotte ME 'carrot'. ?Carrot Bank; f.n. *Carret Close* **4** 253
carriage ModE. Carriage Drive, Carriage Plot
carrion eModE 'garbage'. *Carrion Lane* **1** 155; f.ns. Carrion Pit **1** 270, 302
carthouse ModE, f.ns. *Carthous* **1** 354, Carthouse Grd **1** 151
cast ModE 'what is thrown, a heap'. ?East Hays; f.n. Castway **3** 293
castel(l) OE, ME 'castle, camp'. †Abbotsbury Castle, Black Castle, †Brownsea Castle, *La Casteldych* **3** 247, (The) Castle (× 7), Castle Barn, †Castle Close, †*Castle Close* 1291 (Corfe C.), Castle Clump,

~ Copse, ~ Cove, ~ Fm (× 2), ~ Hill (× 7 of which 3 have †), Castle Lake Spring, Castle Lane (× 4 of which 2 have †), †Castle Rings, †Castleton, Castletown, Castlewell, Castle Wood Fm, Castlewoods Fm, †Coneys Castle, †Corfe Castle (par. and village), †Corfe Castle (remains of), †Crawford Castle, Drummer's Castle, †*Eggarton Castle* **4** 47, Horse Castles, †Lambert's Castle, Lulworth Castle, †Maiden Castle, *Mohun Castle* **4** 333, Pat's Castle, Pats Castle Cottages, Pennsylvania Castle, †Portland Castle, †Rufus Castle, †Sandsfoot Castle, †Sherborne Castle, †Stephen's Castle, Stockland Great & Little Castle, †*Studland Castle*, †Sturminster Newton, †Weatherby Castle, †Woodsford Castle (Fm), Wyke Castle; f.ns. †*Castelwode* **2** 180, †*Castell* **2** 242, The Castle **4** 359, Castles **1** 226. In AS charters, *castel* **3** 206

cat(t) OE 'wild cat'. Catemwood Coppice, *Cathille* **4** 289, Cathole Barn & Copse, Catholes Withy Bed, Catley Copse, ?Catmoor, Catsley Fm, *Cattesdene* **4** 289; f.ns. Cat Hamm(s) **2** 171, *Catmore* **4** 19, Catsbarrow Flg **4** 45, *Catshave* **3** 79, Cats Holes **2** 199, *Cattestich* **4** 270, *Catts Clif Coppice* **3** 226

cat(t)el ME 'cattle, livestock', f.n. Cattlebury **4** 450

cattes-brayn ME 'cat's brain' (applied to mottled soil), f.ns. Catsbrain **3** 161, *Catsbrayne* **2** 13

cauc(i)e ME 'raised way across marshy ground'. The Causeway, Causeway Bridge, Causeway Withy Beds, North Causeway, Sherborne Causeway, South Causeway. Also in f.ns., e.g. *le Caucewey* **4** 43, *le Cauxey* **2** 52, *le Cawsewey* **1** 137, *Cawsey close* **4** 19

caudle ModE dial. 'thick muddy fluid', ?Cuddle

cave ME 'cave, underground hollow'. Cave Hole, Limekiln Cave, Smugglers' Cave, Tilly Whim Caves

***Cawel** pre-English r.n. of unknown origin and meaning. ?R. Cale (*v.* RNs)

***ceacce** OE 'lump, hill'. ?*chece* **4** 16 (*v.* RNs), ?Checkridge Dairy & Fm

ceaf OE 'chaff'. ?Chaffeigh Fm, ?Chaffeymoor Fm; f.ns. ?Chaffcrofts **4** 334, ?Chaff hay Cl **2** 98, *Cheveacre* **4** 384

c(e)alc OE 'chalk'. ?*Celberge* Hundred, Chalk Corner, Chalklands Quarries, Chalkley Coppice, ?Challow Fm & Hill, ?E & W Chelborough, ?Chilbury Plant.; f.ns. ?*Chelbargh* **1** 313, *chelcputte* **4** 266, *Chelkhull(e)* **1** 260, 3 226, *Chelkweye* **1** 345. In AS charters, *cealc pyt* (*v.* Addenda **3** 267), *chealc seðas* **3** 110

c(e)alf OE 'calf'. *Calf* **4** 259, The Calf, Calf Close Coppice, Calf Close Lane, Calfhay Fm, Calf's Horn Lane, Chaldon Herring, Chaw Meadow, *Cow and Calf* **4** 413, Gough's Shoot. Also in f.ns., e.g. Calves Cl **2** 253, Chaffey **3** 161, *Chalfcombe* **3** 226, *Chalfecroft* **2** 70, *Chalvecroft* **2** 215. In AS charters, *chelfgraue* **2** 82

c(e)ald OE 'cold, exposed'. Belchalwell, Calcots Wood, ?Caldron Barn Fm, ?Chalcott's Coppice, *Chaldecot*, *Chaldecots*, *Chaldwel* **4** 94, ?Challow Fm & Hill, ?Charlock Hill, Cold Barrow, *Cold Crosse* **3** 183, *Cold Harbour* **4** 64, Cold Harbour (× 3), Coldharbour, ?Cold Park Wood, Cole Hill Wood, Gold Point, Harbour Bridge & Hill; f.ns. Chalwell **3** 63, *Cheldwlle* **4** 367, Chilwels **4** 380, *Coldeherb(o)urgh* **1** 237, Cold Harbo(u)r **3** 55, 360, **4** 151, 253, 281. In AS charters, *cealdan broc* **2** 161

cēap OE 'trade, market, purchase'. Cheap St. **3** 360; f.ns. *Chapehey* **1** 237, *Cheapside* **2** 185, *Chipstreete* **3** 58

cēap-mann OE 'merchant', f.ns. *Chepemanrewe* **3** 365, *Chepmanhegh* **4** 500, *Chepmans Standinges* **3** 146, *Chepman Way* **3** 237

***cearr** OE 'a turn, a bend'. ?Charborough Ho

cēast OE 'strife, contention'. Chesters Coppice & Hill

ceastel, **ci(e)stel** OE 'heap of stones', f.ns. *Chestell* **4** 10, †Chissels **4** 490. In AS charters, *stancastel*, *stancysten* **3** 208

ceaster OE '(Roman) city', also 'earthwork'. †Dorchester; f.ns. †Chesters Well **3** 72, *Chestreforlang* **4** 235, Chestus **1** 41, †Chinchester **1** 304, *Langh'sester* **3** 199, Staunchester **4** 203

***cęd** PrWelsh, PrCorn 'a wood'. Chetterwood, Lytchett Matravers & Minster, East & West Orchard **3** 133

cedar ModE. The Cedars

***cędiǫg** PrWelsh 'wooded (estate)'. Chideock

cefer, **ceafor** OE 'beetle', f.ns. ?Chaverlands **4** 475, Chiverick Md **3** 161. In AS charters, *cinerget* (for *ciuerget*) *mor*, *cuterget mores*, *seuerȝate* (*v*. Chivrick's Brook in RNs. *supra*)

***ceg** OE 'stump', f.n. *la Chemere* **2** 180

cegel** OE (ci(e)gel** WSax) 'pole, post'. ?*Celberge* Hundred, ?E & W Chelborough, ?*Chelwell Water* (*v*. RNs), ?Chilbury Plant., ?Chillmark. In AS charters, ?*cylberge* **3** 237

celde OE (**cielde** WSax) 'a spring'. ?Childhay (Fm), ?Child Okeford

celer ME 'cellar, store-room for provisions'. *Celer de Cerne* **4** 136, *atte Celer* **1** 87, **4** 136

celle ME 'small religious house'. ?Seldown

***cemde** OE 'having a comb or crest'. ?Hr & Lr Kingcombe

cemetery ModE. Cemetery Fm

cēod(e) OE 'bag'. ?Chedington

ceole OE 'throat, channel, gorge'. ?*Celberge* Hundred, ?*Chelwell Water* (*v*. RNs), ?Chilbridge, ?Chilbury Plant.

ceolor OE 'throat, channel', in AS charter, ?*to cellor* **2** 57

ceorl OE 'peasant'. †Chalbury, Charlton Dairy Fm, Charlton Hr Down, Charlton Marshall; f.ns. *Charelton* **2** 78, *Cherleberge* **2** 183

***ceotol** (variant of **cetel**, **cytel**) OE 'a deep valley surrounded by hills'. Chettle, Chettle Head Copse

cēping OE (**cī(e)ping** WSax) 'market'. Blandford Forum; f.n. *Chupyncliue* **3** 145

cerr OE (**cierr** WSax) 'a turn, a bend'. Jordan Hill; f.n. *le Chur Aker* **2** 52

cēse, **cīese** OE 'cheese'. Cheese Lane, Veiny Cheese Pond; f.ns. *Chese Crosse* **3** 364, Chiwick Md **3** 155

chace, **chase** ME 'chase, tract of ground for breeding and hunting wild animals'. Ashley Chase Ho, Branksome Chase, Chetterwood, Chettle Chase Coppice, Cranborne Chase, Holt Forest

chalet ModE. White Horse Chalet

chalk-pit ModE. (Old) Chalk Pit(s) (freq), Chalk Pit Copse (× 2), ~ Ctg, ~ Fm, ~ Lane, Common Chalk Pit Wood, Coombe Chalk Pit, Dryclose Chalk Pit

chalky ModE. Chalky Close Copse; f.ns. Chalky Fd **2** 74, Chalky Grd **3** 339

chambre OFr, ME 'room', *la Baillifeschaumbre* **3** 365, *Brittoneschambre* **3** 31, Priest's Chamber, *Whitt chamber* **4** 93

champaign ModE 'open, level country', f.n. ?Champions **1** 93

channel ModE 'bed of running water'. Back Water Channel, Main Channel, Middle Channel, North Channel, Quay Channel, E, N & S Ship Channel, Wareham Channel, Wych Channel (*v*. RNs)

chanterie ME 'chantry'. *Abbesse Seynt John is Chauntrey* **3** 152, Brembre's Chantry **2** 188, The Chantry, Chantry Fm, ~ House, *Seint Katerine Chauntrye* **3** 152; f.ns. Chantry Coppice **2** 32, Chantry Md **3** 271, *le Chauntrey strete* **2** 186

chapel(e) ME 'chapel'. Anchoret's Chapel, Chapel (freq), Chapel Cl (× 2), ~ Coppice (× 2), ~ Copse (× 2), ~ Court, ~ Down Fm, ~ Fm, ~ Hill (× 3), ~ Lane, ~ Marsh, ~ Plant., ~ Wood, Chapelhay St., *Chapyllan'* **4** 158, Lady Chapel, *Martynschapell* **3** 21, St Aldhelm's Chapel, St Catherine's Chapel (× 2), St Leonard's Chapel, Venn Chapel, Wilkswood Fm; f.ns. Chapel Acre **4** 159, Chapel Orchd **3** 386

charity ModE. Charity Barn, ~ Bottom, ~ Fm (× 3), ~ Wood; f.ns. Charity Fd **2** 214, Charity Ld **3** 55
charnel ModE 'cemetery, charnel house'. Charnel; f.n. *la Charnel* **2** 192
chartrouse OFr 'a house of Carthusian monks', *le Charterhous* **3** 365
cheker ME 'ground of chequered appearance, land with variegated soil'. ?Chequer (Fm); f.ns. *le Checker* **1** 254, Chequer Cl **1** 323
chequer ModE dial. 'the wild service tree'. ?Chequer (Fm)
cherme lME 'clamour, cry', f.n. ?*Cherma'byrye* **4** 384
chestnut ModE. Chestnut Plant.
chick ModE, f.n. Chick Md **1** 99
chilver Do Dial. 'ewe-lamb' (OE **cilfor**), f.ns. Chilve Acres **3** 297, Chilveracres **3** 314, Chilver Knowls **4** 342
chimney ModE. Old Chimney Ctg, Round Chimneys Fm; f.n. Chimney Croft **4** 221
chiri ME 'cherry-tree'. Cherry Close Coppice, Cherry Ctg, Cherry Orchard Fm, Cherry Tree Copse, Cherrytree Lane; f.n. Cherry hays **1** 51
chopped ME 'cut off', f.n. ?*Choppudehulle* **4** 381
christmas ModE. Christmas Rock
cīcen OE 'chicken', f.ns. Chickens Close **1** 93, *Chicunham mede* **3** 294
***cielce** OE 'chalky ground', f.n. Lt Chilkill **2** 288
cild OE 'son of a royal or noble family, young nobleman'. ?Childhay (Fm), ?Child Okeford, Chilfrome, Chilmore. Also in f.ns., e.g. *Childbergh* **2** 13, *Childebergh* **4** 10, *Childeford* **2** 87, Chill Croft **4** 42
***cille** OE 'spring'. ?Childhay (Fm)
***ciltā** pre-Brit 'steep hill, hill-slope'. Chilcombe
cinnamon ModE, f.n. *Cinnamon Lane* **2** 43
cinu OE 'chine, deep valley'. Branksome Chine, Canford Cliffs Chine, *Chine Head* **3** 126, ?Chine Hill, *Chine Pier* **1** 222, Flag Head Chine, Smuggler's Chine; f.ns. Beacon Chine **2** 41, Bittman Chine **2** 41, †Chinchester **1** 304, Chine Church **4** 42, *Chyne* **4** 384
cipp OE 'log', f.n. ?*Chypylthorn* **2** 52
circle ModE. The Circle
cirice OE 'church'. Abbey Church, *Cherchyde* **4** 21, Chescombe Fm & Lane, *Chircheset* **3** 204, Church (freq), Church Barrow, ~ Bottom (× 2), ~ Bridge, ~ Cliffs, ~ Close, ~ Coppice (× 2), ~ Covert, Church Dairy, Church Fm (freq), Church Hayes, Church Ho, *Churchewey* **3** 221, **4** 21, *Churche(y)nde* **4** 158, Churchfoot Lane, *Churchford* **4** 158, Church Green, Church Ground Belt, Church Grounds, *Church*

Hay **4** 326, Church Hill (× 2), Churchill, Church Hill Fm, ~ Lane, Church Hope Cove, (*the*) *Churchehouse* **3** 365, **4** 264, Church Ho (Fm), Church Knap, Church Knowle, Church Lane (freq), Church Moor Copse, Church Road, Church Rock, *Church(s)ton* **1** 306, *Churchstile* **4** 342, *Church'style* **4** 326, Churchstyle Fm, *Churchulle* **4** 158, Church Walk, *Churichwey* **4** 32, *bithechurcchweye* **4** 216, *Fore Church* **4** 195, Frome Vauchurch, Hawkchurch, Margaret Marsh, *Netherburychirch* **4** 311, Whitchurch Canonicorum, Wild Church Bottom, Winterborne Whitchurch. Also in f.ns., e.g. Church Close **4** 434, *Churchfurlang* **4** 10, *the Churchway* **1** 179. In AS charters, ?*ciric hylle* **3** 281

ciric-stede OE 'church site'. *Church Steeds* **4** 114

***cis**, ***ceos** OE 'gravel', f.ns. *Chesberghe* **1** 326, *Cheslane* **3** 187, Chislands **4** 423

cisel, **ceosol** OE 'shingle, gravel'. Cheselbourne, *Little Cheselbourne*, Chesil, Chesil Beach, *le Chesyll* 1 256, *Chisel* **3** 62, Chiswell; f.ns. *Cheselyngfurlong* **3** 374, Chessel Fd **3** 132, *Chesyll hyll* **2** 173, Chisel **4** 248. In AS charters, *cyselburnan*, *Ceosol burnan*, *chiselburne* **3** 202, *v*. RNs.

cisterne ME 'reservoir', f.n. Cistern Fd **1** 206

city ModE. The City, City Fm, Holy City

clacke ME 'din of speech', ?*Clackestret* **2** 186

***clad-** Brit 'to dig', ***clādo-** 'that which is dug, a ditch'. *Vindocladia* **2** 177

clǣfre, **clāfre** OE 'clover'. Cleverlawns, Clover (Coppices); f.ns. Claver Lds **3** 212, Clover Grd **3** 26

clǣg OE 'clay'. Clay Coppice, Clayford Ave & Rd, Clayford Fm, Clayhanger (× 3), Clay Hard Point, *Clay Hill*, Clay Hill (Ho), Clay Hope, Clay Lake (*v*. RNs), Claylake, Clayland Coppice, Claylands Lane, Clay Lane (× 3), Clay Pit(s) (freq), Clay Pit Plant. (× 2), Claypits Fm, Claypits Lane (× 2), Clay Street, Clay Washing Pits, Claywell, Claywood Coppice, Clinger Fm, Old Claygrounds. Also in f.ns., e.g. *Claye Sharde* **4** 26, Clipper Close **2** 51

clǣgig OE 'clayey', in AS charters, *clæian lane* **4** 34

clǣne OE 'clean, clear of unwanted growth'. Clanden Hill, Clandon, Clanfield **3** 360, Clean Hollow Plant., Clenham Coppice & Mill; f.ns. Clan **1** 198, *Clanebourgh* **4** 476, *Clanfield* **3** 231, Gt Clanfield **3** 308

clām OE 'mud, clay', f.n. Clampits Copse **4** 248

claper ME 'rabbit burrow, warren'. Clapper Close; f.ns. ?Clapper **2** 157, 228, ?Clapper Close **4** 423, 462

clap-gate ME 'swing-gate'. Clapgate, Clapper Gate; f.ns. Clapgate Cl **4** 328, Clap Gate Grd **2** 32, Lr Clap Gate **4** 248, ?Clapper Gate Orchd **2** 199

clapper ModE 'rough bridge'. Clapper Plant.; f.ns. ?Clapper **2** 157, 228, ?Clapper Close **4** 423, 462

clāte OE 'burdock'. Clatcombe Ho & Lane

clawu OE 'claw, river-fork', f.ns. *Clauhulle* **1** 326, *Clawpittes* **1** 229, *Lanclawes* **1** 298

cleavers ModE dial. 'goose-grass', f.n. Cleaver(s) **4** 115

cleft ModE 'crack, fissure', f.n. The Cleft **2** 66

clerc ME 'cleric', f.n. *Clerkesplace* **1** 326

clere ME 'forest glade'. *Clear-walk* **3** 52

clif OE 'cliff, bank, steep slope of a hill-side, escarpment'. Bartlett's Cliff, Bell Cliff, Bincleaves, *Breme(l)clift* **4** 90, Brickells Cliff, Broom Cliff, Burning Cliff, Burton Cliff, *Bynetheclyft* **3** 143, Canford Cliffs, Catsley Fm, Church Cliffs, *Cleave* **2** 246, Cleave Ham, Cleevehill Lane, Lt Cleeve Wood, The Cliff (freq), Cliff Coppice (× 2), Cliff Ctg(s) (× 3), Cliff End, Cliff Gardens, Cliff Ho, Cliff Wood, (The) Clift, Clifton Maybank, Clyffe Fm & Ho, Duncliffe Hill, East Cliff, Eccliffe, Furzy Cliff, Gad Cliff, Grove Cliff, Gun Cliff, Hen Cliff, High Cliff, Hog Cliff Bottom & Hill, Holme Cleve, Hyde Clift Plant., Kitwells Cliff, Lewcombe Woods Cliff, Little Cliff, Priors Cleave, Redcliff Fm, Redcliff Point, Ridge Cliff, Rock Lea Point, Shorne Cliff, ?Stockley Fm, Thorny Cleeve, Toller Cliff Firs, *Under-*, *Upestikeleclif* **3** 246, Wear Cliffs, Westcliff, West Cliff (× 2), Whitecliff Fm, Yew Cliff Coppice. Freq in f.ns., e.g. *a boue cliue* **4** 10, Cle(a)ve(s) **1** 235, **3** 70, Clifts **3** 25, *Clyvemede* **4** 411, *Southcleeves* **3** 6. In AS charters, *clif*, *eficlif* **1** 31, *cliveleghe* **3** 255, *clyfœ* (*v*. Addenda **3** 267)

clock ModE 'clock'. Clock Ho

clock ModE dial. 'beetle', f.n. ?*Clockborow* **4** 58

(ge)clofen OE 'cloven, split', f.n. Clam Barrow **1** 260, *la clouenbergh* **1** 326, *Cloveneburgh* **3** 255

***clopp(a)** OE 'lump, hillock'. Clapton Bridge

clos(e) ME 'enclosure'. Bagwood Close, Barn Close (freq), Birch Close, Boat Close, Broadclose (Fm), Broom Close (× 3), Calf Close Coppice, Calf Close Lane, Castle Close, Chalky Close Copse, Chapel Close (× 2), Cherry Close Coppice, Christmas Close, Close Copse, Copse Close, Dryclose Chalk Pit, Dry Close Plant., Durrant's Close, Furzey Close Row, Furzey Close Wood, Greenclose, Green Close Dairy, Greenclose Fm, Green Close Lane,

Heaveclose Barn, Hilly Close Copse, Homeclose Lane, Horse Close (freq), James's Close Pit, Long Close, Long Close Fm (× 2), Marl Close, Martin's Close Copse, Morris Close, New Close (Coppice), Oatclose Wood, Parson's Close, Pear Close Copse, *Pease Leaze*, Pepperclose Trees, Pit Close Copses, Pond Close (× 2), Pound Close (× 2), Rough Close Barn, Ryeclose, Rye Close, Rye Close Fm, Sing Close Coppice, Spring Close Lake, Thorny Close Copse, Vines Close, Warren Close Plant., Wellclose Plant., Wheat Close, *Whitecliff Mill*, Woodclose (× 2). Freq in f.ns., e.g. *Buryclose* **2** 13, *Courtclos* **1** 249, *Crosse close* **4** 34, *Edge Closse* **3** 6

clott OE 'lump of earth, clod'. ? Clotfurlang Lane

clotty ModE 'lumpy', f.n. Clotty Grd **3** 293

club ModE, f.n. Club Close **3** 388 (*v.* NDEFN 80)

clubman eModE 'one of a mob of local farmers driven to defend their homes in the English Civil War'. Clubmen's Down (*v.* Addenda **3** 100)

clūd OE 'a rock, a rocky hill'. ?Clouds Hill; f.n. Clouds **4** 42

clump ModE 'cluster of trees'. Badbury Clump, Batcombe Hill Clump, Beech Clump (× 2), Blackhill Clump, Black Hill Clump, Boat Ho Clump, Bottlebush Clump, Cannon Clump, Castle Clump, Cherett's Clump, *Clump*, The Clump(s) (× 4), Clump Hill, Con(e)ygar Clump, Cuckoo Clump, Elderton Clump, Ellen Clump, Farringdon Clump, Folly Clump, Gore Clump, Great Down Clump, Green Knap Lane, Grimstone Clumps, Gussage Clump, Half Moon Clump (× 2), Haydon Clump, Heart Clump, Holly Clump, Holton Clump, Horse Clump, Horseshoe Clump, Horse Shoe Clump, Houghton Down Clump, Kidney Clump, Knoll Clump, Lady Caroline's Clump, Legg's Clump, Little Clump, Locust Clump, Maggot Clump, Memory Clump, Milton Mill Clump, Mountain Clump, Mount Pleasant Clump, Nutley Clump, Oakey Clump, Old Clump, Pallington Clump, Pigeon Clump, The Pine Clump, Round Clump (× 2), St John's Clump, Sandhill Clump, Sea Mark Clump, Square Clump (× 2), Starve Acre Clump, Sutton Clump, Telegraph Clump, Three Barrow Clump, The Three Clumps, Three Cornered Clump, Throop Clump, Turnworth Clump, Vary Clump, Windmill Clump, Wolfeton Clump, Wood Hill Clump, Zareba Clump; freq in f.ns. e.g. Clump **1** 371, Clump Grd **2** 22

cnæpp OE 'hill-top, hillock'. Ashey Knapp, Ballast Knap, Beacon Knap, *Bell Knappe* **3** 183, Benville Knap, Bere Knap, Berry Knap, Brunsell's Knap, Burrow Knap, Butt's Knap, Chilly Knapp, Church Knap, Cowherd's Knap, Cowleaze Knap, Crocker's Knap, Dirty

Knaps, Ferny Knap Coppice, Golly Nap Fm, Grandfather's Knap, Green Pit Knapp, Hilling Knap, Hobb Knapp, *Holm Bush Knap* **4** 397, Holmes Knapp, Hornshayes Knap Cross, Horsey Knap, Knap Coppice, Knap Corner, Knap Cottages & Dairy, Knap Fm, Knap Hill, (The) Knapp (× 3), Knapp Bridge & Fm, ~ Ho, *Load Nap* **4** 207, Newnham Knap Fm, *Park's Knap* **4** 39, Pince's Knapp, Rocky Knap, Sandy Knap, Sansom's Knap, Searley's Knap, Seawell Knap, Sheppard's Knap, South Sleight Knap, Stoke Knap, Tiley Knap, Tizard's Knap, Town's Knap, Upper Knapps Fm, *Wacheknappe* **4** 397, Walditch Knap, Windmill Knap, Winnowing Knap. Freq in f.ns., e.g. *Bromyknapp* **4** 10, *Knapfurlang* **1** 260, Pressnaps **4** 410. In AS charters, *cinep*, *cnap* **3** 206, *þorndunes cnep* **3** 208

cnafa OE 'boy, servant'. Knaveswell Copse & Fm; f.n. *Knave acr'* **2** 53

cniht OE 'thane, retainer of a high personage, knight'. East Knighton, Knighton (× 2), Knighton Ho, *Knighton Street*, *Knyghthold* **3** 279, West Knighton; f.ns. Knightney **3** 44, *Knyght brydge* **4** 16

cnoll OE 'hill-top, hillock'. Blacknoll, Boat Knoll, Bucknowle Ho, Chetnole, Chart Knolle, Church Knowle, *Fern Knowle* **4** 324, Hartland Moor, Hinknoll Fm & Hill, Knoll, The Knoll (× 6), Knoll Copse, Knoll Ho, Knolls, ?Knowle Cross, Knowl(e) Fm, Knowl(e) Hill (× 6), Knowlton, Little Knoll Copse, Old Knowle, †Penbury-Knoll, Pins Knoll, Puncknowle, Round Knoll, Squibb's Knoll. Freq in f.ns., e.g. *Dunknoll* **4** 19, *Knollwalle fylde* **1** 264, Knowle Hill **2** 62, *Stuntecnoll* **1** 185

***cnollisc** OE 'hilly', f.n. *Cnollisschefurlang* **3** 255

***cnyll(e)** OE 'knoll, hillock'. Knell Coppice, ?Knowle Cross; f.ns. *la Cnylle* **1** 313, *Knulleweye* **3** 315, Nell **3** 186, **4** 300, Nill **1** 227, **3** 186,

coach ModE. Coach Road

coarse ModE 'rough', f.ns. Coarse Md **4** 115, **4** 423

coastguard ModE. Coastguard Station (Chaldon Herring, West Lulworth, Tyneham)

***cobb(e)** OE 'a round lump, a rounded mass'. (The) Cobb, ?Cobley Fm; f.ns. Cob **4** 266, *Cobham* **4** 437, Cob Wood **4** 115

cobeler ME 'cobbler'. ?Cobblers Pit

cōc OE 'cook', Cooks Row **2** 184

cocc[1] OE 'heap, hillock'. ?Cockett Hill, ?Cockhill Fm; f.ns ?Cockhill **2** 104, *lez Cokkes* **1** 236. In AS charters, ?*cockes þorne* **2** 122

cocc[2] OE 'cock (of wild bird), woodcock'. ?Coaxdon Hall, ?Cockhill Fm, ?Cocks' Moor Pond, Cockwell Coppice, ~ Fm, ?Coxditch,

?Cocknowle, Cock Walk, Cogdean Elms, ~ Fm, ~ Hundred; f.ns. Cocklands **1** 343, *Cokeden* **2** 173. In AS charters, ?*cockes þorne* **2** 122

coccel OE 'cockle, darnel', f.ns. *Cockelhaie* **1** 229, Cockerland **4** 233, Cockleton **1** 236. In AS charters, *coccelmæres dene* **4** 302

cocc-geset OE 'place where woodcock are found', f.n. Cocks Heart **3** 235

cocc-rodu OE 'clearing where woodcock are netted'. ?Cockcrow Copse, ?Cock Crowing Corner, Cock-Crowing Stone, Cockroad (× 2), ~ Copse, ~ Lane, ?Cock Row, ?Cockrow Fm. Also in f.ns., e.g. Cock Crow **3** 287, Cockroad **2** 262, **3** 225, Cock Road **4** 58, Cockroads **2** 22. In AS charters, *coc rodæ* (*v*. Addenda **3** 267)

***cocc-scyte** OE 'glade where woodcock are netted', f.n. *Coxshid Wood* **4** 463

***cōc(e)** OE 'lump, hillock', f.n. ?*Cooke* **1** 229

cock ME 'chief', f.n. Cock Millers **3** 213

cocked hat ModE. Cocked Hat Coppice

cockerell lME 'cockerel', f.n. ?*Cokerelshay* **4** 384

cock-pit ModE 'enclosure for cock-fighting'. Cockpit (Head), Cockpits (rocks); f.ns. (The) Cockpit **1** 283, **2** 139, 234

codd OE 'bag, hollow', f.n. ?*Cod Meadow* **4** 302

coffin ModE, f.n. Coffin Acre **2** 119

cokaygne ME 'an imaginary land of abundance and bliss'. *Cocaygne* (a tower) 1280–5 (at Corfe C. **1** 11), Coking Fm

coke ME 'core, heart of a wood', f.n. ?Coke **4** 328

col[1] OE 'charcoal'. ?Coalhill Drove, Coal Pit Copse, ?Colber Crib Ho, Colcombe Fm, ?Cole Combe, ?Colehill, Cole Street Fm & Lane, ?Cole Wood, ?Coll Wood Ctg/Gt Coll, Coll Wood, ?Colly Fm, Colway Gate & Lane; f.ns. Cold Fd **3** 55, *Colpyttes* **1** 71. In AS charters, *col pytte* **4** 302

cōl[2] OE 'cool'. ?Coalhill Drove, ?Colber Crib Ho, *Colebrok'* (*v*. RNs), ?Cole Combe, Colmer Copse, ~ Fm

cold harbour ME 'cold shelter', *v*. **c(e)ald, here-beorg**

coliere ME 'maker or seller of charcoal'. Collier's Barn/Lane

***colig** OE 'pertaining to charcoal', f.n. ?*Colicroft* **4** 302

***coll** OE 'hill'. ?Colehill, ?Coll Wood Ctg/Gt Coll Wood; Court Koll **2** 78

college ModE. College Down

colt OE 'colt'. ?Colston Fm, ?Colthay Coppice, ?Coltsclose Corner; f.n. *Coltclose* **3** 79, *Coltcroft* **4** 34, *Coltemede* **4** 270

column ModE. Jubilee Column

common ModE, **comun** ME adj. 'common', sb. 'tract of land used in common'. The Common (× 2), Common Barn, Common Bushes, Common Chalk Pit Wood, Common Coppice, Common Drove (freq), Common Fm (× 3), Common Gate, Common Hill, Common Ho, Common Lane (freq), Common Plant. (× 3), Common Pond, Lr Common Rd, Common Water, Common Wood (× 3), Locust Common Plant., Poor Common (× 2), *Qeenes Commen* **3** 265, Stalbridge Common, Stoke Common **3** 229, Weam Common Hill. Freq in modern minor names as ~ Common, e.g. Alderholt Common, *Ashmore Common*, Blackven Common, Bonsley Common, Breach Common, Broadridge Common, Cann Common, Chalbury Common, Chettle Common, Corfe Common, Cowgrove Common, Cranborne Common, Crendell Common, Deadmoor Common, Dewlands Common, East Stour Common, Farnham Common Wood, Fifehead Common, Furze Common Copse, Furze Hill Common, Furzy Common, Gaunt's Common, Gotham Common, Ham Common, Handley Common, *Haselbury Common*, Haythorn Common, Hookswood Common, Horton Common, Leigh Common, Little Common, Lower Common, Lulworth Common, Marsh Common, Monkton Common, *Owre Common*, Parley Common, Rushton Common, St James's Common, St Leonard's Common, South Common, Stoke Common, Stokeford Common, Stroud Common, Sutton Common, Tolpuddle Common, Verwood Common, Wareham Common, Whatcombe Common, White Hill Common, *Witchampton Common,* Woodcutts Common, Woodlands Common. Also in f.ns., e.g. *Comon Crofte* **2** 215, Common Grd **2** 234, The Commons **3** 145

coni ME 'rabbit'. Coneygar Copse, Coney Lane, Coney's Castle; f.ns. Coney Crate **4** 381, Coney Croft **1** 226, **4** 434, *Conye layne* **4** 218

coninger ME 'rabbit warren'. Conegar Hill (× 2), Con(e)ygar Clump, ~ Coppice (× 2), ~ Copse (× 2), ~ Fm, ~ Hill (× 3), ~ Lane, ~ Meadow Coppice, ~ Wood, Coneygore Hill, *Coniger Lane* **1** 251, Conigre Coppice, Connegar Fm, Conniger Lane, Cunniger Copse. Freq in f.ns., e.g. Coneygear **2** 124, *Conyger* **4** 10, *le Conynger* **2** 66, *Cunygere Bottome* **2** 215

coning-erth ME 'rabbit warren'. Coneygar Copse; f.ns. Coney Garths **4** 42, Conygarth **1** 69

conner ME 'a tester'. ?*Conners* **4** 394

***cönǭg** PrWelsh, ***conǭg** PrCorn of doubtful meaning. Combs Ditch (Hundred)

constitution ModE. Constitution Hill

contasse ME 'countess'. Countess Coppice, Countess Point

controversy ModE 'contention', f.n. Controversy **1** 281

cony-burrow ModE 'rabbit burrow', f.ns. Cunbury **4** 381, Cunnerberry **2** 199

copis ME 'coppice'. Ashy Coppice, *Coppes Woode* **2** 94, The Coppice (freq), Coppice Ho, Copse Barn, Copse Close, Copse Gate Dairy, Copsehayes, Copse House Fm, Copse Stile, Greatstone & Littlestone Coppice. Freq in modern minor names as ~ Coppice, ~ Copse, e.g. Ash Coppice (× 2), Badgers Copse, Bagwood Coppice, Ball Coppice, Barkus Coppice, Barn Coppice, Bartlett's Coppice, Beach Coppice, Belhuish Coppice, Binnegar Coppice, Bower's Coppice, Bramble Coppice, Brockhill Coppice, Broomhill Coppice, Brown's Coppice, Buckshill Coppice, Bunker's Hill Coppice, Calf Close Coppice, Cherry Close Coppice, Clayland Coppice, Clyffe Copse, The Coppice, Cowleaze Coppice, Culeaze Coppice, Damerhill ~, Eight Acre ~, Farm ~, Fir Coppice, Fossil Coppice, French Grass Coppice, Furze ~, Furzy Coppice, Gillard's Coppice, Gould's Coppice, Great Coppice, ~ Copse, Grey's Coppice, Gully ~, Half Moon ~, Ham Coppice, Hazel Coppice (× 2), ~ Copse, Heave Coppice, Hewish Coppice, Hill Coppice, Hill's Copse, Holly Coppice, Homemeadow Coppice, Horse Coppice, Hurpston Coppice, Hurst Copse, Hyde Coppice, Jackman's Coppice, Jordan's ~, Kimmeridge Coppice, Kit Hill Coppice, Knap Coppice, Landshare Coppice, Lays Coppice, Limekiln Copse, Little Coppice, Little Copse, Little Hurst Coppice, Little Knoll Copse, Little Puddle Coppice, Long Coppice (freq), Longcroft ~, Long Coppice (freq), Long Copse, Long Mead Coppice, Mate's ~, Moor Coppice, Muston Copse, Muddox Barrow Coppice, Napier's Copse, Nathan's Coppice, New Coppice (freq), Oakley Wood, Orchard Coppice & Fm, Pallington ~ , Payne's Coppice, Peak Coppice, Perry Coppice, Philliols Coppice, Pickard's Coppice, Pigeon Ho Copse, Ploughman's Coppice, Pole Coppice (× 2), Puddle Coppice, Shailes Copse, Six Acre Coppice, Skinner's Coppice, Spear's Coppice, Spring Coppices, Spring Garden Coppice, Square Coppice (× 2), Stafford Park Copse, Tanpits ~, Tatton Coppice, Three Cornered Coppice, Troy Town Copse, Twenty Acre Coppice, Two Mile Coppice, Vicarage Coppice, Vineyard Coppice, Waddock Copse, Walls Coppice, Wrackleford Coppice. Freq in f.ns., e.g. *Greene Coppice* **2** 205, *Hammecopice* **2** 143, *Taile coppice close* **2** 216

copp OE, **cop** ModE 'hill-top, summit'. ?*Copcourt Poynt* **1** 125, ?Golden Cap, ?Tyneham Cap; f.n. ?*Coppehey* **3** 6, *coppehulla* **3** 143, *Coppeland* **3** 280, *Cop(p)ham* **3** 340, Coppice St. **3** 145

coppede OE 'peaked, pointed'. Coppet Hill; f.ns. *Copped close* **1** 42, Coppert Hill **4** 381

copper ModE 'copper (money)'. ?Copper Bridge Md **1** 27

copperas ModE 'copperas', f.n. *coperis mynes* **1** 51

***(ge)coppod** OE 'polled'. Coppleridge; f.ns. Cop Crouch **2** 104, Coped Bush Fd **2** 282, Copythorne **2** 77

copyhold ModE 'land held at the will of the lord of the manor'. Copyhold Lane; f.n. (By) Copyhold **2** 69

***corf** OE 'a cutting, a gap, a pass'. Corfe Castle, Corfe Fm, Corfe Hill/Corfehill Fm, Corfe Mullen, Corton Fm, Coryates. In AS charters, *corfȝet* **4** 302, *corfgetes westran cotan* **4** 27, *miclan corf* **4** 302

***Cori** pre-English r.n. of unknown meaning and origin. ?Corry Brook (*v.* RNs), ?Corscombe

cork ME 'cork-oak tree'. ?Sedge Oak Coppice

cormorant ModE, f.n. *Carmarantes Lane* **3** 178

corn[1] OE 'corn, grain'. *le Cornemarkett* **3** 364, Cornhill **1** 350, **3** 360, *Corn Hill or Corn Market* **3** 146, Cornhill Ctg, Corn Mill (freq); f.ns. Corn Croft **3** 293, *Corne feilde* **4** 10, *les Cornefeilds* **1** 326, Corn Fd **1** 136

corn[2] OE 'crane', f.n. ?*Cornebarrowe* **1** 136, ?*Cornebary* **1** 324

corner ME 'corner, nook'. Amen Corner, Bartlett's Corner, Barn Corner, Beacon Corner, Blackpool Corner, *Brodemede Corner* **4** 376, *Brokecorner* **3** 332, Burrow Corner, Cave's Corner, Chalk Corner, Chantell's Corner, Charleston Corner, Cock Crowing Corner, Coltsclose Corner, The Corner (× 2), Corner Fm, (*the*) *Cornerhouse* **3** 365, Cornerway(s), Cow Corner, Cuckolds's Corner, Darby's Corner, Fleet's Corner, Gallows Corner (× 3), Gipsies' Corner, Grange Corner, Gudgeon Corner, Gunning's Corner, Haydon Corner, Hell Corner, Holmes's Corner, Hope Corner, Hyde Corner, Jack's Hedge Corner, Knap Corner, Miller's Corner Plant., Nine Corners Close, Park Corner, Peaked Corner, Pompey's Corner, Pools Corner, Quarry Corner, Rivers' Corner, Rough Corner Copse, Scadden's Corner, Shepherd's Corner Fm, Sloven's Corner Coppice, Squirrel's Corner, Three Corners, Totnell Corner, Trevett's Corner, Weston Corner Plant.; f.ns. *Brags Corner* **3** 226, *Cornere* **4** 10, Corner Piece **2** 223

cornered ModE 'with corners'. Nine Cornered Copse, Three Cornered Clump, ~ ~ Coppice, ~ ~ Plant.
coronation ModE. Coronation Plant.
cors Welsh 'marsh, bog'. Cosmore; f.ns. *Cosbere* **3** 340, Costly Md **3** 162
cortil(l)e OFr 'garden, enclosure', f.n. ?*Curtell* **2** 10
cosh ME 'cottage, hut', f.n. *le Coysh* **3** 219
cost OE 'trial, dispute', f.ns. Cosloe **1** 79, ?Costell **3** 231, Costlow **2** 157
cost OE 'tansy', f.n. ?Cos(t)ley **4** 347
cot OE 'cottage'. Calcots Wood, ?Chalcott's Coppice, Chaldecot, *Chaldecots*, Cothayes Dairy, Cotness (Wd), *Cot(t)cote* **4** 146, Cott Ho, ?Cottonfields Quarries, ?Court Ho Dairy, Eastcot, Furzecott, Longcutts Fm, Northcote, Southcot, Upcot, Woodcote, Woodcutts, Woodcutts Fm. Freq in f.ns., e.g. *Cot(t)haies* **2** 52, *Geldyngescote* **2** 57, *le Hencote* **1** 264. In AS charters, *corfgetes westran cotan* **4** 27
cotage ModE 'cottage'. The Cottage (freq), Cottage Fm (× 2), ~ Green, The Old Cottage. Freq in modern minor names as ~ Cottage(s), e.g. Florence Ctg, Pentre Ctgs, Waterloo Ctgs (all in Corfe Mullen), Heathfield Ctg, Holly Ctg, Myrtle Ctg, New Gate Ctgs, Summer Ctgs (all in Owermoigne), Alma Ctgs, Cove Ctgs, Lloyd's Ctg (all in Portland). Also freq in f.ns., e.g. Cottage Leaze **2** 51
cotere ME 'cottager', f.ns. ?*Cotiers Drove* **3** 187, *Kotterynpoul* **1** 184
coth(e) ME 'disease', **cothy** eModE 'diseased', f.n. Cothy Plot **1** 93
cot-stōw OE 'collection of cottages'. ?Custard Hill
c(o)unte ME 'female private parts', f.ns. ?*Contelane* **1** 216, *Countedone* **4** 182, *the Narrow Cont* **1** 259
counte-hous ME 'counting-house', *Contehowse* **3** 365
course ModE 'practice of hunting hares', f.n. Course Fd **1** 99
court ME 'large house, manor, hall of justice '. Abbots Court, Abbot's Court Fm, Belmont Court, Bere Fm, *Blunt's Court* **4** 309, Cannings Court, Chapel Court, Chedington Court, Coombe Court, *Copcourt Poynt* **1** 125, Corscombe Court, *Court* **4** 388, The Court, Court Ash Ctg, Courtfield, Court Fm (freq), *Court Fm or Carrant's Court Fm*, Court Hill, Court Ho (freq), Court Orchard, Court Place, Court Pound, Courtwood, *Damory Court*, Dean's Court, Duntish Court, Frampton Court, Gable Court, *Gold Court*, Holditch Court, Hooke Court, King's Court Palace, Little Court, Loders Court, Marsh Court, Melplash Court, Moorcourt Fm, Moor Court Fm, Nash Court, Netherbury Court, Old Court Copse, Ridge Court, Seaborough Court, Somerleigh Court, South Court, Stanley Court, Stretton

Court, Wake Court, Wild/Wyld Court, Wykes Court. Also in f.ns., e.g. *Courtclose* **4** 40, *Courte meade* **4** 19, *le Courtewyke* **1** 313, *Court Koll* **2** 78, *Courtlynch* **4** 10

courthouse ModE. Courthouse Shoot

cove eModE 'small bay'. *Boat Cove*, Bowleaze Coveway, Buck's Cove, Castle Cove, Chesil Cove, Church Hope Cove, Conner Cove, The Cove, Cove Ctgs, Cove Rocks, East Ebb Cove, Gore Cove, Headbury Big Cove, Lulworth Cove, Man o' War Cove, Mike's & Willy's Coves, Mutton Cove, Turner's Cove, Wallsend Cove; f.n. Bucks Cove **2** 36

cover-hole eModE 'hole with a lid or cover'. ?Cover Hole

covert ME 'woody shelter for game'. Badger Covert, Belstone Covert, Black Covert (× 2), The Covert, Big Covert, Deadman's Covert, Dole's Covert, Fook's Covert, Fox Covert, Greenland Covert, Guy's Covert, Hanging Covert, Hatch Covert, Home Covert (× 3), Laurel Covert, Little Covert, Lower Covert, Lox Lane Covert, Luccombe Covert, Main Covert, New Covert (× 2), Nyland Thorn Covert, Pennington Covert, Potwell Covert, Pound Covert, Sixpenny Covert, Soggy Covert, Square Covert, Westhay Covert, Withy Slade Covert, Wood (Ho) Covert

cowherde ME 'cow herd'. Cowherd's Knap; f.n. ?*Cowardes Drove* **2** 94

cow-house eModE. *Cowehowse* **4** 378

cowslip ModE 'species of primrose'. Cowslip Fm

coy ModE 'a decoy'. Coy Pond

cozenage ModE 'deception, fraud'. Cuzenage Coppice; f.n. Cozenage Md **2** 37

crabba OE 'crab'. *Crab* (*v.* Addenda **1** 220), Crab Fm, ~ Hole

crabbe OE 'crab-apple'. Crab Coppice, Crab Orchard; f.ns. *Crabbelond* **4** 10, Crabbe Md **4** 423, Crabb Mdw **1** 151, *Crabelane* **2** 10

crabbetre ME 'crab-apple tree'. Crabtree Coppice (× 2), ~ Copse; f.n. Crabb Tree (Cl) **3** 213, **4** 281

cradeler ME 'one who makes cradles', f.n. Cradlers **1** 41

***cramb** OE 'bend', f.n. *Wood Crampe* **2** 217

cran OE 'crane'. Cranborne, *Craneburn* **3** 32 (*v.* RNs), ?Crane Hill Plant., Cranes Moor; f.ns. *Cranbornella* 2 173, *Craneforlong* **3** 255, *Cranwille* **4** 476. In AS charters, *cranemere* **3** 41, *cranmere* **3** 128, *cran mor* **2** 158

crāwe OE 'crow'. *Gt Crawford*, Crawthorne Fm, Crow Copse, Crowcroft, Crow Hill, ?Crowthorne Lane; f.ns. *Crowenest* **4** 10, *Crowthorne* **2** 52. In AS charters, *crawan forda* **2** 136

creamery ModE. Yonder Hill Creamery

creeper ModE '?bird, insect, plant', f.n. Creeper Linch **1** 343

crest ModE. Hill Crest

crewel Do dial. 'cowslip', f.ns. Cruel Md **4** 463, Cruel Mow Plot **4** 451

cribbe ME (**crib(b)** OE) 'a crib, a manger, an oxstall'. Colber Crib Ho, Cripton Barn

cribhouse eModE 'manger'. Crib House (Fm) (× 2); f.ns. Cribb Houses **2** 100, Crib House **3** 44, 46

crike ME 'creek'. ?Crack Lane, Creekmoor Bridge, Lulworth Cove, Worbarrow; f.n. Fern Hill Creek **1** 242

***crikeled** ME 'indented', f.n. *Krykelydehull* **4** 182

crinkle-crankle ModE dial. 'a zigzag', f.n. Crinkle Crankle **4** 267

cristel-mǣl OE 'cross', in AS charters, *cristemeleighe* **3** 255

Cristen OE 'Christian', f.n. *Cristenecros* **3** 32

***crōc** OE, **crok** ME 'a crook, a bend'. (Baby) Hyde Crook, ?Crockway Fm, ?Crook Hill, ?Jacolines Crock, Knobcrook, Knob's Crook; f.ns. Crook **2** 66, ?Crook Hill **2** 171, *atte Crouke* **3** 64, ?*Cruclemede* **4** 348, Nob Crook Orchd **3** 197

crocc OE 'crock, pot'. ?*Crockern Stoke*, ?Crockway Fm, ?Cromlech Crock Lane, ?Jacolines Crock; f.ns. *Crokewell'hey* **1** 120, *Crokkerne fyldys* **2** 192

***crocc-ærn** OE 'pottery', f.ns. *Crockernewey* **4** 182, ?*Crokerne rewe* **4** 156

***croccere** OE 'potter'. ?*Crockern Stoke*, *Crokerford*; f.n. *Crockeresrewe* **4** 156

croft OE 'small enclosed field'. Abbeycroft Down, Alcroft Copse, Ashcroft, Beaucroft, Blackcroft Coppice, Bowcroft Hill, Broad Croft, Broadcroft Quarries, Coutman's Croft, ?Crack Lane, Crate Coppice, Crate Wood, Crete Bottom & Hill, Croft, The Croft (× 2), The Crofts Down, ?Crofts Fm, Crowcroft, Dencroft Lane, Duncecroft Quarries, Ewecroft Lane, Ferncroft Fm, Goldcroft Fm, Haycraft, James Cross Lane, Jennycroft Lane, *Knap Craft* **4** 294, Longcroft Coppice, Loomcroft, *Ludcroft*, Mancroft, *Maydenecroft* **3** 174, Nutcrack Lane, Oatcroft Lane, *Pease Leaze*, Pettycrate Lane, *Phil(l)ips*, Pit Croft Copse, Sedgecroft Common, Sheepcroft Row, Snailscroft Fm, Staple Croft, Thorncroft, Waycroft Quarries, Whin Croft, Withies Croft Quarries, Wolvercroft, Woodcroft, Wych Croft. Freq in f.ns., e.g. *Chalfecroft* **2** 70, Coney Crate **4** 381, East

Croat **3** 121, *Eliotescroft* **4** 10, *Kingescrofte* **2** 161, Odd Crates **2** 19, *Otecroft* **3** 33

***crōh**[2] OE 'nook, corner', f.n. *Wibbancrowe* **3** 259

crois OFr, ME 'cross', f.ns. *Atte Croyz* **4** 343, *Bolecroys* **3** 226

croked ME 'crooked'. Crooked Oak Hill, Crooked Withies; f.ns. (*the*) *Croked Close* **2** 192, *Croketaysshe* **1** 172, *quercum Crocat* **4** 454

cromlech ModE. Cromlech Crock Lane

crookle eModE 'to bend', f.n. ?*Cruclemede* **4** 348

cros lOE, **cross** ME, ModE 'cross', often 'cross-roads' in modern minor names. Adber Cross, Ansty Cross, Baker's Cross, Bare Cross, Barnes Cross, Beckford Cross, Bellows Cross, Bishop's Cross, Bockhampton Cross, Boniford Cross, Bowood Cross, *Boyscrose* **3** 336, Brimley Cross, Broadhayes Cross, *Butter Crosse* **3** 148, Carters Cross, Champernhayes Cross, Charing Cross, *Chese Crosse* **3** 364, Clarkham Cross, Cockwell Cross, *Cold Crosse* **3** 183, Cole's Cross, *Coles Cross* **4** 323, Corscombe Cross, Cox's Cross, Crandons Cross, (The) Cross (freq), Cross & Hand, Cross Fm (× 2), ?Cross Fm **4** 447, Cross Hill, Cross Ho, Crossleaze Coppice, Cross Ways, *Cutlers Cross* **4** 327, David's Cross, Denhay Cross, Dodge Cross, Drimpton Cross, Duntish Cross, Furleigh Cross, Gain's Cross, Giles Cross, Gore Cross, Green's Cross, Grexy Cross, Ham Cross, Harman's Cross (Rds), Hawley Cross, Headless Cross, Headstock Cross, *High Cross*, Hill Cross, Ivy Cross, Jackman's Cross, Knowle Cross, Lancombes Cross, Langlands Cross, Langton Cross, Leigh Cross, Long Cross, Long Cross Pit, Main Cross, Market Cross, Marlpit Cross, Marr's Cross, Marshwood Cross, Maudlin Cross, Miles Cross, Minchington Cross, Monkton Wyld Cross, Naked Cross, Newcross Ctgs, New Cross Gate, Northay Cross, Paddock's Cross, Penn Cross, *Perry Cross* **4** 319, Pickaxe Cross, *le Pultrey Cross* **3** 148, Quarry Cross, Red Cross (× 2), Ridge Cross, Rodway Cross, Royal Oak Cross, Ryalls Cross, St Catherine's Cross, Sansom's Cross, Shave Cross, Shearing Cross, Short Cross, Shortmoor Cross, Six Cross Ways, Stake Ford Cross, *Stoyes crosse* **4** 163, Three Ash Cross, ?Three Legged Cross, Tincleton Cross, Trickett's Cross, Tucker's Cross, Under Cross Plant., V Cross, Vere Cross, Vinney Cross, Waddock Cross, *Warmwell Cross*, Watton Cross, Westhay Cross, Whetham Mill Cross, Whetley Cross, Whitchurch Cross **4** 449, White Cross, Whitecross, Whitmore Cross, Woodhouse Cross, Woodman's Cross, Wootton Cross. Freq in f.ns., e.g. *Baylyscroce* **4** 9, *Cristenecros* **3** 32, *Crosse close* **4** 34, *Faber Cross* **2** 52

cross ModE adj. 'lying across'. Cross Dyke, Cross Lane(s); f.n. Cross Drove **2** 179

crossing ModE. Axe Crossing, Revel's Crossing

cross-roads ModE. Cross Roads Fm (× 2), Kingcombe Cross Roads

cross-tree eModE 'gallows'. Cross Tree Fm; f.n. *le crosse tree* **4** 10

crūc OE, **crouche** ME 'cross'. Cox's Cross, Crooch Fm, Cross **3** 295, Cross Hill, *la Cruche* **4** 359; f.ns. Crouch **4** 343, *Crouchehey* **3** 32, ?Crouch Hill **2** 145, *Nethere Crouchouse* **3** 58, *Palmecruch* **2** 143, *Wymarecruch* **2** 11

crud(de), **crod(de)** ME 'curd', f.n. *Croddehay* **1** 183

***crūg** PrWelsh, PrCorn 'mound, hill, barrow'. Church Hill **4** 123, ?Churchill **4** 277, ?Churchill Drove, †East Creech, Creech Hill, West Creech, Long & More Crichel, Crook Furze, Crook Hill (× 2), Crouch Lane; f.ns. ?*Brimecherche* **1** 113, ?Churchill **2** 77, Creech **1** 292, **4** 286, ?*Creekes Hill* **1** 229, ?Critch Hill **3** 355, ?Crouch Hill **2** 145. In AS charters, *cruc* **4** 290, *miclan cruc* **4** 290

crumb, **crump** OE 'crooked'. Crumpet's Fm; f.ns. ?Croam Bags **1** 236, Crumpley Md **2** 32

crundel OE 'a pit, a quarry'. Crendell, Crendle; f.ns. *Crundelford* **3** 32, *Depecrundale* **1** 143. In AS charters, *crundel* **1** 31, **2** 57, **3** 128, **3** 206, *crundil* **3** 255, *depe crundel* **2** 82, *ellene crundil* **3** 256, *foczan-*, *fozza crundel* **4** 303, *hlosstedes crundles suðecge* **4** 173, *stanegan crundel* **2** 122

crune ME 'crown'. Crown Fm, ~ Point; f.n. *atte Crune* **4** 152

crux Lat 'cross'. *Kingston Crucis* **2** 167

***crymel** OE 'small piece', f.ns. *Crumbyll* **3** 300, *Crympulforlonge* **3** 255

***crype** OE 'a narrow passage'. Crepe Fm, ?Snail Creep; f.n. *the Westekryppe* **3** 228

crypel OE 'a burrow, a place for creeping through'. Cripplestyle; f.ns. Cripple Cl **4** 80, Cripple Croft **4** 367, *Ivescripple* **3** 6

cū OE 'cow'. The Blind Cow, *Cow* **4** 259, The Cow, *Cow and Calf* **4** 413, Cow Bridge, Cow Corner, Cowdon, Cow Drove, *Cowghhowse* **4** 3, Cowgrove, Cow Grove Fm, Cow Lane (× 2), Cowleaze (× 2), Cowleaze Bottom, ~ Coppice (× 2), ~ Copse (× 2), ~ Knap, Cow Leaze Stall, Cowleaze Withy Bed, Cowpool Fm, Cowpound Wood, Cowridge Copse, Culeaze Coppice/Fm/Ho, Red Cow Dairy. Freq in f.ns., e.g. *Coubere* **3** 32, *Coushete* **4** 454, *Cowbroke* **1** 264, *Cow Close* **2** 205, *Cowemede* **4** 10 [*cf.* **cyn(e)**]

cucking-stool eModE, *le Cockyngstole* **4** 255, *le Cokelynstole* **2** 185

cuckold ModE. Cuckolds's Corner, ~ Pit. Also in f.ns., e.g. ?Cuckolds Md **2** 220, Cuckolds Parlor **1** 27, Cuckolds Pit Moor **4** 423

cuckoo ModE, **cuccu** ME 'cuckoo'. Cuckoo Clump, Cuckoo Coppice, Cuckoo Lane (× 3), Cuckoo Hill, Cuckoo Pen, Cuckoo Pound (× 5); f.ns. *Cuccowe ocke* **3** 341, Cuckoo Lane **3** 225, Cuckoo Pound **3** 287, *Guckow hill* **3** 310

cucumber ModE, f.n. Cucumber Gdn **2** 85

culfre OE 'dove'. *Coluerhegh* **4** 8, Coverwell Coppice, Culverhayes, Culverhayes Rd, Culver Hole, ?Culverlake Lane, Culvers Fm & Lane, Culver Well, *Culverwell*. Freq in f.ns., e.g. *Culuerham* **2** 10, Culverhays **3** 8, Culverlands **3** 204

culfre-hūs OE 'dove-cot'. Culverhayes, Culverhouse Fm, Culvers Fm & Lane; f.ns. *Colverhowse* **4** 173, *the Culuerhouse* **4** 376

***Culm** pre-English r.n. of uncertain origin and meaning. ?*Culmelake* (*v*. RNs)

culture ME 'cultivated ground', f.ns. Culter, Culture **3** 378

cumb OE 'valley' (typically one that is short, broad or bowl-shaped). Abbascombe, After Coombe Plant., *Alrecumeslake* **4** 377, Ashcombe (Copse, ~ Fm), Balmers Coombe Bottom, Barcombe (Fm), Batcombe, Bettiscombe, Bickham Fm, ?Bickham Wood, Bidcombe Mill, Bilcombe Coppice, Bincombe, Bonscombe, Bookham, Bottom Coombe Quarries, ?Bracham Lane, Bramblecombe, Bramble Combe Copse, Brenscombe Fm, Brimbley Coombe Fm, Brimscombe Fm, Bronkham Hill, Burcombe Fm, Chescombe Fm, Chilcombe, Chitcombe Fm, Clatcombe Ho, Colcombe Fm, Cole Combe, (The) Combe, Combe Almer, Combe Barn, Combe Bottom (× 6), Combe Close Batch, *Combe Deverell* **1** 311, Combehays, Combehead Fm, Combe Hill, *Combrock* **3** 191, Compton Abbas, Compton Acres, Nether & Over Compton, Compton Valence, West Compton, Coombe (× 4), Coombe Barn & Plant., Coombe Barton & Valley, Coombe Bottom (× 7), Coombe Court, Coombe Fm (× 4), Coombefield Quarries, Coombe Hill, Coombe Keynes, Coombe Lake & Water Lane, Coombe Lane (× 2), Coombe Plant., Coombe Villa, Coombe Wood (× 3), Coombs, Corscombe, Crincombe Bottom, *Cumbesheued* **4** 345, Delcombe Bottom & Fm, Eastcombe Bottom, ~ Ctg & Wood, Edmund Coombe Coppice, Encombe, Fernycombe Coppice, Filcombe Fm & Wood, Francombe Bottom & Fm, Gorcombe Ctgs & Wood, Great Coombe, Halcombe Vale, Harcombe Coppice, *Hawcombe*, Hawkcombe Lane, Hillcombe Bottom, Hinkhams, *Holcombe* **1** 131, Holcombe Barn, ~ Bottom (× 2), ~ Fm, ~ Pit, ~ Wood, Honeycombe

Fm, Honeycomb Wood, ?Howcombe Ledge, *Hywyshcombyshede* **4** 176, Ilcombe Fm, Incombe (Wood), (Hr & Lr) Kingcombe, Lancombes Cross, Landscombe Lane, Langcombe Bottom, ~ Coppice, Lankham Bottom, ~ Coppice, Lewcombe, Littlecombe Bottom, Longcombe Bottom, Loscombe, Loscombe Barn, ~ Bottom & Hill, ~ Plant. (× 2), ~ Wood, Loscombe's Well, Luccombe Covert & Fm, Luccombe Fms, Luscombe (× 2), Lyscombe Bottom, Manscombe, Mappercombe Fm & Manor, Melcombe Horsey, Melcombe Regis, Moigne Combe, Morcombe Fm, Morcombelake, Motcombe, Muddicombe Coppice, *Mystecombe*, Nether Coombe, Nettlecombe (× 3), Nutcombe Coppice, Old Coombe, Overcombe Beach, Overcoombe, Oxencombe Bottom, Peascombe Copse, Pitcombe Dairy, Purcombe (Fm) (× 3), ?Ransom Coppice, Ratcombe Barn & Wood, Renscombe Fm, Riscombe Coppice, *Rocombe*, Ruscombe Lane, Rushcombe Bottom, Seacombe Bottom, Shatcombe (Coppice, ~ Fm), ?Smacam Down, Smacombe, *Southcombe*, Southcombe Fm, ~ Plant., Stancombe Barn & Coppice, Stonecombe Fm, Stutcombe Bottom, Suckthumb Quarry, Sutcombe Wood, Swainscombe Fm, Tatcombe Wood, Thorncombe (× 4), Thorncombe Beacon, ~ Wood, ?Tillicum, Totcombe Hundred, *Trendlecombe*, Uggescombe Hundred, Wakeham, Wancombe Bottom, Watcombe Bottom, ~ Plain & Wood, Watercombe, Welcome Hill, *Wellcome* **4** 485, Westcombe Coppice, Whatcombe Down, Whatcombe Fm & Ho, Whitcombe (× 2), ~ Bottom, ~ Fm, ~ Hill (× 2), Whitecombe, Woolcombe (Fm), Yelcombe Bottom. Freq in f.ns., e.g. *Combemede* **4** 10, *Gatcombe* **4** 10, *Glidecombe* **4** 14, Whitcombe **2** 10. In AS charters, *cumb* **1** 10, *budencumbe hracan* **3** 128, *flescumbe*, *flexcumbes heuede* **3** 206, *herebrittescombe* **3** 256, *hloscumbes heafud* **3** 328, *holencumb* **3** 99, ~ *cumbe* **3** 102, 111, *hþete cumb* **2** 58, *langen cumbes hauede* **3** 104, *riȝ cumb* **4** 304, *sealtera cumb* **4** 374, *Smalencumbe* **1** 32, *struthgeardes cum* **1** 10, *thorncombe* **3** 246, *tymbercumb* (*v.* Addenda **4** 193), *westrene cumbe* **1** 33, *whetecombe* **4** 123, *wibbelescumb* (*v.* Addenda **4** 193)

cundite, **condit** ME 'water-channel, conduit', f.ns. *the Condit close* **4** 43, *Conditeslake* **4** 10, *Condydeslane* **3** 273, Conduit Fd **3** 372

cunner, **conner** ModE 'species of fish'. Conner Cove

Cupid ModE 'god of love', f.n. Cupid's Bower **1** 371

cupola ModE 'rounded dome', *the Cupola* **1** 355

cuppe OE 'cup', f.n. *Coppeford* **4** 370

curfew ME 'curfew'. Curfew Plot; f.n. Curfew Grd **3** 216

curre ME 'dog'. ?Cowdown Hill

custum(**e**) ME 'customary rent or service, f.n. ?Custom Md **3** 77

***cut**, **cutte** ME 'cut, water-channel'. Cut Mill; f.ns. ?Cut Croft **4** 281, ?Cut Furze **2** 77, *le Cuthouse* **4** 167, *Cutteforlang* **1** 260, *Cutteham* **1** 95, Rough Cut **4** 300

cutel ME 'water-channel', f.n. Cutell Hill **3** 297

***cutted** ME 'provided with a water-channel, f.n. *le Cutted Hill* **3** 297

cutter eModE 'crack in rock', f.n. *Cutterbeds* **3** 26

cutting ME 'piece of land cut off, or cut through'. Charmouth Cutting; f.n. *Kuttyng* **3** 392

cwēad OE 'dirt, mud', f.n. *North-, Southquidlond* **4** 183

***cwelle**, ***cwylle** 'spring, well', f.ns. *la Cuelle* **4** 70, ?Cull **1** 79

cwēn OE 'queen'. *Queene Farme* **3** 18, Queen Oak, Queen's Coppice (× 2), ~ Copse; f.n. *Quenegore* **2** 53

cweorn, **cwyrn** OE 'hand-mill'. *Coringdon*, Cornford Bridge, ?Cowdown Hill, Currendon; f.ns. Quarrendon Fd **3** 109, Quarrington **4** 25

cwicu, **cucu** OE 'live, living'. ?Cookwell Brook (*v*. RNs), also in AS charter, ?*cucan* (*dene*) **2** 158 (?stream-name **Cuce* 'the lively one', *v*. RNs)

***cylfweard** OE 'mace-keeper'. ?Culliford Tree (Hundred)

cycene OE 'kitchen'. Devil's Kitchen, Earl's Kitchen, John o' Gaunt's Kitchen; f.ns. *Kichin Mead* **2** 106, Long Kitchen **1** 236

cyln OE 'kiln'. Bishops Kiln, Killwood, Old Kiln (× 3), Old Pottery Kiln, Smiths Kilns [*cf.* **brickkiln**, **limekiln**, **tilekiln**]

cȳme OE 'convenient, splendid'. ?Kimmeridge

cymed OE 'wall-germander'. ?Hr & Lr Kingcombe

cyne- OE 'royal'. ?The Kendalls, ?Kimerley Wood, King Grove Fm, King's Grove Wood, Kington Magna

cyning OE 'king'. Bere Regis, King Barrow (× 2), King Barrow Quarries, King Down, *Kingesbrigge* **3** 11, *Kingeslyme* **4** 392, King Rock, (The) King's Barrow, ?Kings Barton, King's Bridge, King's Court Palace, Kingsdown, King's Mill, Kingsettle Fm, King's Fm, *King's Mead*, King's Pier, ?King's Standing, Kingstag, Kingston (× 2), Kingston Lacy Hall, Kingston (Maurward), Kingston Russell, King's Wood (× 2), Kington Magna, ?*Kinsworth* **3** 54, *the Kyngges Stere* **3** 365, Winterborne Kingston, Wyke Regis. Also in f.ns., e.g. *By Kinges Barrowe* **3** 263, *Kingescrofte* **2** 161, King Lane Coppice **2** 288, *Kings Doune* **1** 31, Kingway Fd **2** 57. In AS charters, *cing hille* **3** 128, *kinghes imare* **3** 136

cȳpe OE 'osier basket', f.n. ?*Kype-Cross* **1** 81

cyst, cest OE 'chest, coffin', f.n. *stanchust* **3** 343 (*v.* Addenda)
cȳta OE 'kite'. Kitbridge, Kite Hill (Coppice), Kitehill Plant., Lr & Upr Kitemoor, Kite's Nest Lane, Kitford, Kitford Bridge, Kit Hill (× 2), Kithill Plant., ?Kittle Barrow, Kittwhistle, ?Kitwells Cliff, Kit Whistle Barn, Kitwhistle Gorse; f.ns. ?Keet Bridge **2** 105, Kite Corner **2** 205, Kite Hill **4** 267. In AS charters, *ketan lege* **4** 454
cyte OE 'hut', f.n. ?Keet Bridge **2** 105
dæl[1] OE 'valley'. Allendale [ModE dale], Dalwood, Swindale
dǣl[2] OE 'part, share', f.n. Halfendale **4** 281
daffodil ModE. Daffodil Copse (× 2); Daffyclose **2** 220
dagg(e) ME 'shred, strip', f.ns. *Dagelond* **4** 182, ?Dag Md **3** 372
dairy (house) ModE, **deierie** ME 'dairy'. Dairy Ho Coppice, *dayerie* **4** 18. Freq in modern minor names as ~ Dairy (House), e.g. Bincombe Dairy, Bladen Dairy, Bockhampton Dairy Ho, Came Dairy, Church Dairy, (The) Dairy (freq), Dairy Ctgs (× 2), ~ Fm (× 2), ~ Ho (freq), *Dairy Ho*, Dairy Lodge, Dairy Walk, Duck Dairy Ho, East Fm Dairy, Field Dairy, Frome Dairy, Fryer Mayne Dairy, Hewish Dairy, Holworth Dairy, The Hut Dairy, Keynston Dairy Ho, Middle Hyde Dairy, Icen Dairy, Ilsington Dairy Ho, Kingston Dairy Ho, Laycock Dairy Fm, Moreton Dairy, Muckleford Dairy Ho, Northground Dairy, Poxwell Bottom Dairy, Priorslawn Dairy, Red Cow Dairy, Ringstead Dairy, Stafford Dairy, Tincleton Dairy, Winterborne Dairy, Wolfeton Dairy, Woodsford Dairy
daisy ModE. Daisy Fm; f.ns. Daisey Grd **3** 77, Daisey Hill Fd **2** 171
dai-wine ME 'a day's earnings', f.n. *Dawynlonde* **2** 173
dāl OE, **dole** ME 'a share, a portion'. Dole Copse, Doleham Coppice, Dole's Covert, ?Dolway Ctgs; f.ns. Dole Mdw **2** 104, (*la*) *Dolhous* **3** 150, *Larkedole* **1** 238
dāl-land OE 'land held in common', f.n. ?Dalling Grd **3** 197
dam ModE. Mill Dam; f.n. Dam **3** 26
damask ModE 'marked resembling damask'. Damas Lawn Bridge; f.ns. Damas Lawn **4** 151, Dames House Grd **2** 157
dancing ModE. Dancing Drove, Dancing Hill, Dancing Ledge; f.ns. Dancing Cross Rd **2** 204, *Dawnsinge Plotte* **4** 465
dayne ME ?'measure of land', f.n. *la Dayne* **1** 364
dēad OE 'dead, disused, infertile'. Dead Man, Deadman's Coppice, Deadman's Covert & Stile, Deadmoor Common, Dead Well Water (*v.* RNs), ?Dedley Fm; f.ns. Dead Croft **3** 5, Dead Hill **3** 225, Dead Maids' Close **2** 157, *Dedefolkefurlonge* **2** 13, *Dedwill* **2** 10
dēaw OE 'dew'. Dewlands Hill

decoy ModE 'pool with netted approaches for the capture of wildfowl'. The Decoy (× 2), Decoy Heath, Decoy Pond (× 2), Old Decoy Pond; f.ns. Decoy Pond **2** 214

deerfold ModE 'deer fold', f.n. Deer Fold **1** 28

(ge)delf OE 'a pit, a quarry'. The Delph, Poor Delf; f.ns. *delfstichche* **4** 218, Delph Fd **2** 191, *Depdelf* **1** 101, *Standolfurlang* **4** 271

dell OE 'a pit, a dell, a valley', f.n. *Dell close* **1** 152. In OE charters, *mes delle* **3** 117

dene ME 'dean'. Dean's Court, Deans Grove

denn OE 'den, lair', f.ns. *Theuerdenne* **3** 112, *Warledenne* **1** 230

denu OE 'valley'. Beasden Coppice, Blake Dene, Bossleton Coppice, Bugden Bottom, *Cattesdene* **4** 289, Coaxdon Hall, Cogdean Elms, ~ Fm, ~ Hundred, Dean, Deanend, Dean Fm, Dean Hill, Deanland, Dean Lane, Hr & Lr Denhay Fm, Den Wood, Dogdean Fm, Elder Dean Coppice, ?Hawksdean Fm, Haydon Lane & Wood, ?*Hewdon*, Lines Dean Bottom, Meriden Wood, ?*pinguedine* **4** 54, Pribdean Wood, Rowden Coppice, Russon, Stone Down Coppice, Stony Bottom, Tincleton. Also in f.ns., e.g. Dean Cl **2** 51, Dean Hill Flg **2** 148, Holdins **2** 214, *Maleckesdene* **2** 216. In AS charters, *aerne-, earne dene* **4** 302, *ættandene* **3** 310, *cotta dene* **4** 303, 306, *cucan dene* **2** 158, *fideriches dene* **3** 128, *risshdene* **3** 257, *standene* **2** 58, *terente dene* **3** 129, *waddene* **3** 129

dēofol OE 'devil'. *Deuelepole* **3** 317, Devil's Bridge, Devil's Copse, Devil's Den, Devil's Kitchen, *The Devil's Night-Cap* **1** 46, Devil's Staircase, *The Devil's Trencher and Spoon* **3** 81, ?Devils Wood Lane; f.ns. *Dewelepole* **4** 250, *Kill Devil* **4** 398

dēop OE 'deep'. Deep Lane, *Depalders* **3** 233; f.ns. *Depdelf* **1** 101, *Depecrundale* **1** 143, *Depemore* **2** 215, *Deptly* **3** 5. In AS charters, *depe crundel* **2** 82

dēope OE 'deep place'. South Deep

dēor OE 'animal, deer'. Dear Hay Lane **2** 42, Deer Park (× 3), Deer Park Fm. Delcombe Bottom & Fm; f.ns. *Der(r)e Holte* **3** 341, Durcroft **3** 88

deorc OE 'dark'. Dark Hill Plant., Dark Hole, Darkhole Fm, Dark Lane (× 7), Darknoll Fm, The Dark Walk

derne, **dierne**, **dyrne** OE 'secret, hidden'. *Dernepath* **1** 126, Durnford, Durngate St.; f.ns. Dun Water **3** 55, *Durnecote* **4** 188, *Durnelane* **1** 350, *Durnelee* **2** 10

desert ModE. The Desert

dew-berry ModE dial. 'blackberry', f.n. Dewberries **1** 142

dey(e) ME 'dairy maid', or ***dey** ME 'dairy', f.ns. Day Cl **2** 77, *lez dayhay* **4** 58, ?Deehay **4** 159

deyhus ME 'dairy'. ?Dyehouse; f.n. *lez dayhowse* **4** 58

dial ModE 'dial', f.ns. Dial Cl **1** 28, ?Dyle Corner **1** 312

dīc OE 'ditch'. †Bokerley Ditch, *Borysdych* **2** 280, Bowditch or Burridge, *Bystaweidich* **4** 163, †Combs Ditch (Hundred), Coxditch, Cross Dyke, †Ditchey Coppice, †Dyke, Dyke Head, Fox Ditch Coppice, *gildenedich* **3** 194, Great Ditch, †Grim's Ditch, Hawkhill Ditch, †High Ditch Coppice, High Ditch Copse, Hodge Ditch, Holditch, Hollow Ditch, *Myldyche* **4** 379, Rowditch Wood, Walditch. Freq in f.ns., e.g. *Biridiche* **4** 34, *Dychemore* **4** 10, *le dychewey* **2** 14, *Soredich* **1** 238, *Southedich* **3** 208. In AS charters, *alde dich* **2** 57, *chelbrichtes dich* **3** 163, *cinninces dic* **2** 71, *dic* **3** 310, **4** 193 (*v*. Addenda), *dich* **1** 31, **2** 57, **2** 122, **3** 41, 99, 237, 255, *dice hyrnan* **4** 34, 454, *diche* **3** 123, 163, *diches hirne* **3** 206, *ealdan dic* **3** 298, *ealde diches heued* **3** 206, *ealdene dich* **1** 31, *Elde Waldich* **3** 256, *horsedich* **2** 122, *hwitendich* **2** 122, *live dic*, *lyf dic* **4** 304 (*v*. Addenda **4** 193), *muledich* **1** 272, *ruwe dich* **3** 257, *stod dic* **1** 30, *þwers dich* **1** 32, *twifelde dich* **3** 208, *Walle dich* **3** 238, *weal dic* (*v*. Addenda **3** 267), *wulfrices dice hyrnan* **4** 454, *wyndrede dic* **3** 159

dikede ME 'with a ditch', f.n. *Dikedemede* **3** 263

dile OE 'dill', f.n. *Dylemour* **3** 157

dip ModE 'hollow', f.n. Dips **1** 28

dirty ModE 'dirty'. Dirty Gate (× 2), Dirty Knaps, Dirty Lane Coppice

dish ModE 'dish-shaped hollow'. Cull-peppers Dish

docce OE 'dock, water-lily'. ?Dogdean Fm, Doghill Coppice; f.ns. Dock Lake **3** 26, Docklands **3** 204, *Dokham* **2** 52

dock ModE 'pool where a boat can rest'. ?Puddle Dock Dairy Ho; f.n. Paddle Dock **1** 29

doctor ModE, f.ns. Doctors **2** 199, Doctor's Grd **3** 121

dodde ME 'a rounded hill-top'. ?Dodding's Fm, Dodpen Fm & Hill, ?Duddle Fm; f.n. *Doddehulle* **4** 411

dodgy ModE 'difficult'. ?Dodgy Close

dogga lOE 'dog'. Dogbury, ?Dogdean Fm, Dogholes, Dog Kennel Copse, Dog Plant., ?Dogshole Ctgs, ?Dogwell Fm; f.ns. Dogs Close **1** 127, *Dokthorn'* **3** 206. In AS charters, *doggeneberwe* **4** 161, *doggeneford* **3** 255

dogh, dow(e) ME 'dough', f.n. *le Dowe* **2** 14

dog-house ModE 'kennel'. Doghouse Fm, Doghouse Lane Gate (st.n., *v*. Addenda **3** 360)

dog-kennel ModE, f.n. Dogkennel Leaze **2** 72
dohtor OE 'daughter', f.n. *Helle doughter* **4** 10
(ge)**dolfen** OE 'dug', f.n. *Nywedolveneford* **2** 216
***dolu**, ***dulu** OE 'valley'. ?Dullar Fm
dōmere OE 'a judge'. ?Damer Hill
donkey ModE. Donkey Moor; f.n. Donkey Plot **2** 109
dor, **duru** OE 'door, gate, gap'. Durdle Door; f.ns. Dometts Door **4** 281, Dicks' Door **3** 176, *le Twine Dores* **1** 314
dote eModE 'dowry', f.n. Dote **4** 281
double ModE, **duble** ME 'double'. Double Barrow, ~ Bow, Satchel's Double; f.ns. Double Close **3** 272, *Dowbell'putte* **3** 206
dour Corn 'water'. ?Loders
dovecot ModE. Culverhayes, Dovecot (× 2)
dovehouse ModE. Culverhayes
dowere ME 'dowry'. Dowerfield Dairy Fm & Ho, The Dower House; f.ns. Dower **4** 409, *Dowerclos* **3** 32, *the Dowerfield* **4** 471, *la Dowre Oke* **3** 156
draca OE 'dragon'. Drakenorth; f.n. ?*Drakecroft* **1** 345
drǣf OE 'drove'. The Drives, The Drove **1** 178, Haydon Drove; f.n. *le Dreve* **1** 120
drāf OE, **drove** ME 'herd, drove', also 'road on which cattle are driven'. Ashy Drove, Back Drove (Copse), Bailey Drove, Barnsley Drove, Batterley Drove, Bolter's Drove, Bourne Drove, Boys Hill Drove, Brick Kiln Drove, Broad Stone Drove, Broomhill Drove, Burleston Drove, Butterwick Drove, Cancer Drove, Caphays Drove, Coalhill Drove, Combehead Droveway, Common Drove (freq), Cow Drove, Daggles Drove, Dancing Drove, The Drove (× 3), Drove Barn, ~ Coppice, ~ End, ~ Hill, ~ Lane, Eastfield Drove, East Morden Drove, Green Down Drove, Haydon Drove, Hayesend Drove, Hewlett's Drove, Hr & Lr Drove, Lackington Drove, Littleton Drove, Long Bridge Drove, Luton Drove, Monkton Drove, Morgan's Drove, Mullett's Drove, Old Drove, Osmington Drove, Pitt's Drove, Pound Drove Copse, Poxwell Drove, Ridge Drove, St Mary's Drove, Sandy Drove, Sheephouse Drove, Slough Drove, South Drove, Stent Drove, Stockbridge Drove, Stonerush Drove, Sweetbrier Drove, Two Droves, Waddock Drove, Warmwell Drove, Whitmore Drove, Winfrith Drove, Wire Drove, Yard Drove, Yearlings' Drove, Yelcombe Drove. Freq in f.ns., e.g. Anvils Drove **2** 36, *drafaker* **2** 52, Drove (Lane) **2** 13, *Droveyate* **3** 32
dragon ModE. Dragons Hill
drake ME 'male of the duck', f.n. ?*Drakecroft* **1** 345

drash ModE dial. 'to thresh', f.n. ?Dratch Hay **4** 434
draught ModE '(animals used for) drawing or pulling', f.ns. Draught Moor **4** 151, Draughts **4** 33
draw-bridge ModE, *(the) Drawbridge* **1** 258
dreche ME 'trouble', f.ns. ?Dratch Hay **4** 434, *Drecchethornesforlang* **1** 345
driblet ModE 'small quantity', f.n. Dribbett **4** 201
drift ME 'drove'. Drift Plant., ~ Road; f.ns. Driftway **3** 67, *Dryftclos* **1** 203
drinking ModE 'ditch'. Drinking Barrow; f.n. Drinking Plot **4** 298
drit OE 'dirt', f.ns. ?*Derthehilla* **3** 64, Dirt Croft **3** 88, Thirtoop **1** 228
drive ModE. Carriage Drive, Drive End, Drive Lane, (The) Drive Plant. (freq), Lady Caroline's Drive (× 2), Moreton Drive, Pony Drive
drong ModE dial. 'a narrow way between hedges'. ?Coombe Drain, Drong, Furzy Drong; f.ns. Drang **1** 331, **4** 131, Drong **4** 475, Drong Md **1** 99, Drong Way **2** 98, Throng **1** 331
drowned eModE 'submerged'. *Drowned Landes* (*v.* Addenda **1** 232)
droy eModE 'a drudge', f.n. Droy **4** 286
druerie ME 'love-token, sweetheart'. Drury Lane
***drust** Brittonic 'door'. ?Owermoigne (*v.* Addenda)
drȳge OE 'dry, dried up'. Dryclose Chalk Pit, Dry Close Plant., Dry Hill Moor, Dry Wood; f.ns. Dry Cl, Dry Md **3** 55. In AS charters, *drian hyrstes lace* **4** 454
drying ModE 'drying of cloth or clothing', f.n. Drying Plot **2** 109
***duß, dü** PrWelsh 'dark, black'. Devil's Brook, R. Divelish (*v.* RNs), Dewlish
***dußr** PrWelsh 'water'. ?*Toller* (*v.* RNs), giving name to Tollerford, Toller Fratrum and Porcorum, Toller Whelme
duble ME 'double', f.n. *Dowbell'putte* **3** 206
duck ModE. Duck Dairy Ho, Duck Lane, Duckpool, Duck St (× 2), ?Duck St; f.ns. Duck Moor **2** 148, Duck's Nest **3** 251
duckpond ModE. Duck Pond, Duckpond Plant.
dūfe OE 'dove'. Dovehayes Fm; f.ns. *Dovelond* **4** 160, *Dufurlang'* **2** 241
duke ME 'duke'. Duke's Copse, ?Duck St
dūn OE 'hill, down' Abbeycroft Down, Ailwood Down, Ash Down, Ashmore Down, Askerswell Down, Bareden Down, *Barn Down*, Batcombe Down, Bazon Hill, Beaminster Down, Bere Down, *Bestedon*, Bewley Down, Bincombe Down, Bindon Abbey, Bindon Hill, Little Bindon, ?Bingledon Wood, Bishop's Down, Blackbush Down, Black Down (× 5), Blackdown (Hill), Blagdon Barn, ~

Copse, ~ Fm (× 2) & Hill (× 2), *Blandford Down*, Bloxworth Down, Bokerly Down, Bottlebush Down, Bowden, Bowden Hill, Bowdens, Boyden Wood, Bradford Down, Bramdon Lane, Brockington Down, Brown Down Plant., Bubb Down Fm & Hill, Buckham Down, Buckland Down, Burleston Down, ?Burnt Bottom, Bussey's Down, Caldron Barn Fm, Came Down, Camp Down, *Canendone Hundred*, Cayles Down Copse, Chaldon Down, Chaldon Herring, Chapel Down Fm, *Chappell Downe* **4** 242, Chardown Hill, Charlton Down, Charlton Hr Down, Charminster Down, ?Chedington, Cheselbourne West Down, Chettle Down, Chilfrome Down, Chitcombe Down, Clanden Hill, Clandon, College Down, Compton Down, Coombe Down Hill, *Coringdon*, Corscombe Down Barn, Corton Down, Cowdon, Cowdown Hill, Crichel Down, The Crofts Down, Cullington, Currendon, *Dalwood Down* **4** 389, Deverel Down, Dimstone Hill, Dogbury Down, Donedge Lodge Fm, *Down*, The Down, Downash Lane, Down Barn (freq), ~ Bldgs (× 2), Downend Coppice (× 2), Down Fm (× 3), The Down Ho, Down Ho (Fm), Downlands, Down Lane, Downley Ctg & Coppice, ?Downshay Fm & Wd, Downside, Down(')s Lane, (The) Down Wood, Dunbury, ?Duncliffe Hill, Dungrove Fm, Duntish, East Down Ctgs, †Eggardon Hill & Hundred, *Egleston Down*, Eldon Hill, ?Elsdon's Batch, Everley Down, Ewern Down Plant., Eype Down, Farm Down Barn, Farringdon Clump, Farrington, Fern Down, Ferndown Barn, Ferney Down Wood, Fontmell Down, Ford Down Lane, Fordington Down, Forked Down End, Foxdown, Fox Ground Down, France Down, *Fredownes* **3** 125, Free Down (Border), Furze Down, Furzedown Fm, Garston Down, Gold Down, Great Down, Great Down Clump, Green Down (Drove), Grimstone Down, Gunville Down, Gussage Down, Haddon Copse, ?Haddon Lodge, †Hambledon Hill, Ham Down Copse, Hammiton Hill, ?Hampton Barn, Hampton Hill, Handley Down, Hardown, Harley Down, ?Hawksdean Fm, Haydon (× 2), Haydon Clump & Hill, Haydon Down, Haydon Plant., *Hemsworth Down*, Henning Hill, ?*Hewdon*, Higher Down, *Higher Down*, Higher & Old Down Barn, Hingsdon Fm, Hinton Lane, Hoggen Down, Honey Down Plant., Horse Down, Houghton Down Clump, Houghton N & S Down, Hungry Down, Ibberton Long Down, Jordan Hill, Keynston Down, Kimmeridge Down, King Down, Kingsdown, Langdon Fm, ~ Hill, Launceston Down, Lewesdon Hill, ?Linen Hill Fm, Linton Barn & Hill, Little Down (freq), Little Piddle Down, Lord's Down, Low Down Fm, Lower Down, Luccombe Down, Luton Down, Main

Down, Maine Down, Martin's Down, Melbury Down, Mildron Copse, Mill Down, Minchington Down, Moigns Down, Monkton Down, Morden, Mosterton Down, Muston Down, Necklace, Nine Barrow Down, Norden Fm, Norden Hill (× 2), ?Nordon, Nordon Hill, Nordons Estate, North Down, *North Down*, Northdown Barn & Fm, ?Norton Coppice, ?Norton Ctgs/Lane/Wood, Norton Hill, Norton Lane, Oddens Wood, Ogden Down Fm, Old Down Ctgs, Owermoigne Down Barn, Patson Hill (Fm), Payne's Down, Pentridge Down, †Pilsdon, Pimperne Down, Pistle Down, Poyntington Down, Prior's Down, Profits Down, Puddletown Down, Quarleston Down, Race Down, Racedown Fm, *Ralph Down*, Rampisham Down, Ramsden Coppice, Rawston Down, Rodden, Rough Down Coppice, ~ Lane, Round Down (× 2), Rowden Bridge, Rowden Hill, Rowden Mill Fm, Row Down Coppice, Seldon Hill, Seldown, Shapwick Down, Sheep Down, *Le Shepe Downe* **3** 125, Smacam Down, ?Smitten Corner, Snow's Down, ?Sodern, *Soddon*, Sodom Lane, *South Down*, South Down, South Down Ctg, South Down Fm, South Field Down, Sovell Down, Stone Down (Coppice, ~ Lane & Stall), Stourpaine Down, Stratton Down, Stubhampton Down, Sutton **1** 68, Sutton Fm & Hill, Tarrant Hinton Down, Tenantry Down, Thickthorn Down, Thorney Down Fm, Toller Down, Turnworth Down, Underdown, ~ Hollow, ~ Lane, *Undermildone* **4** 130, Vine's Down Bldgs, (Little) Waddon, Waddon Fm, ~ Hill (× 2), Wallis Down, Walls Down, Wardon Hill, Warmwell Down Barn, ?Watton Hill, Week Street Down, Well Bottom Down, West Down (Fm), Westdowne, Whatcombe Down (× 2), White Down, Whitelands Down, Winfrith Down Bldgs, Woodsdown, Wreden Plant., Wyke Down, Yeldon. Freq in f.ns., e.g. *Bromdun* **4** 10, Down Lands **4** 42, Waddon **2** 14, *Wyllesdon* **4** 411. In AS charters, *alden doune* **3** 254, *blacan dune* **2** 161, *brugh doune* **3** 255, *brumdun* **4** 302, *dounen tit* **3** 242, *dunæ* (*v*. Addenda **3** 267), *dungete* **2** 121, *dydelingdune* **2** 135, *Gretindune* **3** 206, *hamelendune* **3** 106, *heandene beorg*, *hendune beorch* **3** 218, *lindune* **2** 151, *meleberig dune* **3** 131, *safundune*, *saþendune* **1** 32, *sanddoune* **3** 200, *swindone*, *-dune* **3** 208, *þorndunes cnep* **3** 208, *trildoune* **3** 191–2

dūne OE 'below', f.n. *Dounlangelond* **1** 313

dung OE 'dung, f.ns. Dung Cl **4** 87, *Dunghaie Close* **3** 55

dungeon ME 'a deep dell or hollow'. The Dungeon, Dungeon Close, †Dungeon Hill, ~ Pit (× 2); f.ns. Dungeon Close **1** 211, Dungeon Copse **3** 55, Dungeon Mdw **2** 157

dunghill ME 'dunghill', f.ns. *Dunghill* **2** 174, *Dunghilles* **2** 106, *le Dunghyll'* **3** 237
dunn OE 'dun, dull brown, dark'. ?Duncliffe Hill; f.n. Dun Willow **3** 26
***dunnoc** OE 'hedge sparrow', f.n. Dunocks **2** 220
***dūno-** Brit 'a fort'. *Dunium* **1** 377
***durno-** Brit 'fist'. Dorchester
dusty ModE, f.n. Dusty Fd **4** 475
dutch ModE, f.n. Dutch Cl **4** 267
dutch cap ModE, f.ns. Dutch Cap **1** 336, **4** 91
dyehouse ModE. *Dyehouse*
dyncge OE 'manured land'. ?Dunch Lane, Dunch Mouth, ?Dungy Beach & Head; f.ns. *Culuerdyngge* **1** 313, Dungey **3** 101,
***dȳne** OE 'slope', f.n. ?*Dynecum* **4** 182
dȳnige OE obscure plant-name. ?Denhay Fm **4** 427, Denny Bridge **3** 308; f.n. ?*la Dunye* **1** 313
ēa OE 'stream'. Pucksey Brook (*v*. RNs), ?Ryeclose; f.ns. ?*Everlong* **2** 180, *le Re* **2** 38
eagle ModE 'eagle'. Eagle Plant.
eahta OE 'eight'. Eight Acre Coppice/Copse (× 5), ~ ~ Plant., Eightacres, Eight-Hatch Plant.; f.ns. 8 Acres **2** 51, **3** 204
ēa-land OE 'island', f.ns. ?*Iland Moore* **1** 194, ?Ryeland(s) **3** 162, 187, Yeolands **1** 229
(e)ald OE 'old, long used'. *Eldeportwey* **3** 148, Eldon Hill, ?Elwibber, Little Bindon, Old Butts (× 2), Old Chalk Pits, Old Claygrounds, Old Clump, Old Coombe, The Old Cottage, Old Court Copse, Old Drove, (*la*) *Oldehall* **3** 366, Old Fm, Old Ford, Old Hill, *Old Hill*, Old Kiln (× 3), Old Knowle, Old Lane, Old Lawn Coppice & Fm, Old Limekiln (× 3), Old Lodge, Old Manor Ho, Old Marl Plant., (The) Old Mill, Old Orchard, Old Park (× 4), Old Portisham, Old Pottery Kiln, Old Quarr(y), Old Ram (Plant.), Old Read's Copse, Old Rectory (× 3), Old Road Wood, Old Turnpike, Old Warren (Ctgs, –Hill), Old Wood (× 2), ?Yelcombe Bottom, Yeldon. Freq in f.ns., e.g. *Eldelond* **3** 32, Old Down **3** 5, *Oldffold* **4** 10, Old Hayes 2 51, *yelde Barrowe* **4** 43. In AS charters, *alde dich* **2** 57, *alden doune* **3** 254, *ealdan dic* **3** 298, *ealdan-, alde stodfald* **1** 30, *ealdan hagan* **3** 289, *ealdan herpað* (*v*. Addenda **4** 193), *ealdan wale* **4** 34, *ealdan weg(e)* **2** 136, **3** 289, **4** 35, *ealde ad* **3** 102, 131, *ealde berig* **3** 206, *ealde diches heued* **3** 206, *ealdene dich* **1** 31, *ealdene hage* **3** 237, *ealden forð* **3** 128, *ealden hege rewe* **1** 31, *ealden herepaþe* **3** 111, *ealden reshbed* **3** 206, *ealde paþe* **3** 206, *ealde rode* **1** 31, *ealde seale*

2 82, *ealde weg* **1** 31, *elde lidyate* **3** 256, *Elde Waldich* **3** 256, *elþen stret* **2** 57, *olde wested'* **3** 201

early ModE, Early Md **4** 399 (*v*. NDEFN 123)

earm OE 'wretched', f.n. *Armemede* **4** 384

earn OE 'eagle'. *Earnley*, Yon Barrow; f.ns. *Ærnelai* **1** 198, *Erneborge* **2** 271. In AS charters, *aerne-*, *earne dene* **4** 302, *earna lea weʒe* **4** 290, *earnes hlin(g)c* **4** 374

ears OE 'fundament, buttock'. *Arish* **1** 133, Arish Mell, Blackers Hole, Scratch Arse. In AS charters, *ears mores heaued* **3** 128

ēast OE 'east'. East Almer Fm & Lodge, East Barn, East Bottom, East Bridge, Eastbrook Fm, ~ Ho, East Burton, Eastbury, Eastbury Ho, East Chelborough, East Cliff, Eastcombe Bottom, ~ Ctg & Wood, East Compton, East Coppice, Eastcot, East Creech, East Down Ctgs, East Ebb, East End, East Fm (freq), East Field Fm, Eastfield Fm, ~ Hill, ~ Plant., Easthay Fm, East Hemsworth, East Hewstock, East Hill (freq), East or Ice Hill, East Holme, East Ho, ?Eastington, East Knighton, East Lane, East Lodge, East Lulworth, East Melbury, East Mill, *Eastmore*, Easton, East Orchard (× 2), East Over, East Park, East Plant. (× 2), East Point, East Road, *East Stafford*, East Stoke, East Stour, East Street, Eastward, Eastwell Lane, Eastwood, East Wood, Eastworth Fm, *estbroke* **4** 395, *Esthamme* **2** 226, *Eston*, *Estyep* **4** 427, ?Ice Drove. Freq in f.ns., e.g. Icely Moor **3** 44. In AS charters, *æst Sæ* **1** 257, *eastgeat*, *east gete* **2** 134, *eastlea* **2** 161

ēastan OE 'east, east of'. *Bestedon*, Bestwall Fm & Ho, East Brook, ?Eastington; f.n. *Byestoune* **2** 52

ēasterra OE 'more easterly'. Easterly Copse, *Estretuna* **3** 131, *Estrhep* **4** 427, Litchfield Copse; f.ns. *Easter Close* **3** 32, *the Easter Grove* **4** 19

ēastmest OE 'furthest east', in AS charters, *eastemestan holan weg* **3** 380

eastward ModE 'easterly', f.n. Eastward Whetcombe **2** 32

easy ModE, f.n. Easy Go **1** 112

ebba OE 'ebb, low tide'. East Ebb, Great Ebb, Long Ebb; f.ns. *Ebbemor* **1** 101, Ebb's Fleet **1** 271

ecclesia Lat 'church'. *Netherbury Ecclesia* **4** 311

ecg OE 'edge'. Corfe Edge, Donedge Lodge Fm, River's Edge; f.ns. *Edge Closse* **3** 6, *Shortegge* **2** 90. In AS charters, *blacan dune ecge* **2** 161, *ecge* **3** 310, 380, *ecge ford* **1** 31, ?*norðeage* **2** 161, *north ecge* **3** 106, *sledes Northecge* **3** 123, *suðeke* **3** 104

-ede OE adj. suffix. ?Knap Barrow; f.ns. *Dikedemede* **3** 263, *Goredaker* **3** 32, Napperty Hall **3** 373

edisc OE 'enclosed park', surviving as ModE dial. **eddish** 'aftermath, stubble'. ?Bowldish Pond, ?Duntish; f.ns. Barley Eddish **4** 280, Eddish **4** 463, ?Raddish **4** 217

ēdre OE 'stream', f.n. ?*Ederley* **1** 305

eegrass ME 'aftermath', f.ns. *le Egras* **1** 326, Hay Grass **3** 387, *le Yegrasse* **1** 129

efen, **efn**, **emn** OE 'even, smooth, level'. ?Emley Lane, ?Evenings Ford; f.n. ?*Theven'howe* **2** 230

efisc OE 'edge, border', in AS charters, *euisc* **1** 31

ēg OE (**ī(e)g** WSax) 'island, land partly surrounded by water, dry ground in marsh, well-watered land'. Brownsea Island, Chaffeymoor Fm, Eye Bridge & Mead, ?Fishey Lane, Furzey Island, Green Island, High Mead Lane, Lowness Coppice, Pipsford Fm, Rye Water Coppice & Lane, Rye Water Fm. Freq in f.ns., e.g. *La Cresseye* **1** 249, *Efurlang* **2** 52, Knightney **3** 44. In AS charters, *mæd (æcras)...on ige* **2** 136

ēgeð OE (**īgeð** WSax) 'small island'. ?*Horsyth*, f.ns. ?Eights Close **1** 236, Night Pool **3** 379

ēg-land OE (**īeg-land** WSax) 'island'. Brownsea Island, Furzey Island, Isle of Purbeck, Hr & Lr Nyland; f.ns. ?*Iland Moore* **1** 194, *Place Ilande* **2** 181, ?Ryeland(s) **3** 162, 187. In AS charters, *igland* **3** 257

*__eglẹ̄s__ PrWelsh 'church'. ?*Eglysham*

-el³ OE noun suffix. ?Duddle Fm

elf OE 'elf, fairy', *Vlnene-*, *Ilnenelane* (? for *Vluene-*, *Iluenelane*) **1** 350; f.n. *Elfhede* **1** 313

elle(r)n OE 'elder-tree'. ?Elder Dean Coppice, Elderen Coppice & Row, Ellen Clump, ?Elmore Fm, ?Elwell Fm; f.ns. *Elderland* **2** 204, *El(le)forlang* **2** 52, Yelder Close **3** 80. In AS charters, *ellene crundil* **3** 256, *ellenstubbe* **1** 237, *ellen þirnen* **3** 206

elm OE 'elm-tree'. Cogdean Elms, Cokers Elm, Duntish Elms Fm, The Elm, The Elms (× 3), Elm Fm, Elm Grove, Elmside Coppice, Elm Tree Fm, Elmwood, Five Elms, Gussage Elms, Longelm Fm, *Martyns Elme* **3** 21, One Elm, *Strowde Elme* **3** 182, Three Elms Inn, Walton Elm; f.ns. *Elme Close* **1** 372, *les Elmys* **2** 70

*__elmen__ OE 'made of elm', f.n. *Elmon Stubbe* **2** 52

el(n)boga OE 'elbow', f.ns. Elbow **4** 499, Elbow Colbourne **3** 196, *Nelbowe* **1** 327

*__elre__ OE 'alder'. In AS charters, *þurh þo aelres* **3** 106

elvisch ME 'elvish', f.n. *la Haluesch'thorne* **3** 225

embankment ModE 'embankment'. Embankment Bridge **1** 252

emn OE 'even, flat, level, smooth'. ?Emlands Bridge, ?Empool Bottom & Heath

emnet OE 'plain, flat ground', f.n. *la Empnet* **3** 210

-en[2] OE adj. suffix. ?Hatherington, ?Holm & Ivy Fm

enclosed ModE, f.n. Enclosed Heath & Lands **1** 51

enclosure ModE. Gundrys Inclosure, Hr & Lr Inclosure, The Inclosure; f.ns. Enclosure Plot **1** 28, Inclosure **3** 27

encroachment ModE 'common land taken into private tenure', f.ns. Encroachment **2** 157, **4** 201

ende[1] OE, ME 'end, district or quarter of a village or town'. Brister End, *Cheyne End* **3** 126, *Churche(y)nde* **4** 158, Cliff End, Deanend, Downend Coppice (× 2), Down End Fm Bldgs, Drive End, Drove End, East End, End Barrow, Farmer's End, Forked Down End, Hedge End Fm, Kimbert's End, Lane End (Fm) (× 2), Loders West End, North End (× 2), ~ ~ Fm, Peak End Hill, Redend Point, Reeds End, Three Lanes End, Town's End (Fm) (× 5), Townsend, Wall Lane End, Wallsend Cove, West End (× 2), West End Barn, Westend Ctgs, West End (Ho), Woodsend Fm. Freq in f.ns., e.g. *Touneshende* **2** 63, *Westend* **2** 11. In AS charters, *higeweges ande* **3** 137, *leas ende* **4** 304, *stanhecheres ande* **2** 58, *seuen diche suðende* **3** 123

en(d)leofan OE 'eleven'. Eleven Acre Coppice; f.n. *Enleveacres* **2** 72

endless ModE 'deep'. Endless Pit

ened, **æned** OE 'duck'. ?Enford Fm, Enmore Green; f.n. *andforde* **1** 293

engine ModE 'engine'. Engine Stream or Back Brook (*v*. RNs)

enterclose ME 'partition', f.n. *le Enterclose* **1** 62

entrance ModE. Entrance Copse, New Entrance; f.ns. Entrance Plant. **1** 167

envy ModE, f.n. Envy Close **3** 97

eofor OE 'wild boar'. ?Everley (Hill) Fm, Evershot; f.ns. Everley **2** 98, *Everlond* **4** 10, ?*Theuerdenne* **3** 112

eorl OE 'nobleman'. Earl's Fm, Earl's Hill, Earl's Kitchen, *Earl's Mead*, ?Early's Fm; f.n. ?*Hurlescruche* **3** 33

eorð-burh OE 'earthwork'. ?Herbury (Fm); f.ns. Arberry **3** 250, *Erthburie* **1** 333, ?*Yerdbury* **3** 259

eorðe OE 'earth, soil'. Earthpit, Earthpits Lane

eowestre OE 'sheepfold', f.ns. ?Ister **2** 220, ?Oustry Close **4** 156

eowu OE 'ewe'. Ewecroft Lane, ?Ewern Down Plant., Eweyards Coppice; f.ns. Eweleaze **3** 225, Ewing Down **4** 17, ?Young Creech **1** 194

episcopus Lat 'bishop'. Bishop's Caundle, Bishop's Down
epistle ModE 'epistle'. ?Pistle Down & Hill
erber ME 'garden, orchard', f.ns. *the Arbor garden* **1** 194, ?Norber **3** 186, *Southerber* **4** 80
ere- ME 'ploughing, arable'. ?Early's Fm, ?Yarlands, ?Yearlings' Bottom & Drove; f.n. Yearlings **1** 293
ermite ME 'hermit', f.n. *hermyteswode* **4** 204
ersc OE 'ploughed field', surviving as **arrish** ModE dial. 'stubble field, aftermath'. Downash Lane, Oakridge; f.ns. Barley Arish **4** 370, *Errishe field* **1** 62, Oaterish **4** 282, *Pokershe* **3** 207, Urrish **4** 352, *Wheate Earesh* **4** 58 . In AS charters, *bealtunes ersc* **3** 288–9
erð, **ierð** OE 'ploughing, ploughed land', f.ns. *Gavelerthe* **3** 199, *Gowerth*, *Vrthe* **1** 229. In AS charters, *Ierðlonde* **3** 207
***Escā** Brit 'water'. R. Axe (*v.* RNs)
escheated ModE 'lapsed to the lord of the manor', f.n. *Escheated Peece* **1** 229
estate ModE. King's Head Estate
ete OE 'grazing, pasture', f.ns. *Ettelonde* **2** 52
***etisc** OE 'pasture, plot'. ?Duntish, *dounen tit* **3** 242 (AS charter)
-et(t)(e) OE, ME noun suffix, f.ns. *Hyllett* **3** 334, Shrubbets **3** 253, Spanyard **3** 198
ever-grass Do dial. 'rye grass', f.ns. Evergrass **4** 409, 267
everlasting ModE, f.n. Everlasting Close **1** 324
eweleaze ModE 'pasture for ewes'. Bushy Eweleaze, Cerne Ewe-leaze, Eweleaze Barn (× 3), ~ Coppice (× 2), ~ Fm, ~ Spinneys, *Ewe-lees*, Higher ~, Home ~ (× 2), Lower ~, Pond Eweleaze, Lankham Eweleaze, North Ewe Leaze, South Ewe Leaze, Thorncombe Eweleaze, Wolfeton Eweleaze, Yewleaze Barn
eweyard ModE 'yard for ewes'. Eweyards Coppice
extrinseca Lat 'outside'. Beaminster Forum & Redhone Hd, *Ryme Extrinseca*
factory ModE. Factory Hill & Pond, Factory Lane **3** 360, Manufactory **4** 499
fæger OE 'beautiful'. Fair Barn, Fair Meadow, Fair Mile Rd, Fair Water (*v.* RNs), Verwood
fær OE 'ford, stream-crossing'. Hr & Lr Farway Fm
fær-weg OE 'cartway', f.ns. ?*Foreweye* **1** 128, ?*Varewaye* **3** 129
fæsten OE 'stronghold'. The Foreland or Handfast Point
fāh OE 'coloured, multi-coloured'. Frome Vauchurch
fald OE 'fold, enclosure for animals'. *Fold Bridge* **4** 195, Openfold Copse, ?Woodville Lane; f.ns. *Oldffold* **4** 10, *Therssheföldes* **2** 78,

Vacombe **4** 181, *Wydefoldessherde* **2** 175. In AS charters, *fald heʒe* (*v.* Addenda **4** 193)

faleise OFr 'cliff'. The Cliff **2** 255, Cliff Coppice **3** 36

falwe ME 'fallow', f.n. *lez Vallofeld'* **4** 335

farmery ModE 'farm buildings', f.n. Farmery **4** 298

farm-house eModE, *the fearme Howse* **1** 62

farthing, *v.* **fēorðung**

fat ME 'rich, fertile', f.ns. Fat Acre **3** 353, Fatlands **1** 142, Fat ox **2** 104, *Vaytelond* **1** 138

fatting ModE 'fattening', *Fatting Park* **2** 91; f.n. Fatting Grd **2** 114

***(ge)f(e)all** OE 'a felling of trees, a clearing'. ?Woodfalls Coppice, ?Woodville Lane

fearh OE 'pig'. ?Vary Clump

fearn OE 'fern', sometimes in a collective sense 'ferny place'. Blackfern Plant., Farnham, Farringdon Clump, Farrington, *Fearne Hill* **4** 324, Fern Barrow, Fern Brook (*v.* RNs), Fernbrook Fm, Fern Coppice, ?Fern Down, Ferndown Barn, ?*Ferne*, *Ferngore*, Fernham, Fernhayes Copse, Fern Hill (× 3), *Fern Hill*, Fernside, Ferns Plant., Fern Wood, Ferny Barrows, Fernycombe Coppice, ?Furleigh Fm, ?*Hawneferne*, ?Verne Yeates (*v.* Addenda). Freq in f.ns., e.g. *Fernhill* **3** 6, *Frizferne* **2** 158, *Verne Hill* **3** 35, *Vernolde* **2** 66. In AS charters, *fearngaren* **3** 206, *fern helle* **3** 102

fearnig OE 'ferny'. Ferncroft Fm, Ferney Down Wood, Ferney Way, *Ferngore*, Ferny Hole Plant., Ferny Knap Coppice; f.ns. Ferney Grd **2** 199, *Fernie Close Meade* **3** 32, *Vernigore* **2** 37, *Virniforlang* **2** 86

fearr OE 'bull'. ?Vary Clump

fee ME 'estate held on condition of feudal homage'. *Abbot's Fee*, *Vinter's Fee*

feire ME 'fair, market'. Fair Field (× 5), Fairfield, Pamphill Green, Vere Cross

feld OE 'open country, tract of land', later 'enclosed plot of land'. *Ashmore Fields*, *Baresfeld* **2** 279, Barrow Field, Belfield Fm, Benville Fm & Manor, Blynfield Fm, Broadfield (Coppice), Brockfield, Brookfield, *Bushfield Fm* **3** 46, Cannfield Fm, Clanfield **3** 360, Coombefield Quarries, Dowerfield Dairy Ho & Fm, East Field Fm, Eastfield Plant., ~ Hill, Farm Field Barn, Fair Field (× 5), Fairfield, Field Barn (× 5), ~ Fm, Field Dairy, Field Grove, Field Lane (× 2), Fordington Field, Forston Field Bottom, Gore Fields, Greenfield Coppice, Grimstone Fields Plant., Heathfield Ctg, Heathy Field Coppice, Hethfelton, Highfield Ho (× 2), Highfields Fm & Wood, Hillfield, Hog Field, Ladies' Walking Field, *Layfield*,

Litchfield Copse, Lower Field Stall, Merryfield (× 2), Merry Field Hill, Middle Field Lane, Muston Field Barn, New Field Coppice & Lane, Newfield Fm, Northfield Barn, Northfield Fm (× 2), ~ Ho & Point, North Field Hill, Perryfield, Planefield Ho, Smithfield Coppice, South Field Down & Hill, Southfield, Ho, ~ Lane, ?Sovell Down, Springfield (freq), Stockfield Fm, Stourfield, Sturminster Fields, Tomson Field Barn, Truckle Field Lane, Waddon Field Barn, Westfield (Fm), West Field Coppice, ~ ~ Plant., ~ ~ Wood, Westfields, Whitefield Bottom, ~ Coppice, *Whitfield*, Whitfield Fm, Winfrith Fields Fm, Woodville. Freq in f.ns., e.g. *Dollysfeld* **4** 10, *Estfeld* **4** 40, Heath Fd **2** 199, *le Stubble Fildes* **4** 26. In AS charters, *blinchesfeld*, *blinnesfeld* **3** 90–1, *feld* **3** 289

felh OE 'ploughed'. ?*Felback*

fell eModE 'cutting down of timber', ?*Velle bridge* **1** 333

fence ME 'fence', f.ns. *Bartlett's Fence* **4** 43, Fencewood **4** 328

fenn OE 'fen'. Blackven Fm, Black Venn (× 2), Bowerswain Fm, Vale Acre Copse & Fm, Vembury Dairy, Venn Chapel & Hill, Venn Dairy, Venn Fm

fennig OE 'fenny', f.ns. *Netherfennywey* **3** 226, *le Vennyeclause* **1** 237, *Vennyford'* **3** 274, Venus Style **4** 249

fente ME 'rift in the ground', f.n. Fent Md **2** 229

feoffee ME 'person invested with a feoff', f.n. Feoffee Grd **3** 26

fēorðan-dǣl OE 'a fourth part', f.n. *Hudeleforundele* **2** 130, Vernal **4** 125

fēorðung OE, **farthing** ME 'fourth part, quarter', also 'rent of a farthing' . Farthing Gate, Farthing Lane, Farthings Brake Coppice, ?Farthing's Coppice, Penny Farthing Wood; f.ns. Farthing Breach **4** 423, Farthing Fd **4** 434, Farthingsland **3** 97, *Ferthyng* **3** 266

fēower OE 'four'. Four Acre Coppice, Four Ashes, Four Barrow Hill, Four Gates

fēowertēne OE 'fourteen', f.ns. (The) 14 Acres **1** 167, **2** 214

fergen OE (**fiergen** WSax) 'mountain, wooded hill'. ?Fern Down, ?*Ferne*, ?*Hawneferne*, ?Verne Yeates (*v.* Addenda); f.n. ?*Furnedon* **2** 57

ferme ME 'rent', later 'farm'. Farm Down Barn, Farm Coppice & Heath, Farm Field Barn, Farm Hill Bottom, ~ ~ Copse, Farm Wood, Green Fm **2** 154, Long Bredy Fm, *Myddelmersh Ferme* **4** 165, *Queene Farme* **3** 18, Upton Fm, North Wootton Fm. Freq in modern minor names as ~ Farm, e.g. Candys Fm, Clapcott's Fm, Court Fm, Flag Fm, Glebe Fm, God's Blessing Fm, Hill Fm, Peaton's Fm, Pitt's Fm, Plainfield Fm, Poplartree Fm, Slade's Fm, Sparrowbush

Fm, Violet Fm, White Fm, Whitehall Fm; f.ns. *le farme* **2** 215, *the Ferme* **2** 14, *the ferme close* **1** 95, *Ferme lands* **3** 266

ferry ME 'ferry'. Ferry (× 2), Ferry Bridge, Mintern's Ferry; f.n. Ferry Cl **1** 51

fersc OE 'fresh'. Freshwater Bay, Freshwater Steps; f.n. Freshwater lake **2** 38

feðer OE 'feather'. Featherbed Lane, Featherlake; f.n. Feather Md **4** 131

feudary ME 'feudal tenure'. *Feodary lands* **3** 273

fever ME 'fever' or '?fishpond', f.n. ?*la Vivere* **3** 311

***(ge)fierne** OE 'fern brake'. ?Fern Down; f.n. ?Furne **4** 17, ?*Furnedon* **2** 57

fiddle ModE 'fiddle'. Fiddle Wood

fīf OE 'five'. *Fifash* **2** 71, Lower Fifehead or Fifehead St Quintin, Fifehead Magdalen, Fifehead Neville, Fifehead Sydling, Five Barrow Hill, Five Barrows, Five Bridges, Five Elms, Five Marys, Five Ways; f.ns. *Five Acres* **3** 226, 5 Acres **4** 34, *Vifacres* **3** 208, *la Vyfhoke* **1** 298

fīftēne OE 'fifteen'. Fifteen Trees; f.n. *Fifteneacre* **2** 72

fīftig OE 'fifty', f.n. *Fiftie acre feylde* **2** 215

filbert ModE 'hazelnut', f.n. Filbert Gdn **4** 140

***filde**[2] OE 'dwellers in open country', f.ns. *Feldyngfordesbrigge* **3** 263. In AS charters, *fildene lane* **2** 122, *fyldena wylle* **3** 289

fileðe OE 'hay'. Feltham Fm, Filcombe Fm & Wood, Filford, ?Fillymead. In AS charters, *filithleighe* **3** 199

fīn OE 'heap (of wood)', f.n. *Fynslade* **3** 256

fīna OE 'woodpecker'. Philleyholme Coppice & Mill; f.ns. Fineham **1** 171

finc OE 'finch', f.n. *Finchcombe* **2** 192

***fīning** OE 'place where wood is heaped', f.n. Gt & Lt Vining **3** 30, *Vynynges* **3** 238

fir(r) ME 'fir-tree'. Fir Coppice, Fir Hill, Fir Mount, Fir Plant., Fir Pound, Fir Tree Ctg, Firtree Ctg, Fir Tree Inn, ?Ragged Firs, Wytch Fir Pound; f.ns. Fir Close **4** 347, Fir Plant. **1** 242 [some of the names under **fyrs** may rather belong here]

fisc OE 'fish'. Fishway Lane, Fishweir Lane, ?Fishy Mead Copse; f.ns. Fish Pool Md **4** 347, *Fyshey* **3** 231, *Fysshemere* **3** 256, *Fysshewer* **2** 8

fiscere OE 'fisherman', f.ns. *Fyssherisbrigge* **1** 237, *Visghernehevede* **3** 258

fishery ModE 'fishery'. The Fishery

fishhous ME 'fish-house', f.n. *Fychehowsse* **4** 376
fishing ModE 'a fishery'. Fishing Barrow, Fishing Ctg
fishpond ModE 'fishpond'. Fishpond Bottom, Fish Pond(s) (freq), Fishponds Coppice; f.n. Fishpond Pce **2** 199
fitt OE 'fight, contest, dispute'. ?Fitzworth
fives court ModE. Fives Court Row
flagge ME 'rush, iris'. ?Flag Fm, ?Flaghead
flary ModE 'gaudy, showy', f.n. ?Flary Pins **4** 381
flasshe ME 'swampy grassland', f.n. Flash Lots **1** 324
***flasshett** ME 'place characterised by swampy grassland'. Flashet Plant., f.ns. Flaishet **3** 386, Flashett **1** 60, **3** 251
flat ModE 'flat or level ground'. The Flats; f.n. Lewell Flat **1** 211
fleax OE 'flax'. Flaxland Lane, Flaxley Coppice, Flax Mill; f.ns. Flax Cl **3** 391, *Flaxepoke* **1** 229, Flax Grd **3** 26, *Flexlond* **1** 137. In AS charters, *flescumbe*, *flexcumbes heuede* **3** 206
***flēd**, ***flēde** OE 'intermittent stream'. ?Arfleet Mill; f.n. *Flaydlane* **4** 384
flēot OE 'inlet, creek, estuary', also 'small stream'. Fleet, East & West Fleet, Fleet Fm, Fleet's Corner, Fleet St. **4** 240, Longfleet, West Fleet; f.ns. Fleet **1** 293, *Fleet Craft* **4** 332, *Lee Fleet* **4** 452
flint OE 'flint'. Flint Pit; f.ns. Flint Close **3** 387, ?*Flyntesfeld* **4** 222
(ge)flit OE 'dispute', f.n. *Flitegore* **4** 82
flōc OE 'fluke', f.n. ?*Fluxford Bridge* **1** 81
flocc OE 'flock (of sheep)', f.n. Flockshard **2** 85
flōd OE 'flood, tide'. Blood-way Coppice, Flood Ho & Lane, ?Shaggs Flood; f.ns. Floodhatch Md **4** 423, Floodlands **2** 104
flod-yate ME 'flood gate', f.ns. *Les Flodeyates* **1** 365, *Flude yate* **3** 341
flooding ModE '(land) irrigated by flooding', f.n. Floodings **3** 40
floody lME 'liable to flood', f.n. *Flodyput* **4** 284
flop Do dial. 'mass of thin mud', f.n. Flop Yard **4** 381
flōr OE 'a floor'. ?Flower's Barrow; f.n. ?Flower Lds **1** 343
flote ME 'flooded', f.ns. *Floteham* **1** 372, **3** 110, *Floteshorte* **1** 381
flour ME 'flower', f.ns. *Flo(u)r(e)lond* **1** 183, 216, 298, ?Flower Lds **1** 343, *Flowrelane* **4** 22
flowen ME 'flooded', f.n. *Flowenham* **1** 372
fodder ModE 'food for cattle', f.ns. *Castle fodders* **1** 62, Foder Ho Grd **1** 312
fogge ME 'aftermath', f.n. Fogberry **4** 143
foggy ModE 'marshy', f.n. ?Woggy Cl **4** 311
fola OE 'foal', f.n. *Folmore* **4** 19

folc OE 'folk, the people of a tribe or family, the common people'. Folke; f.n. *Dedefolkefurlonge* **2** 13

folie ME 'folly'. Baker's Folly, Cain's Folly, Folly, The Folly (× 3), Folly Barn & Clump, Folly Barrow, Folly Ctg, Folly Fm (× 3), Folly Hill, Folly Ho, Folly Mill, Folly Pier, Folly Plant., The Folly Wood, Frog's Folly Plant., Grange Arch, Piper's Folly, Stagg's Folly, Woolcombe Folly; f.ns. Brook's Folly **2** 22, Folly Lane **4** 486, Folly Plot **3** 213, Volley Mdw **4** 344

font ME 'fountain, well-spring', *Bouelemmanescumbefonde* **3** 244

fontein ME 'fountain, spring', *de la funtaine* **4** 265, St Edward's Bridge; f.n. The Fountain Gdn **4** 399

footpath ModE, f.ns. Foot path Fd **2** 22, Footpath Mdw **2** 157

forca OE 'fork'. ?Stivvicks Bridge [*v.* Addenda **2** 103]

ford OE 'ford'. Abbey Ford Bridge, Ameysford, Axe Ford, Barford Fm/Old Barford, Battleford Fm, *Blachford* **4** 406, Blackmore Ford Bridge, Blackwater Ford, Blandford Forum/St Mary, Bradford Abbas, Bradford Fm, Bradford Lane, Bradford Peverell, Canford, *Cheselbourneforde* **1** 316, Child Okeford, *Churchford* **4** 158, Clayford Fm, Cornford Bridge, *Gt Crawford*, *Crokerford*, Dollerford **3** 367, Elsford Bridge, Enford Fm, Evenings Ford, Fiddleford, Filford, Filling Ford, Ford (freq), Ford Heath, Fordington, Ford Coppice & Fm, Forde Abbey, Ford Fm (× 32), Ford House (× 2), Ford Lane (× 2), *Fordmere* **3** 146, Ford Plants., Fordwater, Garvey Hill, Gozleford, Greenford Lane, Hanford, Hayward Bridge, Heniford Fm & Mill, High Ford Lane, *Hogford*, *Horyford*, Hunters Bridge, Huntingford, Hyford Ctg, *Hymindesforde* **4** 496, *Joppysforde* **3** 165, Kitford, Kitford Bridge, Langford Fm, Langton Long Blandford, Lawford Sheard Gate, *Leftisford*, Longford Bridge, *Lovard*, Luckford Coppice, Lydford Fm, Maidenford Weir, *Milborne Mamford*, Metford Mill, *Milford Ctg*, Mount Radford, Muckleford, Muddyford Lane, Nutford Fm, Oakford Fm, Okeford Fitzpaine, Old Ford, Organ Ford, Pilford, Pinford, Povert Bridge, Redford Fm, Rodford Withy Bed, Romford, *Ruddlesford*, *Saltforde* **4** 395, Sandford (Ctg & Fm), Sandford Orcas, Sharford Bridge, Sherford, Shillingstone, *Smallsford Lane*, Snatford Bridge, *Spadeforde Bridge* 1381 **1** 18, Stafford Park Copse/Fm, Stafford Row, West Stafford, Stake Ford Cross, Stanford Point, *Stapleford*, Stinsford, Stintford Lane, Stockford Ctg, Stokeford, *Stratford*, *Stretford* **1** 275, Synderford, Thornford, Tollerford, Turford, Twofords Bridge, Twyford, Walford Bridge & Fm, Watford Bridge & Fm, Westford Fm, *Wikenforde* 948 **1** 18,

Williford, Wizard Bridge, Woodsford, Woolford's Water, Hr & Lr Wrackleford. Freq in f.ns., e.g. *fordetune* **4** 17, Longford **3** 28, *Rotherford* **2** 11. Also in AS charters, *chirchelford* **2** 274–5, *crawan forda* **2** 136, *doggeneford* **3** 255, *ealden forð* **3** 128, *ford* **2** 122, **4** 193 (*v*. Addenda), *forda* **3** 206 (now **1** 316, *v*. Addenda), *funtemel forð* **3** 136, *fyres forid*, *fyres hylleford*, *fyrs ford* **4** 303 (*v*. Addenda **4** 193), *horsford* (*v*. Addenda **4** 193), *hweoul forda* **4** 454, *landscher' ford* **3** 199, *sandford* **3** 129, *scyleford, Scylenford* **1** 94, *staford* **4** 304, *sticelan ford* **4** 454, *stræt forda* **4** 454, *wicanforð*, *Wichenford*, *Wickenford*, *Wikenforde* **1** 18, *winterburne ford* (*v*. RNs.)

fore OE 'in front of, before', also 'prominent, jutting forward'. Ford Down Lane, *Fore Church* **4** 195, Fore Hill (Plant.), Fore Place **4** 240, Fore Street (× 2), Fore Top; f.ns. Fore Door Cl **4** 423, *Foremede* **2** 52, *Forlond* **1** 260, *Ye Longe Fore pitt* **3** 80

fore-burg OE 'an outwork', f.ns. *Forbury* **4** 3, *Vorburycroft* **4** 183

forehead eModE, f.ns. Forehead Cloth **2** 124, Goodforward Flg **1** 362

forein ME 'outside a boundary', f.n. ?*La Forengore* **1** 273

foreland ModE 'cape, headland'. The Foreland or Handfast Point, St Alban's or St Aldhelm's Head

forest ME 'forest'. Blackmoor Forest, Charmouth Forest, Forest Bourne Fm, Forest Fm (× 3), Forest Hill, Forest Side, Fossil Forest, Gardiner Forest, Holt Forest, *le old Forest* **2** 233, Poorstock Forest, Wareham Forest; f.ns. Forest **3** 262, Forrest Croft **4** 139

forester ME 'officer in charge of a forest'. *Forstershay*

forge ME 'forge'. Forge Lane, *Le Smithyes Forge* **3** 184; f.n. Gt Forge **4** 486

for(i)erð, -yrð OE 'projecting piece of ploughland', f.ns. *la Forurde* **4** 10, *le Langefeherth* **2** 224, *Langeforurde* **1** 215, *la Rydeforhurde* **3** 34. In AS charters, *forerð akere* **3** 206

forinseca Lat 'outside'. Beaminster Forum & Redhone Hd

forked ModE 'forked'. Forked Down End; f.ns. Forked Bridge Fd **3** 101, Forked Cl **2** 41

forneis ME 'furnace', f.n. ?Furnace Md **3** 339 (*v*. NDEFN 161)

forsacen OE 'abandoned', f.n. *la Forsakenegore* **3** 199

***forst** OE 'ridge'. ?East & West Fossil Dairy & Fm; f.n. *Forstedune* **4** 411

fort ME 'fortress'. The Fort, *the Nothe Fort* **1** 256

forð OE 'in front, before', f.ns. *Forthdon'* **2** 271, Forward Lawn **3** 108

***forð-īeg** OE 'island in marshland'. Forty Pond; f.ns. Forty Hedge **4** 42, Gt Varty **3** 30

fortelet or **fortalyce** lME 'small fortress', f.n. *Fortell'* **4** 376

fortune eModE 'luck, fortune'. Fortuneswell
forty ME 'forty'. Forty Acre Bottom, ~ ~ Plant.; f.ns. 40 Acres Mead **3** 197, Gt Fourty Acres **1** 83
forum Lat 'market'. Blandford Forum [*v.* Addenda **2** 87 for a different explanation of this affix]
***foss** OE 'ditch'. *la vosse* **4** 198; f.n. *la Fosse* **1** 143
fossil ModE 'fossil'. Fossil Forest
fōt OE 'foot (of a hill)'. Churchfoot Lane, Footlands Coppice, ~ Lane, Godlandfoot, Greenhays Foot, High Stoy, Rowdenfoot Coppice, Sandsfoot Castle; f.ns. *Cumbesfote* **4** 475, *Footecrofte* **4** 465, *Foteacr'* **3** 79, *la Fotlond* **3** 32
fote-bridge ME 'footbridge', f.n. *Fotebridge* **1** 346
founder ModE 'to sink'. ?Founder Hole
foundering eModE 'collapsing', f.n. Foundering Bank Mdw **2** 74
foundry ModE 'foundry'. Foundry Ctgs
fount eModE 'spring', f.n. *le Founte* **1** 287
fox OE 'fox'. Fox Allers Copse, Fox Barrow Fm, Foxbrake, Foxbury (Coppice), Foxbury Lower Hedge, Fox Covert, Fox Ditch Coppice, Foxdown, Fox Fm, Fox Ground (Down), Fox Hill (× 2), Foxhill, Foxleaze Withy Bed, Foxmoor Coppice, Fox Pound (× 4), Pimperne Fox Warren & Wood; f.ns. Fox Cover **1** 286, Foxen hills **3** 162, *Foxfforlang* **4** 10, *Voxle* **3** 82, Voxley **2** 172. In AS charters, *foxlee* **3** 128
fox-hol OE 'fox hole'. Fox Hole Goyle, Fox Holes (Ctgs & Wood), Foxholes Coppice, Dairy & Wood, ~ Ctgs & Plant.; f.ns. *Foxe Holes* **3** 32, Foxholes **3** 251
frank(e) ME 'pig-sty', f.n. ?*Frankemere* **2** 174
frary ME 'brotherhood, fraternity'. Friary Hill & Lane, Friary Saw Mill; f.n. Freary Barton **3** 26
frater Lat 'brother, member of a religious order'. Toller Fratrum
fraunchise ME 'district over which the privilege of a corporation extends'. Little Francis
freehold ModE, f.n. Freehold Orchd **3** 272
french grass ModE 'sainfoin'. French Grass Coppice; f.ns. (Old) French Grass **3** 40, **2** 199, French Grass Fd **2** 85
Frensche ME adj 'French'. French Mill; f.ns. French Fds **1** 136, French Grd **3** 26
frēo OE 'free from service or charge'. *Fredownes* **3** 125, Free Down (Border), *Free Tithing* **3** 10; f.ns. *Frelease* **3** 178, *la Fremede* **1** 333
***frēo-land** OE 'estate that is free from certain dues and obligations'. *Fryland*; f.n. *la Fryelandshor* **1** 326

frēo-mann OE 'free-man', f.ns. *Freeman Meade* **2** 192, *Fremancroft* **1** 70

frere OFr, ME 'friar'. Friar Waddon, Friary Lane **1** 351, Fryer Mayne; f.ns. *Fryarnebridge* **2** 78, Frying Croft **4** 55

fresh ModE 'fresh'. Burton Freshwater

frīgedæg OE 'Friday'. Long Friday; f.ns. Friday Street Plot **4** 55, *Good Friday* **2** 109, Long Friday **1** 312

frogga OE 'frog'. Frog Hole, Froghole Fm, Froghole Withy Bed, Frog House Fm, Frog Lane (× 3), *Frogmoor*, Frogmoor, Frogmore Ctgs, ~ Dairy House, ~ Fm (× 2), ~ Hill, ~ Ho, ~ Pool, Frog's Hole, Froxen Copse; f.ns. *Froggegore* **3** 256, Frogmore Fd **2** 146, *Vroggemere* **3** 59

***frǭm** PrWelsh 'running water, stream'. R. Frome (*v*. RNs), giving name to: Chilfrome, Frampton, Frome St Quintin, Frome Vauchurch

front OFr, ME 'face, forepart'. Front Beach, Front Lane, Front St; f.ns. *Beawffrount* **4** 399, Front **4** 381, Front Grd **3** 287

frontage ModE 'land in front', Frontage **4** 39

frosc OE 'frog'. Breach Plant.

frost ModE '(subject to) frost', f.ns. Frost Fd **2** 270, Frost Lands **2** 237

***frud** PrWelsh 'stream'. Winfrith Newburgh, R. Winniford (*v*. RNs), Wynford Eagle

fryingpanne ME 'frying pan'. Fryingpan; f.n. *le Fryingpanne* **2** 14

fugol OE 'bird'. ?Foul Hill, ?Foul Moor Lane; f.n. Vulscombs **4** 283. In AS charters, *fugelan pæþ* **2** 136

fūl OE 'foul, dirty'. Filley Brook (*v*. RNs); f.n. Fullfords **4** 110. In AS charters, *fulan lake* **2** 158, *fulan slo* **4** 303

***funtǭn** PrWelsh, **funta** OE 'spring, stream'. Fontmell Brook (*v*. RNs) giving name to Fontmell Magna & Parva; f.n. Long Fount **1** 362 [cf. **fount**]

furh OE 'furrow, trench'. Firch Lane; f.ns. *Furchee* **4** 434, Furches **4** 381, *la Heldevorugh* **3** 226, *Waterfury* **1** 199. In AS charters, *furch* **3** 206, *fures* **2** 122, *furuh* **3** 122

furlang OE 'furlong'. Church Coppice, *Churcheforlang* **4** 264, Clotfurlong Lane, Furlong, Furlong's Homestead, Lamplands Copse, *Northdunesfurlange* **4** 259, Waverlawns, *Widecombefurlang* **3** 246. Freq in f.ns., e.g. *Brokfurlang* **1** 260, *Churchfurlang* **4** 10, *Efurlang* **2** 52, *Watfurlong* **4** 35.

furthor OE 'further', **furthra** OE 'more distant'. Further Ho; f.ns. *Furthemustetwyste* **3** 238, Furtherlands **4** 409

fyniht OE 'fenny, marshy', f.n. Fent Md **2** 229

fyrhð, **(ge)fyrhðe** OE 'a wood, woodland'. Frith Fm & Wood, Writh Ctg, Writh Fm & Copse; f.ns. Frith Md **2** 98, *le Frythe* **2** 174, Parks Frith **4** 25, *Pinfrithe* **3** 34

fyrs OE 'furze'. ?Barrow Firs, ?Burnt Firs, *Chapple Furses* **4** 242, Crook Furze, ?France Firs, ?Frizzel's Hill, Furze Brach Lane, Furze Brake (Plant.), Furzebrook, Furze Common Copse, Furze Coppice, Furzecott, Furze Down, Furzedown Fm, Furze Hill (freq), Furzehill Coppice & Fm, Furze Ho, Furzeland Barn, Furze Lane, Furzey Close Row, Furzey Island, Furzhill, Furzy Cliff, Grighay Furze Coppice, ?Gulliver's Firs, Heckford Rd, ?Lily Bank Firs, ?Ragged Firs, ?Stone Firs, ?Strouds Firs, ?Woolland Firs. Freq in f.ns., e.g. *Furshull* **1** 264, *lez Fursys* **2** 70, Furze Hill **4** 22, *Longe Furze* **4** 43. In AS charters, *firesrewe* **3** 256, *fyres forid*, *fyres hylleford*, *fyrs ford* **4** 303 (*v*. Addenda **4** 193), *snieren fyrsgaran* **3** 207

***fyrsen** OE 'growing with furze', f.ns. Furzen Grd **2** 146, *ffuson Bottom* **4** 82

***fyrset** OE 'place overgrown with furze', f.n. Fuzzards **4** 226

***fyrsig** OE 'growing with furze'. Furzey Close Row, Furzey Close Wood, Furzey Coppice, Furzy Cliff, ~ Common, ~ Coppice, ~ Drong, Furzy Woolly Coppice; f.ns. Furzey Marsh **4** 19, Furzy Cl **3** 251

gable ModE 'gable'. Gable Court, The Gables

gad ModE 'spike, wedge'. Gad Cliff

gærs OE 'grass', f.ns. *Grashay* **4** 182, *Grasslade* **3** 81

***gærsing** OE 'grazing', f.n. ?Gaz(e)ing Fd **1** 206

gærs-tūn OE 'grass enclosure, paddock'. Garston Down & Wood, Gascoyne Bridge & Lane, Gassons Lane, Goschen, Hingarston, *Mauryesgarston* **3** 194, Rix(t)on, *Sugestone*; f.ns. Gascon **2** 69, Gassons **3** 5, Gaston **3** 213. In AS charters, ?*hengersten* **3** 252, *lyngærstun* **3** 380

gærs-yrð OE 'the customary ploughing of grassland', f.n. *Graserth* **2** 28

gafol[2] OE 'tax, rent'. Galton; f.ns. *Gavelerthe* **3** 199, *Gavelmede* **3** 374, *Gawle acr'* **4** 19

gail(l)e ME 'gaol'. *la Gaole* **1** 355, *Gaole Lane* **1** 352; f.n. Gayhole **4** 337

galga OE 'gallows'. Gallica Bridge, Gallows Corner (× 3), Gallows Gore Ctgs, Gallows Hill (× 3), Gallows Plot; f.ns. Gallows Acre **2** 234, Gallows Cl **4** 423, *Gallows Hill* **1** 352, *Gallyhey* **1** 62

gallery ModE 'an underground passage, a level or drift'. Seacombe Gallery

gallows-trams ModE 'the two upright posts of a gallows', f.ns. Gallett Rams **4** 192, Gallytrams **4** 233

game ModE 'game'. Game Copse; f.ns. Games Mead **3** 388, Games Plot **4** 329

gamekeeper ModE. Gamekeeper's Lodge

gander ME 'male of the goose', f.ns. Gander Cl **4** 201, *gander Howsse* **2** 174

gang OE 'track', f.ns. Overgang **3** 353, *la Overgange* **3** 200

gangstede OE 'walk or pasture for cattle', f.n. *Ganstede* **1** 313

gappe ME 'gap'. Dorsetshire Gap, Gaulter Gap, Harley Gap, Winyard's Gap; f.n. *Mesurer's Gap* **2** 45

gāra OE 'triangular plot of land'. Atrim Gore, ?Binnegar, *Ferngore*, Gallows Gore Ctgs, Gallpits Gorse, *Garbroke* (*v*. RNs), ?Gorcombe Ctgs & Wood, Gore, Gore Barn & Hill, Gore Coppice, Gore Cove, Gore Cross, Gore Fm (× 2), Gore Fields, Gore Gate, Gore Hill Plant., Gore Holmes, Gore Ho, Gore Lane (× 2), Goren Fm, Gore Pit, Gore's Hill, Gorewood, Guineagore Lane, Peter's Gore. Freq in f.ns., e.g. Goors **3** 26, Gore Acre Ho **3** 8, *Goredaker* **3** 32, Gt Hazel Gare **2** 37, *Quenegore* **2** 53, Wager **3** 30. In AS charters, *biricgaran* **2** 161, *fearngaren* **3** 206, *garen* **2** 82, *snieren fyrsgaran* **3** 207

garbage ModE 'refuse', f.ns. Garbage Md **2** 199, ?*Gulbedge brooke* **4** 22

garbrede ME 'land in the gore of the common field', f.n. *Gorprode* **1** 327

gardin ME 'garden'. Alexandra Gdns, Barton Gardens, Bee Garden, Broad Garden, Building Garden Coppice, Cliff Gardens, Coy Pond Gdns, Garden Coppice, ~ Copse, ~ Wood (× 2), Mulberry Gardens, Nothe Gdns, Pageant Gardens, Radipole Park Gdns, Sluice Gdns, Spring Garden Barn, Trinidad Gdns Estate; f.ns. The Garden **3** 293, *Estgardin* **2** 52, Garden Tillage **2** 199

gār-lēac OE 'garlic', f.ns. Garlick Md **4** 399, *Garliham* **1** 183

gāt OE 'goat'. *Gatelinche* **4** 470, Gatemerston, Gatmore Fm, Goathill, ?Goathorn, Gotham; f.n. *Gatcombe* **4** 10. In AS charters, *gata ford*, *–forð* **4** 374

g(e)alla OE, ModE dial. **gall** 'wet or barren spot in a field'. Gaul Hill; f.ns. Bitters Gall **1** 78, ?Fourteengals **4** 39, ?Gall Hill **3** 197, Gall Md **4** 423

geant ME 'giant'. The Giant, Giant's Grave & Stones, Giant's Grave & Trencher

***gēap**[1] OE 'steep place'. Eype

gēap[2] OE 'steep', f.n. *Yapehaie* **4** 466

***gear** OE 'yair, weir', f.n. *Yerhede* **4** 454; in OE charters, *miliere* **3** 207

geard OE 'yard'. ?*Bush-head Gdn*, Dodding's Brick Yard, Eweyards Coppice, Hop Yard, (Huish) Lambing Yard (× 3), Manor Yard, Monk's Yard, ?Smetherd Fm

gearn OE 'yarn, spun fibre of flax, etc'. Yarn Barton, *Yarne Markett* **3** 146; f.ns. Yarn Barton **3** 5, *Yarne Closse Streate* **2** 186

geat OE, **yate**, **gate** ME 'an opening, a gap'. Alms Grove (Gate), Ashley Sand Gate, Bailie Gate, Biddlesgate Fm, Birdsmoor Gate, Bloody Shard Gate, Brockhampton Gate, *Brodemede yate* **4** 376, Burngate (Fm), Cobb Gate Beach & Jetty, Cockett Hill Gate, Colway Gate, Common Gate, Coombe Gate, Copse Gate Dairy, Corfe Castle, Coryates, Daggers Gate, Dirty Gate (× 2), Dogbury Gate, Farthing Gate, Four Gates, Gate, Gate Barn, Gate Coppice & Fm, Gate House, Golden Gate, Gore Gate, Gorwell Gate, Grange Gate, Highgate Ctgs, Highgate Lane, Hill Pit Gate, Lakegate Lane, Langford Gate, Lawford Sheard Gate, Leigh Gate, Lob Gate, *Lucombe yate* **4** 197, Lych Gate, Lyon's Gate, Maiden's Grave Gate, Marriotts Fm, Max Gate, Moryatt Fm, New Cross Gate, New Gate (Ctgs), Newlands Gate, Northgate, Park Gate, Picket Fm, *quarr' yate* **4** 379, Remedy Gate, Rempstone Gate, Scutt's Gate, Sessions Gate, *sewes yate* **4** 207, Shermel Gate, Sixpenny Gate, Slaughter Gate, *Sleybrondesgate* **3** 149, Snipe Gate, Somerset Gate Ctg, Stag Gate, Stone Hill Gate, *Stoys gate* **4** 163, Strongate Fm, Ten Acre Gate, Thistlegate Ho, Three Gates, Toller Gate, Two Gates (× 4), Verne Yeates, Wareham Gates, Watergates Ctg & Lane, West Gate, Whitegate Fm, Whitepost Gate, Woodlands Gate, East & West Woodyates, *Wyche gate* **3** 69, *Wythehocke yate* **4** 206. Freq in f.ns., e.g. Chiverick Md **3** 161, *Le Goryate* **4** 40, Peckashite **4** 410, Pound Yate **3** 5, *le yetcroft* **1** 299. In AS charters, *cinerget* (for *ciuerget*) *mor*, *seuerȝate* (*v*. Chivrick's Brook in RNs. supra), *corfȝet* **4** 302, *corfgetes westran cotan* **4** 27, *dungete* **2** 121, *eastgeat*, *east gete* **2** 134, *get* **3** 123, *horgate*, *hornget* **2** 57, 82, *houndesete* **3** 257, *la-*, *lobirihtes gete* **4** 454, *midelgete* **2** 82, *ruwan leges gete* **2** 162, *tor scylget* **3** 102, *westgeat*, *west geate* **2** 134, *widdisȝate* **3** 259, *widesgete* **2** 82

***gefl** PrWelsh 'fork'. R. Yeo (*v*. RNs)

gentilman ME 'gentleman', f.ns. *Gentilmannesmede* **1** 365, Gentleman's Furze **1** 271

geolu OE 'yellow'. Yellowham Hill & Wood, Yellow Ledge, Yellow Withy Bed

geong OE 'young'. Young Copse; f.n. *Yongewode* **3** 159

gerd OE, **gierd** WSax 'a rod or spar' also 'a measure of land consisting of about thirty acres'. ?*Bush-head Gdn*, Halfyard Coppice, Urless Fm, Yard Ctg & Fm, Yard Dairy, Yard Fm **4** 405, Yardgrove Fm, Yard Lane, Yardley, Yards Brake; f.ns. *la Girde* **2** 19, 3 Yards **3** 71, *Yerdlondes* **4** 454. In AS charters, *ȝyrdlea* **4** 292, *gyrd weg* **4** 374, *þreora gyrda* **3** 205 (now **1** 316, *v.* Addenda)
gerner ME 'granary', f.n. *Netelgurner* **1** 184
gib ModE 'hump, hook', f.n. The Gibb **1** 60
gibbet ME 'gallows', f.n. Gibbett Cl **3** 97
gig ModE 'a light, two-wheeled, one-horse carriage'. Gig Lane
gig(ge) ME ' a fool', f.n. ?*Gygehorn* **3** 226
gild-hall OE 'guildhall'. *le gildhall* **3** 180, (The) Guildhall **3** 10, 150, **4** 360, (*the*) *yelde halle* **2** 187, **3** 366
gild-hūs OE 'guild house'. *Yeldehous* **3** 366, *yeldhouse* **3** 10
ginne OE 'wide, spacious', in AS charters, ?*ginum hocum* **3** 138
gipsy ModE. Gipsies' Corner, ~ Pit, Gipsy Lane; f.ns. Gipsey Corner **2** 74, Gipsey Plot **4** 124
girdle ModE 'girdle', f.ns. Ash Girdle **4** 192, Girdle Ring **4** 451
glassen eModE 'made of glass, fitted with glass'. Bradford Abbas
glebe ME 'glebe, land attached to a benefice'. Glebe Ctg(s) (freq), Glebe Dairy Fm, Glebe Fm (freq), Glebe Ho, Glebeland Lane, Glebe Withy Bed; f.ns. *the Gleabelands* **4** 43, *Glebeland* **2** 215
glen ME 'valley'. The Glen (× 2)
***gles**, **gleis** PrWelsh 'stream'. Devil's Book, R. Divelish (*v.* RNs), Dewlish
glida OE 'kite or other bird of prey'. Glyde Path Hill & Rd **1** 351, Lydford Fm; f.n. *Glidecombe* **4** 14
gloriette OFr 'highly decorated chamber'. *Gloriette* 1280 (at Corfe C. **1** 11)
goad ModE 'measure of land', f.n. The Hundred Goad **3** 205
gobelin ME 'goblin', f.n. *Gobelynshole* **4** 35
gōd OE 'good'. Godlandfoot, ?God Nore; f.ns. *Godshepen* **4** 234, *Godstiche* **3** 199, Good Acre **4** 267
***gogge** ME 'a bog', f.n. *Goggford* **3** 321
go(w)ke ME 'cuckoo', f.ns. Cuckooford **4** 216, Go Croft **1** 79, Gookhays Copse **4** 409, Gooks Hill **2** 139
gold OE 'gold, golden-hued'. ?Goldcroft Fm, ?Gold Down, Gold Hill **3** 145, Gold Oak; f.n. Gold Hill **4** 143
golde OE '(marsh) marigold'. ?Goldcroft Fm, ?Gold Down, Gold Hill; f.n. *Goldecrofte* **3** 256

golden ModE. 'golden'. Golden Bowl, Golden Cap, Golden Gate, Golden Square; f.n. Gold Hill **4** 143 [*cf.* **gylden**]

gold-hord OE 'gold hoard, treasury'. Gaulter Ctgs; f.n. Golter **1** 41

goodman ME 'head of a household', f.n. ?*the Goodman* **1** 169

gor OE 'dirt, dung, filth'. Garvey Hill, ?Gorcombe Ctgs & Wood, Gorwell Fm

gorst, **gors** OE 'gorse, furze'. Bide's Gorse, Clarke's Gorse, Gallpits Gorse, Gordon's Gorse, (The) Gorse, Gorse Plant., Gorse Triangle, Holwell Gorse, Kitwhistle Gorse, New Gorse, Pulham Gorse, Ranksborough Gorse

gōs OE 'goose'. Goose Hill, Gooseland Lane; f.ns. *Gesefield* **1** 152, Goose Ham **3** 44, Goosehaum **2** 94, *Gousmede* **3** 32

gospel ModE. Gospel Hall; f.n. *Gosspell Bushe* **2** 174

gotere ME 'gutter', f.ns. *le Goter* **1** 120, Gutter Md **4** 451, 463, *the lawe gutter* **3** 372

***gotu**, ***gote** OE 'water-course, channel, stream'. ?Goathorn, ?Gutch Pool Fm & Hole; f.ns. *Cherchegot* **3** 273, *Gotfurlang* **4** 133, *Gout acr'* **4** 385 [*cf.* **gut**]

gouge eModE 'chisel or spade with concave blade', 'to hollow or scoop out'. ?Gutch Pool Fm & Hole

go(u)le ME 'ditch, channel', f.ns. Gole **4** 487, *Thorne Goale* **2** 54

goulet ME 'a gully, a ravine, channel'. Birch Hill Gully, Gully, Gully Coppice, Long Gully, Newfoundland Gulley, Park Gully; f.ns. *Gullimead* **3** 231, Gully **4** 39, Gully Md **3** 66, *le gullyes* **1** 169

goyle, **goil** ME 'ravine'. Alford's Goil Coppice, Chanters Guile, *Egliston Gwyle*, Fox Hole Goyle, Girt Goyle, Liberty Goyle, Lutton Gwyle, N & S Gwyle, The Gwyle (× 3, *v.* RNs), *Ridge Goyle* **4** 416, Swalland Fm, Tyneham Gwyle; f.ns. Bottom Goil **4** 409, Goil(e) **4** 423, Goyle Orchd **4** 390, *Little Gyoll* **3** 256

***grǣd-tūn** OE, *v.* **grotten**

græf, ***grafa** OE 'grave, pit'. Betty's Grave, Elliott's Grave, Giant's Grave (× 2), ?Graston Ho, Maiden's Grave Gate; f.n. *la Grave* **4** 19

grǣg[1] OE 'grey'. Grey Cliff; f.ns. *grey wethy lane* **3** 273, Greystone Bottom **1** 181

***grǣg**[2] OE 'a badger'. ?*Grayswick* **1** 357

grāf(a) OE 'grove, copse'. Alder Grove (× 2), Alms Grove, Ashgrove, Beech Grove, Bowgrove Fm, Cowgrove, Cow Grove Fm, Deans Grove, Drove Island, Dungrove Fm, Elm Grove, Field Grove, ?Graston Ho, Grove, The Grove (freq), Grove Coppice, ~ Fm (× 3), ~ Hill, ~ Ho (× 2), Grovehouse Fm, Hr & Lr Grove, Grove Lane (× 2), The Groves, Grove Stall, Grove Wood, Hargrove Fm, Hartgrove,

Heron Grove, Heron Grove Coppice, High Grove Fm, Hollow Grove, Holly Grove Coppice & Fm, Honey Grove, Joe's Grove, King Grove Fm, King's Grove Wood, Lover's Grove, *Lussh'groue* **3** 203, Madgrove, Nether Grove, Nunnery Grove, Oak Grove, Poxwell Grove, Ring Grove, Rook Grove, Thorngrove, *Walgrove*, Weathergrove, Yardgrove Fm. Freq in f.ns., e.g. *Cookesgrove* **2** 215, *Wychegrave* **2** 54. In AS charters, *chelfgraue* **2** 82, *graf* **4** 35, *groue* **3** 199, *langengroue* **3** 257, *sulangraf* **3** 112

grāfet OE 'trench', f.n. *La Grauette* **1** 287

grain ModE 'corn', f.n. Grain Grd **1** 28

grammer eModE 'grandmother', f.n. Grammers Grd **3** 97

***grand** OE 'gravel', f.n. *Graund acres cope* **2** 14

grandfer eModE 'grandfather', f.n. Granfers Grd **3** 97

grange ME 'an outlying farm belonging to a religious house where crops were stored'. Abbey Grange (× 3), Balholm Grange, Chaffeymoor Grange, Creech Grange, Forde Grange, Forston Grange, (The) Grange (freq), Grange Arch/Gate/Heath/Hill/Plant., Grange Copse, Grange Corner, Grange Ctgs, Grange Dairy, Grange Fm (freq), Grange Mead, Grange Woods, Merton Grange, Middlemarsh Grange, Newell Grange, Wooth Grange; f.ns. Grange **2** 229, *Homegraunge* **2** 14

gra(u)nger ME 'one in charge of a grange', f.n. *Grangeresbrigge* **3** 32

grant, **grand** OFr, ME 'large'. Kington Magna

grāpian OE 'to grope'. ?*Cropelane* **3** 148, ?*Groop Lane*, *Groupe Street* **1** 155, ?*Grope Lane* **1** 354

grate eModE 'prison', *Grate* **3** 150

gravele ME 'gravel'. Alderholt Gravel Pit, Alderney Gravel Pit, Gravel Hill (× 2), Gravel Lane, (Old) Gravel Pit(s) (freq), Gravel Pit(s) Plant., Gravel Pits Wood; f.n. Gravel Pit Grd **2** 139

grēat OE 'large'. Great Barrow, ~ Batch, ~ Bournes, ~ Common, ~ Coppice (freq), ~ Copse, ~ Ditch, Great Down Barn, Great Down Clump, Great Head, Great High Wood, Great Hill (× 2), Great Ho (× 2), Great House Fm, Great Moor, Great Pit Lane, Great Plant. (× 2), Great Pond (× 2), Greatstone Coppice, Great Wood (× 2), Melbury Great Wood; f.ns. *grete close* **3** 79, *Grete Robyn* **3** 29, *Gretesherde* **3** 32. In AS charters, *graetem beorge* **2** 122, *greatan ac* **3** 289, *greate hlinc* **3** 206, *Gretindune* **3** 206

grēne[1] OE 'green'. Greenclose, Green Close Dairy, Greenclose Fm, Green Close Lane, Green Copse, Green Dale, Green Down (Drove), Green Fm (× 2), Greenfield Coppice, Greenford Lane, ?Greenham, Greenhill, Green Hill (× 6), Green Island, Greenland, Green Lane (freq), Green Leaze, Green Pit Bridge, Green Point, Green Pond, Green Wall Plant., *Greenway*, Greenway (Barn, ~ Head, ~ Lane). Freq in f.ns., e.g. *Greene Coppice* **2** 205, *Greene haies* **3** 6, *Greneclose* **2** 10. In AS charters, *ȝrenan weȝ* **4** 303, *grenan weges* **4** 454, *grenenhille* **2** 57, *grenwai* **3** 128

grēne² OE 'grassy spot, village green'. Adam's Green, Bittles Green, Bonny Green, Brockhampton Green, Camelgreen, Church Green, Cottage Green, Dalwood Green, Eddygreen Fm, Enmore Green, Fidler's Green, Gerrard's Green, God's Blessing Green, The Green (× 2), Harraby Green, Hermitage Green, Higher Green, Holebrook Green, *Howke Grene* **2** 203, King's Green Ctg, Lambs' Green, Leigh Green **4** 194, *Okegreene* **4** 460, Osehill Green, *Parkston Green*, Pamphill Green, Pleck Green, Post Green, Sharnhill Green, Sheepwash Green, Shorts Green Fm, Tollard Green Bottom, Tytherleigh Green, Wingreen, Wytch Green; f.n. *Almer Greene* **2** 57

grēot OE 'gravel', f.ns. Girt **4** 451, ?*Gretebergh* **2** 183

***grēoten** OE 'gravelly', in AS charters, ?*gretenlinkes* **4** 173

gribbele ME, Do dial. 'crab-tree'. (Hr) Gribb, ?Gribb's Fm; f.ns. Gribbles Orchd **4** 9, ?Griblayne **4** 35, *lez gribles* **4** 335

grigge ME 'dwarf', f.n. ? *Griggegore* **1** 211

***Grīm** OE (probably a by-name of Woden). Grim's Ditch

grīma² OE 'goblin', f.ns. Grimbury **1** 28, *Grymbargh* **3** 226

grin ME 'snare for catching fish', f.n. *Tweygryn* **1** 366

***Grinde** OE '?grinder'. ?Greenham (*v.* Addenda **4** 263)

grist ModE 'corn for grinding', f.n. Gristhay **4** 298

grist-mill ME 'mill for grinding grain'. Bindon Mill, *the griste mill* **4** 325, *Gristmyll* **1** 234

grone, **grane** ME 'trap, snare', f.ns. Abbey & Field Groan **3** 235

grōp(e) OE 'ditch', ?*Cropelane* **3** 148, ?*Groop Lane*, *Groupe Street* **1** 155, ?*Grope Lane* **1** 354

grot(a) OE 'particle of sand or gravel', f.n. *Grodbury* **1** 327

***groten** OE 'sandy', f.n. *La Grotene* **2** 85, *Groten Close* **4** 465, ?*Grotehill* **3** 392

grotten, **gratten** Do dial. 'sheep pasture' (OE ***grǣd-tūn** 'stubble field'), f.ns. Crotten Close **3** 323, Gratton Close **1** 336, Grotten **1** 28, **2** 109, Grottens **4** 480, Grotton(s) **3** 314, **4** 39

grotto ModE 'cave', f.n. Grotto Grd **3** 197

grow eModE 'growth', f.n. ?Growlands **1** 227

groyne ModE 'breakwater'. Bincleaves Groyne

grubbed ModE 'dug', f.n. Grublands Md **4** 423

grubbing ModE 'digging up of trees', f.n. Grubbing Cl **4** 470

grund OE 'ground, stretch of land'. Bottom Ground, Church Ground Belt, Fox Ground (Down), Horseground Coppice, Long Ground Plant., Lower Grounds, Middle Ground, Mill Ground Ctgs, *New Ground*, Northground Dairy, Old Claygrounds, Pleasure Ground, Pond Ground, Rough Ground Coppice, South Ground Plant., White Ground Lake, Windmill Barn & Knap. Freq in f.ns. e.g. Furzy Grd **2** 199, Long Grd **2** 200, Lowsy Grd **4** 435, Rough Grds **1** 249

***grundel** OE '?deep place', f.ns. ?*Grundalvesput* **3** 206, ?*Grundelhull* **2** 14

gryppe OE 'ditch, drain'. Grip Heath; f.n. Gripe Grd **4** 79

***grytte** OE 'stony soil or land'. Girt Fm; f.ns. Girt **4** 188, *le Grette* **1** 237, *la grutte* **3** 235

guinea eModE 'guinea'. ?Guineagore Lane, Three Guinea Bridge; f.n. Guinea Fd **2** 109

gull ModE 'sea-bird', f.n. Gull Moor **4** 423

gunne ME 'cannon'. Gun Cliff; f.n. Gunhill **1** 187

gurnard ModE 'fish', (*le*) *Gurnard* **1** 365

gut ModE 'water-channel', f.ns. Gutt Flg **1** 336, Long Gut **2** 200

guttered ModE 'furrowed', f.n. Gutter'd Plot **4** 310

gydig OE 'mad, foolish'. Giddy Green; f.ns. Giddy Lake Plot **2** 191, Giddy Md **2** 8, Giddy Mead Cl **2** 229

gylden OE 'golden, gold-coloured, splendid'. *Gildenock*; f.ns. *gildenedich* **3** 194, *Gilten acre* **1** 313, (*la*) *guldenelonde* **2** 52, **3** 256

***gyre** OE 'dirt, mud', f.ns. ?*Geremede* **2** 10, *Girehull* **2** 78, ?*Guymede* **2** 10

***gyse** OE 'a gush of water'. Gussage Brook (*v*. RNs), giving name to: Gussage All Saints/St Andrew/St Michael (*v*. also Addenda **2** 276)

haca OE 'hook'. Hackridge Fm

***haca-þorn** OE 'thorn-tree'. Hackthorn Hill

hack ModE 'cut by hacking', f.n. Hack Furze **4** 281

***hacket** ME 'cleared ground', f.n. ?Hacket Stile **3** 391

hæc(c) OE 'hatch-gate, wicket' also 'hatch, grating, floodgate, sluice'. Doe's Hatch, Eight-Hatch Plant., Hatch, Hatch Covert, Hatch Hole, Hatchlands, Hatch Pond Rd, Pickaxe Cross; f.ns. Floodhatch Md **4** 423, Hatch Flg **2** 157, Hatch Md **3** 27

***hæcce** OE 'fence', in AS charters, *hæccean* **4** 454

***hæcc-geat** OE 'hatch-gate'. Hatchet, Hatchetclose Coppice; f.ns. *Haccheghyte* **3** 27, Hatch Md **3** 27, Hatchet Acre **2** 204, Hatchet Grd **3** 355, Hatchet Plot **4** 417

***hæddre** OE 'heather'. ?Hatherington

hæfen[1] OE 'haven, harbour'. Brideport *Hafene* **4** 361, Lulworth Cove, North Haven Point, *Poole Haven* **2** 44, South Haven Point, Tatten, *Waymoth haven* **1** 255, Wych Haven

hæfer[1] OE 'he-goat', f.ns. *Hauerheye* **2** 148, ?*Haverlang* **4** 270

***hæfera** OE 'oats', f.ns. *Haverland Field* **1** 152, ?*Haverlang* **4** 270

(ge)hæg OE, **hay** ME 'enclosure, part of a forest fenced off for hunting'. Andrewshays Fm, Ashey Copse, ?Ashy Drove, ?Axen Fm, Baber's Fm, *Beauchin*, ?Benchy Hill, Berehays Fm, Bilshay Fm, Birdshays Fm, Blackney Fm, Bluntshay, Bolehays Copse, Bowhayes, Bowleaze, Brays Fm, Brighthay Fm, Broadhayes Fm & Ho, Bucehayes, Bully Wood, Calfhay Fm, *Castlehay* **4** 359, Causeway, Chaffeigh Fm, Champernhayes Fm, Chasehayes, *Cherry Hay* **4** 245, Childhay (Fm), *Church Hay* **4** 326, Church Hayes, Colly Fm, *Collyshey* **4** 231, Colthay Coppice, *Coluerhegh* **4** 8, Colway Gate & Lane, Combehays (Pit), Cookney, Copsehayes, Cornshay, Cothayes Dairy, Cothays Fm, *Coytesheyes* **3** 54, Culverhayes, Culverhayes Rd, *Cumheigh or Cymeheigh*, Cuthay Fm, Hr & Lr Denhay Fm, Ditchey Coppice, ?Dodseye, Downshay Fm & Wd, *Dudley Moor*, ?Early's Fm, Easthay Fm, East Hays, Fernhayes Copse, ?Fishy Mead Copse, *Forstershay*, Furzy Woolly Coppice, Gashay Fm, Greasehays Fm, Griddleshay Fm, Grighay Fm, Gummershay Fm (× 2), Guppy, Gypshayes Estate, Harmshay Fm, Harper's Fm, Hay Coppice, ?Haydon (× 2), Haye Fm, ?Hayes **4** 263, ?Hayes Lane, Hayne Fm, Hay Wood, Hewood (Fm), Hey Wood, High Wood **2** 209, Hornshayes, Horsey Fm, Huntshayes Fm, Hursey, Inmosthay Quarries, *the Johnes hay* **4** 431, Kershay Fm, Kilderhayes Fm, Larkshayes, Lenthay Common, ?Lilymoor Copse, Long Hayes, Mackey's Copse, *Mandevills* **4** 444, Manshay Coppice, Marshalsea, Meerhay (Fm), Millhayes, Netherhay, Northernhay, Northhay (Fm), North Hayes Fm, Oakey Plot, Oselhay Fm, *Parson's Hay* **4** 334, Phoenix Hayes, Pilmeshay Barn & Coppice, ?Pomice Fm, Popehayne Fm, ?Puxey, Revelshay Fm, Roger Hayes, Rookery Coppice & Fm, Roundhayes, Rushy Ctgs, *Shaphegh* **4** 33, Shipney Lane, Shute Haye, Silkhay Fm, Sminhay Fm, Spearhay, Talbothays, *Thurstanesley*, Trimshayes Fm, Tudhay, *Uppeheye* **4** 107, ?Vary Clump, *Uphay*, Westhay (Fm), Whithay Lane & Moor, Whitty Mead, (?)Winterhays (× 2), *Womeslegh* **4** 456, Woodhays Fm, Woodhouse Lane, Wylmington Hayes. Freq in f.ns. e.g. Brambly **3** 25, *Culuyrhey* **2** 13, *Darelleshay* **4** 10, Hemphay **4** 435, *Magattehayes* **2** 70

***hægen** OE 'enclosure', f.n. *Hayne close* **4** 332, in AS charters, *hegen* (*paðe*) **3** 110
hǣl OE 'omen, good fortune'. ?Holywell **4** 94
hǣle OE 'hale, safe, healthy'. ?Elwell, ?Holywell **4** 94
hǣlu OE 'health, healing'. ?Elwell, ?Holywell **4** 94
hǣme OE 'dwellers'. Ashington, Hr & Lr Bockhampton, *Buchœmatune* **4** 242, ?Brockhampton Fm, Brockington Fm, Shilvinghamton, Stubhampton, Witchampton; f.ns. *Hemecrofte* **3** 256, *hemefild* **2** 216
hænep OE, **hempe** ME 'hemp', f.ns. Hemphay **4** 435, *Hempe land* **4** 223, Hempits **3** 162, *la Henepheye* **4** 11
***hær** OE 'rock, heap of stones , tumulus'. ?Arne, ?Haremere Wood
***hæren** OE 'stony', f.n. ?*Harnehull* **1** 183
***hærig** OE 'stony', f.n. ?*the Harry Leach* **1** 62
hærfest OE 'autumn, harvest', f.ns. Harvest **4** 463, *Harvest Bower* **1** 95
***hǣs** OE 'brushwood'. Hayes Ctgs & Fm, Hayes Fm, ?Hayes Lane
hæsel OE 'hazel'. Haselbury Bryan, Hasler, Hasler Hundred, Hazel Barton, ~ Coppice (× 3), ~ Copse (× 2), ~ Fm, ~ Lane, ~ Wood, The Hazels; f.ns. *Haselley* **2** 216, *Haselstaire* **3** 58, Hazel Hole **3** 27
***hæslett** 'hazel copse', f.ns. *Haselet* **1** 145, *Haselet* (*coppice*) **1** 145, **4** 38, *Hasselet* **4** 227, Hazletts **4** 335
hǣð OE 'heath'. *Alderholt Heath*, Canford Heath, ?Haddon Lodge, Hardown, Heath Bottom, ~ Coppice, ~ Ctg(s), ~ Fm (× 2), Heathfield Ctg, Hethfelton, *Mines heath* **2** 40, *Woolbridge Heath*. Freq in modern minor names as ~ Heath, e.g. Affpuddle Heath, Bhompston Heath, Black Heath (× 2), Bottom Heath Barn, Bovington Heath, Bryants Puddle Heath, Carey Heath, Chamberlayne's Heath, Coombe Heath, Creech Heath, Decoy Heath, Duddle Heath, Farm Heath, Ford Heath, Galton Heath, Godlingston Heath, Grange Heath, ?Hadber Coppice, Highwood Heath, Holt Heath, Holton Heath, Hurst Heath, Hr & Lr Hyde Heath, Ilsington Heath, Knighton Heath, Lulworth Heath, Middle Heath, Morden Heath, Moreton Heath, New Mills Heath, North Heath, Northport Heath, Outer Heath, Owermoigne Heath, Pallington ~, Philliols Heath, Povington Heath, Puddletown Heath, Skippet Heath, South Heath (× 2), Southover Heath, Stoborough Heath, Stoke Heath, Stokeford Heath, Studland Heath, Throop ~, Turners Puddle ~, Warmwell Heath, Watercombe Heath, West Holme Heath, Winfrith Heath, Woodsford Heath, Wool Heath, Woolbridge Heath, Worgret Heath, Wytch Heath. Freq in f.ns., e.g. Heath Plot **2** 8, *la Hethe Fylde* **2** 52

hǣðen OE 'heathy', f.ns. *Ethenwode* **2** 124, Heathern Drove **2** 94. Also 'heathen', in AS charters, *æþenam byrigelsas* (*v*. Addenda **3** 267), *haþene berielese* **4** 173

hæt(t) OE 'hat-shaped hill'. ?Hatts Barn & Copse, Hatts Coppice; f.ns. *Fries hatts* **2** 250, *Furshat* **1** 313, Little Hat **2** 200

hafoc OE 'hawk', ?Halcombe Vale, *Hawcombe*, Hawkhill Ditch & Mill, Hawkmoor Fm, Hill & Marh, Hawksdean Fm, Hawk Wood; f.n. *Haukeshulle* **1** 137

hafocere OE 'hawker'. Hawker's Hill

haga[1] OE 'enclosure', f.n. *les Hawes* **3** 213. In AS charters, *ealdan hagan* **3** 289, *ealdene hage* **3** 237, *hagan* **3** 289, **4** 303, *hagen* **3** 110, 256, *hawen* **3** 256, *lyc haȝan* **4** 304, *mearchagan* **2** 162, *merehawe* **3** 200, *merhawin* **3** 257

haga[2] OE 'hawthorn'. Hawley Fm

***hagen** OE 'enclosure', in AS charters, ?*agen þorn* **3** 128

***hagga** OE 'haw', f.n. *Hagdon* **4** 303, in AS charters, ?*hacggen hamme* **3** 237

hagu-, **hæg-þorn** OE 'hawthorn'. Hatherly Fm, Haythorn; f.n. *Hatherly* **3** 263

hakeney ME 'horse kept for riding or hiring out', f.ns. Hackney Gate **4** 298, *Hakeney londes* **1** 120, *Hakneylond* **1** 298

hāl OE 'whole', f.n. *le Whole Place* **2** 181

hald[2] OE 'sloping', f.n. *Holdeley* **2** 125

half OE 'half'. *Halfhide*, Half Hide Coppice & Down, Half Moon St. **3** 360, Halfyard Coppice; f.ns. Halfendale **4** 281, *Halfeplace* **2** 180, Long-halves **1** 260, Tappers Half **2** 215, Three Halves **2** 19. In AS charters, *healfan treowe* **4** 303

halfcrown ModE, f.n. Half Crown Grd **3** 186

halfpenny ModE, f.ns. *Halfepeny Hay* **4** 218, Half pence **3** 101, Halfpence Pce **2** 204, Halfpenny Pce **2** 204

halfway ModE. *Halfway Hill*, Halfway Ho (× 2)

hālga OE 'a saint'. All Hallows Fm, All Saints; f.n. *Alhalon Well* **1** 168

hālig OE 'holy'. Holyrood Fm, Holy Stream (*v*. RNs), Holywell. In AS charters, *halgan wyl* **4** 35

halh OE (**healh** WSax) 'a nook, a hollow, a corner of land'. ?Cudnell, Hile Fm, ?Hilton, Hyle Fm, *Mulnehale* **4** 277, ?Tadnoll Dairy & Mill, Uphall, Wortheal, Wraxall; f.ns. *Chattoshale* **4** 284, Hale **3** 378, *Hylemede* **2** 52, *Sutton Hales* **3** 82. In AS charters, *amesheal* **3** 254, *browoldes heal* **4** 302, *heal* **2** 158, *merehale* **3** 181, 200, *norðheal*, *suðheal* **2** 162

halig OE 'holy'. Halstock, ?Holly Wood, Holwell, Holy City, Holy Stream, Holy Well, Holywell; f.ns. *Halewell* **4** 284, *Halgewal* **3** 33. In AS charters, *halgan stowe* **3** 359, *halgan weies lake* **3** 110

halke ModE 'a corner, a nook', f.n. *Halkes* **4** 227 [*cf.* **halh**]

halse ME 'narrow neck of land', f.n. Halse Close **4** 423

halter ModE 'rope for leading an animal', f.ns. Halter Path (lane) **3** 165, **4** 409

hām OE 'homestead, village'. ?Attisham, ?Hr & Lr Buckham, ?Edmondsham, ?Farnham, ?Moulham, ?Petersham, ?Pulham, ?Tyneham, ?Wareham [*cf.* **home**]

hamelet OFr, ME 'a group of houses or small village in the country, especially one without a church'. Hamlet, *Stoke Galliard Hamlett* **3** 350; f.n. Hemlet **3** 297

hamm OE 'enclosure, river-meadow, land in a river-bend'. ?Attisham, Bagman's Fm, Blashenwell Fm, ?Boyston Lane, ?Bracham Lane, Brinsham Fm, Broadenham Fm, Brockham (Lane), Brumham Coppice, ?Hr & Lr Buckham, Bulhams, Catenwood Coppice, Clarkham Cross, Cleave Ham, Clenham Coppice & Mill, Densham Fm, Doleham Coppice, ?Edmondsham, *Eglysham*, ?Farnham, Fernham, Feltham Fm, Frankham Fm, Gotham, Greenham, (The) Ham, Ham Common, Ham Coppice, Ham Fm (× 2), Ham Lane, Hammoon, Hampreston, Hams, Ham Waste, Ham Wood Fm, Hamworthy, Hillamsland, ?Lancin Fm, Langham, Longham, ?Lowsome Lane, ?Markham Ho, Marnhull Ham, Milborne Stileham, *la Millehamme* **4** 130, *Millhams* **4** 189, ?Millum Head, ?Milom Lane, ?Moulham, *Newnham*, Hr & Lr Newnham, *Odenham*, Parnham, ?Pelsham Fm, ?Pentsome Coppice, ?Petersham, Portisham, ?Pressham Wood, ?Pulham, *Ramesham*, Rampisham, Rookhams, Roundham Nursery, *Smokeham*, Smokeham Bottom, Stockham Dairy & Lane, Sutton Ham Coppice, Tatnam Fm, Thornham, Thorny Ham, ?Tyneham, ?Wareham, Westham, Whetham Fm, ?Whitham Ball Pit, Witcham Fm, ?*Wrykesham* **2** 39, Wyndlam Fm, Yellowham Hill & Wood. Freq in f.ns., e.g. *Baggardesham* **2** 199, Barley Ham **1** 136, *Dokham* **2** 52, the Hams **4** 39, Newnham's **1** 127, *Ordesham* **2** 10. In AS charters, *archet hamm* **3** 138, *br(e)owoldes ham* **4** 302, *hacggen hamme* **3** 237, *ham(me)* **3** 206 (now **1** 316, *v.* Addenda), *hamme* **3** 256, *med ham* **2** 122, *molenhame* **2** 58, *scearpenhame* **3** 112, *snelles hamme weghe* **3** 112

***hamol**, ***hamel** OE 'scarred, mutilated', also 'winding'. †Hambledon Hill; f.ns. *Hamelton* **1** 71, ?Hamlake **2** 171. In OE charters, *hamelendune* **3** 106

hamor OE 'hammer', f.ns. *Hamerham* **4** 500, Hammer Md **4** 337

hām-stall OE 'homestead', f.ns. The Homestal **1** 100, Homestall **2** 8

hām-stede OE 'homestead'. *Hampstede*; f.n. *Hamstede* **2** 52

hām-tūn OE 'home farm'. Bothenhampton, Brockhampton Fm

hān OE 'rock, stone'. Beaminster Forum & Redhone Hd, ?The Foreland or Handfast Point, Hanford, ?Hen Cliff; f.n. Hone Grd **2** 148. In AS charters, *rede hane* **2** 122

hand ModE. Cross & Hand

handkerchief ModE, f.n. Handkerchief half **2** 119

hang ModE 'slope, bend'. ?The Hang, ?Hankmoor Coppice; f.n. Hang Pce **3** 262

***hangel** OE 'slope', f.n. ?*Hangleyate* **3** 315

hangende OE pres. part. 'hanging'. Druce Hangings, Hanging(s) Coppice (× 3), ?Hanging Piece Belt; f.ns. Hanging Land **3** 27, Hanging Mdw **3** 225, *Hanginge Ryme* **4** 222

hanging eModE 'steep slope'. *Bishop's Hanging* **3** 94, Druce Hangings, The Hanging, Hanging Banks Coppice, Hanging Covert, ~ Fd, ?Hanging Piece Belt, Hussen Hanging, Nineteen Acre Hanging, Sandy Hole Hanging, South Hanging, Tincleton Hang, West Hanging; f.ns. Hanging **3** 27, **4** 39, Hanging Leaze **4** 434, Hangings Hill **3** 27, *Hangynges* **4** 222

hangman eModE, f.ns. Hangmans Cl **4** 434, Hangmans Corner **4** 248, *Hangmans Thorne* **1** 194

hangra OE 'wooded slope'. ?Binegar, ?Binnegar, Clayhanger (× 3), Clinger Fm; f.ns. Bingers **4** 233, *le Hanger* **3** 33, Hangsam **3** 235

happy ModE. Happy Bottom, ~ Valley; f.n. Happy Md **4** 281

hār[2] OE 'grey'. ?Haremere Wood, Hargrove Fm, Harpitts (Fm), ?Hartgrove, *Horeappledore* **3** 13, Hore Stones; f.ns. Hoare **1** 112, *Horewythey* **4** 223, the Wurstone **2** 259. In AS charters, *haran apuldran* **2** 136, ~ *apoldran* **2** 161, *haran stanas* **2** 161, *haran wiðig* **2** 161, *haren torre* **3** 206, *here apeldren* **3** 128

hara OE 'a hare'. Hare Coppice, ?Hare Lane Fm, ?Harely Fm, ?Haremere Wood, Hare Moor, Hargrove Fm, ?Hartgrove; f.n. Hare Castle **3** 388

hard ModE 'firm beach or foreshore, landing-place, jetty'. Clavell's Hard, Clay Hard Point

harlequin ModE. Harlequin Hill

harpin OFr 'harper'. ?Hurpston

hassuc OE 'clump of coarse grass'. The Hassocks; f.ns. *le Hassakes* **4** 227, Hassock(s) **3** 70, 137

haul ModE. ?*Boat Haul*

hawking ModE 'hunting with hawks', f.n. Hawking Down **2** 249

hayward ME 'guardian of the hedge or enclosure, keeper of cattle in the common field'. ?Howard's Lane; f.ns. *the Haywards Ham* **4** 391, Long Hayward Rd **3** 78, *Haywardwyke* **3** 199

hēafod OE 'hill, (end of) a ridge, promontory', also 'upper end of a valley, river-spring'. Gt & Lt Alderwood Coppice, ?Ballard Cliff, Down & Point, Bat's Head, Black Head, Bride Head (Lake), Buck's Head, Chettle Head Copse, *Chine Head* **3** 126, Cockpit Head, Combehead Fm, *Combeshed* **4** 114, *Cumbesheued* **4** 345, Delcombe Head, Devonshire Head, Donedge Lodge Fm, Dungy Head, Durlston Head, Dyke Head, Encombe, Fairwater Head, Flaghead, Gillin's Head Ctgs, Great Head, Greenway Head, Head Lane, *Hewdon*, Holcombe Head Pit, *Hundred head* **4** 30, *Hywyshcombyshede* **4** 176, Incombe Head Plant., *Lemmanescombesheved* **3** 244, *le Milhed* **4** 379, Millum Head, Pond Head, Poole Head, *Quarry Head*, Rew Head, Rope Lake Head, St Alban's or St Aldhelm's Head, Shore Head, Skilling Head, Spring Head, Springhead, Spring Head Hotel, Stair Head, Stony Head, Swyre Head (× 2), Tunnel Head, *Wellhead* **3** 107; f.ns. *ly Hedweyre* **4** 385, *Hefdacre* **2** 52, *le Heuede* **1** 71, *Heuedforlange* **3** 33. In AS charters, *aettan wylles heafde* **3** 310, *akere heueden* **3** 163, *ealde diches heued* **3** 206, *ears mores heaued* **3** 128, *flexcumbes heuede* **3** 206, *hefden* **3** 206, *he(f)deswelle* **3** 162, *heued* 3 256, *heuede(n)* **3** 122, 199, *hlinchesheaved* **3** 207, *hloscumbes heafud* **3** 328, *iheafde* **3** 206, *langen cumbes hauede* **3** 104, *linkes haued* **3** 207, *litiges heuede* **3** 163, 199, *mearc broces heafde* **2** 136, *sucgimade hauede* **3** 136

***hēafod-land** OE, **hevedland** ME 'the head of a strip of land left for turning the plough'. Headland Common, Headlands Coppice, Hr Headlands Quarry, *Nordon Headlons* **4** 325; f.ns. *Esthevedlond* **1** 260, Headlands **4** 234, Yedlands **1** 293

hēafod-lēas OE 'headless'. Headless Cross, Heedless William's Pond

hēafod-stocc OE 'a post on which the head of a beheaded criminal was exposed'. Headstock Cross, East Hewstock/Hewstock Fm, Yewstock; f.n. *Heuedstokk* **1** 260. [For a different interpretation of these names, *v.* Addenda **3** 160]

hēah (wk.obl. **hēan**) OE 'high, chief'. ?The Foreland or Handfast Point, Great High Wood, Hambury Ho & Tout, ?Hampton Barn, Hampton Hill, Hanover Hill, Hembury Fm & Mill, ?Henbury

Manor, ?Hen Cliff, Henning Hill, High Bridge, High Chaldon, High Cliff, *High Cross*, High Ditch Coppice, ?Higher Down Barn, Highfield Ho (× 2), Highfields Fm & Wood, Highgate Ctgs, Highgate Lane, High Grove Fm, High Hall, High Hill, High House Fm, Highlands, High Lea Fm, Highmoor, High Orchard, High Stoy, High Street (freq), Highwood, High Wood (freq), Hinton Lane, Hinton St Mary, Sixpenny Handley; f.n. *Heyghalres* **4** 160. In AS charters, *heandene beorg*, *hendune beorch* **3** 218, *hean wifeles hylle* **3** 288, ?*hengersten* **3** 252

h(e)all OE 'large residence, manor house'. ?Alcroft, Athelhampton Hall, Bear-, Beerhall Fm, Binnegar Hall, Downe Hall, Duncliff Hall Fm, Gisborough Hall, Gospel Hall, *Halle Hais* **3** 173, *Hall(fielde)* **4** 143, High Hall, Jubilee Hall, Kingston Lacy Hall, Lardenhall, Merley Hall Fm, *Middlemarsh hall* **4** 165, Nazdar Hall, New Hall, (*la*) *Oldehall* **3** 366, Salisbury Hall, Swain's Hall, West Hall, West Parley Hall, Whitehall (× 4), White Hall (Barn), Whitehall Ctgs, ~ Fm, *le Whitehaull* **3** 366, Wyke Hall; f.ns. *La Blakehalle* **2** 192, *Hallewode* **2** 216

hēap OE 'heap, mound'. Stone Heaps Plant., Hr & Lr Stonehips; f.n. *Stonyhipe* **1** 238

heard OE 'hard'. Harley Down & Wood; f.n. Hard Labor **1** 119

hearpe OE 'harp, triangular piece of land', f.ns. Harp **3** 314, *Harp acre* **4** 56, *Harpeland* **4** 92, Harpland 4 475

hearpere OE, **harpere** ME 'harper'. ?Harper's Fm, ?Hurpston

***hearpere** OE 'river occasionally dried up, barren place'. ?Hurpston

heart ModE. Heart Clump

hearth ModE 'hearth', f.ns. *Hearth Bay* **1** 184, *Hearth Close* **4** 254

heartsease ModE 'wild pansy', f.n. Heartsease **1** 167

heather ModE. Heatherlands

heathy eModE. Heathy Field Coppice, Heathy How, Heathypiece Bungalow; f.n. *Hethy howe* **2** 230

heave ModE dial. 'a heap, a hillock'. ?Heave Coppice/Fm, Heavland Lane; f.n. Heave **3** 252

heaven ModE, f.ns. Heaven **3** 378, Heaven down **4** 435

hecg, **hecge** OE 'hedge'. Boxen Hedge, Dean Hill, Hedge End Fm, Jack's Hedge Corner, *le Parkhegge* **4** 225, Three Hedges, Vicarage Coppice, *Walhege* **4** 38, West Wood; f.ns. *Heg Acre* **2** 52, *Langgehegge* **2** 53, *Neweheigge* **2** 216, *la Nywhegge* **3** 227, *Twohegyn* **4** 174. In AS charters, *hecgan sled* **1** 31, *langhecgan* **3** 138

***hēcing** OE 'part of field sown whilst the rest lies fallow', f.ns. *Hechyn* **2** 38, *les Hechynges* **3** 256, Hitchen House **4** 253

hedgehog ModE, f.n. Hedgehogs Bank **4** 68

hefding ME 'headland', f.n. *Hydynggrove* **2** 179

hēg OE (**hī(e)g** WSax) 'hay'. Haddon Copse, Hail, Hay Barn Ctgs, Hay Bridge, Haycraft, ?Haydon (× 2), Haydon Clump & Hill, Haydon Down, Haydon Lane & Wood, Haydon Plant., ?Hay Holes, Haymoor Bottom, *Haymore*, Hayward Bridge, ?Higher Down Barn, ?High Ford Lane, High Lea Hill, Hyford Ctg. Freq in f.ns., e.g. Hay Cross **4** 435, *Heycroft* **3** 45, Heywey **2** 130, Highlands **4** 435. In AS charters, *higeweges ande* **3** 135, 137

hege OE 'hedge'. ?Haydon (× 2). In AS charters, *fald heȝe* (*v*. Addenda **4** 193), *hege* **3** 237, *hegen* **3** 163, *heges hirne* **3** 206, *land share hegen* **3** 138, *stigel hege* **3** 139, *wiþiȝheȝe* (*v*. Addenda **4** 193)

hege-rǣw OE 'hedgerow'. Draglens Hedgerow; f.n. Hedgerow 2 57. In AS charters, *ealden hege rewe* **1** 31, *hege rœwe* **3** 206 (now **1** 316, *v*. Addenda), *hege reawe* **2** 57, **3** 111, 122

helde, **hielde** OE 'slope'. Hillway Copse, ?Hilton; f.ns. *Heldehegh* **3** 256, *Heldwater* **4** 40, *Hildeweie* **2** 255, *Hyldeway* **1** 31, *Neldeshende* **4** 167, *Twynelde* **1** 70. In AS charters, *mearhhilde mere* **4** 173

hell OE 'hell' [*cf.* also ME **helle** 'lower room']. Helen Lane **1** 252, ?Hell Corner, Hell Fm, Hell Pit, †Hell Stone, *Hell-Stone* **4** 54, Hill Fm & Wood **2** 30, Hill St. **2** 42; f.ns. *Hell Close* **3** 226, *Helle* **4** 10, *Helleclif* **3** 256, Hell Grd **1** 286, *Hell Meadowes* **4** 88, *helpytt* **4** 167

helm OE 'helmet'. ?Goathill

hemlock ModE. Hemlock

***hemm-stall** OE 'the site of a hem or border', f.n. ?*North-*, *Suthymstal* **1** 216

hengest OE 'stallion'. Hingsdon Fm; f.n. *Hengstelega* **2** 255. In AS charters, ?*Hœngestes ricg*, *hengestes rigc* **4** 374

hen-house eModE, f.n. *Henhouse* **4** 188

henn OE 'wild bird'. Encombe, ?Henbury Manor, ?Hen Cliff, Henley, Henwood (Hill), Hincknoll Fm & Hill; f.ns. ?Hagusson **3** 252, *le Hencote* **1** 264, *Henforlang* **2** 52, *Hen Meade* **3** 33. In AS charters, ?*hengersten* **3** 252

henroost eModE. Henroost Wood

hēopa OE 'dog rose' or **hēope** OE 'hip', f.n. *Hebdychelane* **3** 315

heorde-wīc OE 'a herd farm'. *Hurdick*

heorot OE 'hart'. *Hartescath* **4** 494, Hart Hill Plant., Hartland Moor, Hartley Fm, Hartmoor (Fm); f.n. Hart Hill **2** 199

herbage ModE, f.ns. Herbage **4** 435, Herbage Md **4** 329

here OE '(invading) army'. ?*Herbekne* **4** 371

here-beorg OE, **herberȝe** ME 'place of shelter'. Cold Harbour (× 3), *Cold Harbour* **4** 64, Coldharbour, Harbour (at Weymouth), Harbour Bridge & Hill, Poole Harbour, Portland Harbour, ?Snagharbour Wood; f.ns. High Harbor **2** 157, Harbour Fd **3** 309, *Hereburghcrofte* **4** 374

here-pæð OE 'highway'. ?*Harpydegor* **3** 391, Hartfoot Lane; f.ns. Harpway Flg **2** 148, *Harpydway* **3** 237, *Westharpydwey* **3** 226, *Herpath* **2** 136. In AS charters, *cern hlinces herpaðe* **4** 454, *ealdan herpað* (*v*. Addenda **4** 193), *ealden herepaþe* **3** 111, *hærpaðæs* (*v*. Addenda **3** 267, *herepað*(*e*) **3** 111, 206, 122, *here path* **3** 256, *herepathes* **3** 256, *here paþ*, *richte herepath* **1** 31, *herpaþe*(*s*) **2** 82, 122, **4** 173, 4 303, **4** 454, *herpaþ*, *-poðe* **3** 206 (now **1** 316, *v*. Addenda), *wic herepaþes* **2** 144

here-weg OE 'highway', in AS charters, *her weg* **3** 102

hermitage ME 'hermitage, house of friars hermits'. *Ermitage* **4** 215, Hermitage (× 2), The Hermitage (× 3)

(**ge**)**hērness** OE 'jurisdiction', f.n. ?*Harnes Wey* **2** 14

hete, **hette** ME 'heat', f.n. *Hetwell* **4** 382

hiche ME 'enclosure of hurdles for penning sheep', f.ns. *Gillinghech* **3** 128, Hitchcroft **4** 409, *Hitche* **3** 33

hīd OE, **hide** ME 'hide of land'. *Cherchyde* **4** 21, Lower Fifehead or Fifehead St Quintin, Fifehead Magdalen, Fifehead Neville, Fifehead Sydling, *Halfhide*, Half Hide Coppice & Down, (The) Hyde, *Hyde* (× 2), Hyde Coppice, Hyde Ctg & Wd, Middle Hyde Dairy, Hyde Fm (× 3), Lr Hyde Heath, Hyde Hill, Hyde Hill Plant., Hyde Ho, The Hyde, Hydes, Piddletrenthide, Woodyhyde Ctgs & Fm; f.ns. *Hydelond* **2** 58, *le Hyde londes* **2** 14

hielfe OE 'handle', f.n. *Le Langehilve* **4** 368

higher ModE. Dewlish Higher Fm, Higher Barn (freq), Higher Barton, Higher Bldgs (× 2), Higher Clifton, Higher Common (× 2), Higher Coppice, Higher Down, Higher Fm (freq), Higher Green, Higher Hill, Higher House, Higher Lodge (× 2), Higher Melcombe, Higher Mill, *Higher Mill*, Higher Pond, Higher Street Lane, Higher Warren, Higher Waterston, Higher Wood, Higher Woodsford, Higher Wraxall

hilly ModE. Hilly Close Copse, Hilly Ho & Pond

hind OE 'hind, doe'. Hendover Coppice, Hinkhams

hindan OE 'behind'. Hingarston; f.n. *behind Hayes* **1** 344

hirde OE 'herdsman', f.n. *Hurde acre* **1** 327

hither ModE. Hither Coppice, *Hither Ham* **2** 21; f.n. *The Heather Furchase* **4** 381

hīwan (gen.pl. **hī(g)na**) OE 'a household (of monks), a religious community'. Hinton Martell & Parva, Piddlehinton, Tarrant Hinton; f.n. *(H)ywenemede* **2** 53. In AS charters, *hina gemære* **2** 146, *hina imares* **3** 136

hīwisc OE 'a household, a measure of land that would support a family'. Belhuish Fm, Hewish Dairy House, Hewish Fm, ~ Hill, ~ Plant., *Hiwes* **1** 335, Huish, Huish Barn, Huish Fm (× 2), Huish Lane; f.ns. Hewish 4 481, *Hywisschcroft* **2** 85, *Hywysshhegge* **1** 378. In AS charters, *hiþisc bergh* **3** 206–7

hlǣfdige OE, **ladi** ME 'lady', often 'the Virgin Mary'. Ladies' Copse, Ladies' Lookout, Ladies' Walking Field, Lady Chapel, Lady Mead, Lady's Copse, Lady's Lane, Lady's Well; f.ns. Lady Hole **2** 124, *Lady Mead* **3** 219, *Levidiforhurde* **2** 85

hlāford OE 'lord'. ?Lardenhall, Lord's Barrow, Lord's Down, *the Lordes Quare* **3** 69, Three Lords' Barrow; f.n. *Levidiforhurde* **2** 85, *Lordesclos* **1** 298, *Lordishalle* **2** 43, Lords Bush **3** 309

hlāw OE 'hill, tumulus'. ?Foxbury/Great Lower hedge, ?Lawford Sheard Gate, Low Down Fm, ?Winslow; f.ns. Costlow **2** 157, Hounslow **3** 27

hlēapewince OE 'lapwing', f.n. Leapwinch **4** 329

hlēda, **hlȳda** OE 'seat, ledge'. f.ns. ?*Lydelonde* **1** 199, Norledge **1** 41, ?*Soutlude* **1** 185

***hlenc** OE 'hill-side', f.ns. Lench **3** 136, *le Lenche* **1** 121

hlēo(w) OE (***hlīeg**, ***hlīg**, ***hlīewe** WSax) 'protection, shelter'. Lewcombe, Lewell Fm, ?Luccombe Covert & Fm, ?Luccombe Fms

hlēow OE 'sheltered'. Lewcombe, ?Luccombe Covert & Fm, ?Luccombe Fms

***hlēp**, **hlīep** OE 'a place that can be crossed by leaping'. St Lucas Leap; f.ns. Lype **2** 229, **3** 121, *Lyppehedge* **3** 324

hlēt, **hlīet** OE 'a share, a lot', f.ns. *la Lette* **4** 43, *La Lyete* **3** 200

hlid OE 'lid, door', possibly also 'gate'. Lily Fm

hlid-geat OE 'swing-gate', f.ns. *Lidyeate* **4** 284, Lydgates **3** 252. In AS charters, *elde lidyate* **3** 256

hlīep-geat, **hlȳp-geat** OE 'a leap-gate (for deer and other animals)'. Leap Gate, Lipgate Fm; f.ns. Leap gates **2** 105, *la Lupeghite* **3** 33, *Lupyate* **1** 298, *la Lypyȝate* **4** 454. In AS charters, *hlype ȝete* **4** 303, *lipgete* **3** 136

hlinc OE 'ridge, bank'. Buckland Down & Hill, Fifehead Sydling, Horse Lynch Plant., Links Plant., Long Lynch, Lydlinch, The Lynch

(× 2), Lynch Coppice, ~ Dairy, ~ Fm (× 4), ~ Lane, ~ Lodge, ~ Rd, Stickley Coppice, Sydling St Nicholas, Up Sydling, West Lynch, Woodlynch Plant.; f.ns. *Buglinche* **3** 31, *Courtlynch* **4** 10, Linches **3** 28, Lynchet Stile Fd **2** 220, Lynchy Cl **3** 44, *la Westrelinche* **2** 54. In AS charters, *burch linken* **3** 110, *cern hlinces herpaðe* **4** 454, *earnes hlin*(*g*)*c* **4** 374, *greate hlinc* **3** 206, *gretenlinkes* **4** 173, *hlinc* **3** 207, *hlinc*, *langan hlinc* **2** 136, *hlinchesheaved* **3** 207, *hlinc reawe* **3** 207, *landscar hlinc*, *land scorhlinc* **3** 207, *linc reawe* **1** 15, **3** 128, *linkes haued* **3** 207, *minnanlinche*, *nunnen linc* **3** 163, 200, *Sticholnelinche* **3** 246

hlōse OE 'pig-sty'. Loose Path Coppice, Loscombe, Loscombe Barn, ~ Bottom, ~ Hill (× 2), ~ Plant. (× 2), ~ Wood, Loscombe's Well, Lousley Wood, ?Lowsome Lane, Luscombe (× 2); f.ns. Looselea **4** 424, Loscombe **2** 139, Pigsloose Orchd **4** 410, Pigs Louse **3** 253. In AS charters, *hlosmoc* **4** 374 (*v*. RNs), *hloscumbes heafud* **3** 328

hlōs-stede OE 'site of pigsty or other shelter for animals', in AS charters, *hlosstedes crundles* **4** 73

hlot OE 'lot, share, allotment'. Coombe Lots, Latmore Fm, Lotmoor Hill, Poor Lot(s), Winterborne Poor Lot; f.ns. Lot Barrow **4** 45, *Lot*(*te*)*meade* **3** 64

hlūd OE 'loud'. ?Lodburn; f.ns. Loudwell **4** 343, Ludwell **4** 58

***hlȳde** OE 'loud stream, torrent'. Litton Cheney; f.n. Lyde **3** 78 . In AS charters, *hlidan*, *hlyda*(*n*) *pol* **4** 299

***hlȳding** OE 'loud stream, torrent'. ?Litton Cheney

hnecca OE 'neck', f.n. *Meade necke* **4** 193

hnoll OE 'top of the head'. ?Tadnoll Dairy & Mill

***hnoþ** OE 'knoll, hill'. Nath Point, The Nothe, White Nothe or White Nose

hnott OE 'bare'. ?Nottington (*v*. Addenda); f.ns. *Notayssh* **4** 284, *Nottefurlang* **1** 260. In AS charters, *hnottan stybbe* **4** 454

hnutu OE 'nut'. OE Nutcombe Coppice, Nut Coppice, Nutcrack Lane, Nutford Fm, Nutley Clump, ~ Way; f.ns. *Notecroft* **4** 411, *la Noterewe* **4** 205, Notley **3** 28, Nuthays **2** 37 [*cf.* **nutting**]

hob ME 'hobgoblin', f.n. ?*Oblands* **1** 121

hobb(e) ME 'tussock'. Hobb Knap f.ns. ?Hobbs Mdw **3** 5, ?*Oblands* **1** 121

hobbing ME 'place with tussocks'; f.n. ?*Hobbyngmor* **1** 101

hōc OE 'hook, angle, corner, bend'. Hogboro' Coppice & Ctgs, Hook, Hooke, Hookley Copse, Hookswood Fm, Huck Barrow, ?Mudoak Wood, ?Roke Barn & Fm, *Sciterhuc* (*v*. RNs), Withyhook Mill; f.ns.

Hokebrygge **3** 346, Hook Hood **4** 267, Hook Mdw **3** 44, Out Hooks **2** 8. In AS charters, *ginum hocum* **3** 138

hocc OE, **hock** ME 'hock, mallow', f.ns. *Hockehaye* **4** 40, Hockford **4** 423, Hock Hams **4** 435

hōcede OE 'curved', f.ns. *Hokedelond* **3** 256, *Hokedwale* **1** 345, *Hokydedich* **2** 180, *Howked Walles* **3** 324

***hocer**[1] OE 'a hump, a rounded hill. ?*Hockerill*

hocer[2] OE 'contempt, abuse, dispute'. ?*Hockerill*

hōd[1] OE 'a hood'. ?†Hod Hill, Oddens Wood; f.ns. Gt & Lt Hod **4** 417, ?Hod Moor **3** 186

***hōd**[2] OE 'a shelter'. ?†Hod Hill; f.n. ?Hod Moor **3** 186

hogg OE 'a hog', also 'a young sheep'. ?Hodge Ditch, ?Hogchester, Hog Cliff Bottom & Hill, *Hogen Downe* **2** 123, *Hogenhull* **3** 223, Hog Field, ?*Hogford*, *hoggen lease* **3** 118, Hoggen Down, Hog Hill (× 4), Hog Leaze, Hogleaze Fm, ?Hogshill St. **4** 240, ?Hogstock, Ogden Down Fm. Freq in f.ns., e.g. *Hogge more* **4** 411, *Hoggen Bottom* **2** 216, *Hoghey* **1** 250, Hogleaze **4** 25

hōh OE 'heel or spur of land, ridge end'. ?Hobarrow Bay, Hoe Coppice, ?East & West Howe, *Wytihoug* **2** 240; f.n. Hough Fd **2** 37

hol[1], **holh** OE 'hole, hollow'. Ansty Hollow, Bacon Hole, Bat's Hole, Big Leaze Hole, Blackers Hole, Black Hole, Blow Hole, ?*Boat Haul*, Bowerswain Hollow, Boyne Hollow, Bradford Hollow, Butt Lane Hollow, Cathole Barn & Copse, Catholes Withy Bed, Cave Hole, Clean Hollow Plant., Cover Hole, Crab Hole, Culver Hole, Dark Hole, Dawson's Hole, Dogshole Ctgs, Ferny Hole Plant., Founder Hole, Froghole Fm, Froghole Withy Bed, Frog Hole, Frog's Hole, Gutch Pool Hole, Hand's Hole, Hatch Hole, Haydon Hollow, Hay Holes, Hethfelton Hollow, *The Hole* **1** 255, Hole Batch, Hole Cmn & Ctg, Hole Ho (× 2), Hole Plant., ?Holes Bay, The Hollow, Hollow Grove, ?Hollow Oak, ?E & W Holton, Holway, Holworth, Horton Hollow, Jones's Hole, King's Pier Hollow, *Knacker's Hole* **1** 241, Knacker(')s Hole (× 2), *Maidshole* **3** 174, Minnows Hole, Mutton Hole, *Pexy'shole* **2** 94, Pink Knoll Hollow, Potter's Hole, Pug's Hole, Red Hole, Redhole Lane, Sand Holes, Sandy Hole Hanging, Shark's Hole, Shit Yallery Hole, Smokey Hole, Stair Hole, Throop Hollow, Tolpuddle Hollow, Trunk Hole, Underdown Hollow, Vesey's Hole, Weaver's Hole, White Hole, Wood Fold Hill; f.ns. Bishops Hole **4** 39, Cats Holes **2** 199, *Gobelynshole* **4** 35, *Holweton* **2** 216. In AS charters, *deoh holes hyllæ* (*v.* Addenda **3** 267), *holu* **3** 207, *holun* **4** 35

hol[2] OE 'hollow, deep'. Hawkcombe Lane, Holbrook, *Holcombe* **1** 131, Holcombe Barn, ~ Bottom (× 2), ~ Fm, ~ Pit, ~ Wood, Holditch, Holebrook Fm & Water (*v.* RNs), Holland's Hill Coppice, Hollis Hill, ?*Hollish*, Holloway Lane, Holloway's Wood, Hollow Ditch, Hollow Hill, Hollow Way, Holway Fm, ~ Hill, Holwell, Holwell Fm, ?Howcombe Ledge; f.ns. *Holbrouke* **4** 11, Holdins **2** 214, *la Holeden* **2** 52, *Holemers* **3** 33, Holloway Plot **3** 83. In AS charters, *eastemestan holan weg* **3** 380, *Holambrok'* **3** 348, *holan broc* **2** 161, **4** 193 (*v.* Addenda), **4** 374, *holan lace*, *holan wege* **4** 454, *holebroke* **3** 109, *holencumb* **3** 99, *-cumbe* **3** 102, 111, 207, *holenwelle* **3** 102, 111, *holenwylle* **3** 102, *holewei* **3** 111, *imeren hole weg* **3** 102

hold ME 'tenure, property'. *Kingshold* **1** 279; f.ns. Eight Holes **1** 60, *Ferthyngholde* **1** 362, *Frydayshold* **3** 310, *Hanbyshold* **3** 237

holding ModE 'tenure, property'. White Ho Holding

holegn OE (eModE dial **holm**) 'holly tree'. East Holme, Gore Holmes, Hold Acre Coppice, Hollow Hill Plant., Holly Bank Fm, Holly Brake, Holly Clump, Holly Coppice, Holly Ctg, Holly Fm & Lane, Holly Grove Coppice & Fm, Hollymoor Common Lane, ?Holly Wood, Holm & Ivy Fm, The Holm Bushes/Holm Bushes Fm, Holmbush (Plant.), Holme Cleve & Craig, Holm Fm, ?Holmes Copse, Holmwood, Holnest, ?Home Plant. **4** 346, Philleyholme Coppice & Mill, Sutton Holms, West Holme; f.ns., e.g. *Holme mede* **2** 52, *Holminge Stubbes* **2** 174. In AS charters, *holenbedde*, *holne stoke* **1** 31

holt OE 'wood, especially a single-species wood'. Alderholt, ?Curry Hole Lane, Holt, Holt Coppice & Lane, Holt Ctgs, Hr & Lr Holt Fm, Holt Hill & Woods, Holt Lane, North Holt Fm (× 2), ?East & West Holton; f.ns. Hazel Hole **3** 27, *Northholt* **2** 216, *Presteholte* **3** 236. In AS charters, *holt* **3** 136

home ME, ModE 'manor, residence, dwelling place'. Barnes Homes, *Coleweyishome* **4** 396, Home Barn, Home Bottom Coppice **4** 448, Homeclose Lane, Home Coppice (× 2), ~ Copse (× 2), ~ Covert (× 3), ~ Eweleaze, ~ Fm (freq), Homemeadow Coppice, Home Plant. (× 2), Home Wood, *Person's home* **4** 392, *Sherbo*(*u*)*rn*(*e*) *Holme* **4** 393. Freq in f.ns., e.g. *le Home Close* **3** 33, *Homegraunge* **2** 14, *Homemede* **3** 339, Home Orchd **4** 34 [*cf.* **hām**]

homestead ModE. The Homestead, Furlong's Homestead

honeysuckle ModE, f.n. Honey Suckle Md **4** 329

hop[1] OE 'small enclosed valley'. Hope Bottom & Fm, Hope Fm, Hope Wood; f.ns. *Le Hope* **1** 184, Hope Close **3** 33

hop[2] OE 'a hoop, an inlet, a small bay'. Arne Bay, Hope Corner, Hope Square **1** 252, ?Mupe Bay

***hopet** OE 'small enclosure', f.n. Hoppett **4** 298

***hoppa** OE 'grass-hopper', f.n. *Hoppehull* **2** 216

hoppe ME 'hop'. Hop Yard, Hopyard Coppice; f.ns. Hop Gdn **2** 214, Hopyard **1** 345, *Le Hopyard* **1** 194

hord OE 'hoard, treasure', also 'store-house'. Drakenorth; f.n. Winterhead **3** 392. In AS charters, ?*hordþiuel* **3** 207 [cf. **gold-hord**]

horig OE 'filthy'. *Horyford*; f.n. Horylane **4** 271, *Horyslowe* **3** 334. In AS charters, *horgan sloh* **3** 289

horn OE 'projecting piece of land, headland'. Calf's Horn Lane, Horn Bridge, Horn Hill, Horner Hill, ?Ramshorn Lake, ?Reforne, Rhodehorn Plant.; f.ns. *Gygehorn* **3** 226, Horn **4** 435, *Sherpehorn* **2** 41, White horn Md **2** 37. In AS charters, *horgate*, *hornget* **2** 57, 82, *hornes beorȝ*, *-beorh* **4** 303

hornede OE 'with horn-shaped projections', f.n. Three Horned Hay **3** 136

horny eModE 'with projections' or 'hard', f.n. *Hornymore* **1** 81

hors OE 'horse'. Horse Castles, Horse Close (Copse) (× 6),~ Clump, ~ Coppice (× 2), ~ Copse, ~ Down, ?Horseground Coppice, Horsehill Fm, Horse Hill Lane, Horse Leys Pit, Horse Lynch Plant., Horse Moor (Coppice), Horse Park Plant., Horse Plot (× 2), Horseplot Ctg, Horsepool Lane, *Horsewere* **3** 368, Horsey Fm, ?*Horsyth*, White Horse Hill; f.ns. *Horscombe* **4** 11, *La Horscrofte* **4** 19, *Horse Fourde* **2** 158, Horse Leaze **2** 214, *Horse Poole Close* **3** 6. In AS charters, *horsedich* **2** 122, *horsford* (*v*. Addenda **4** 193)

horseshoe ModE. Horseshoe Clump

hors-mulle ME 'mill driven by a horse'. Horsemill Corner

hors-þegn OE 'groom', f.n. ?*Horsingtons* **3** 27

horu OE 'filth, dirt'. ?*Horsyth*, Horton; *Horsithe* **3** 199. In AS charters, *hore tuning gemære* **2** 159

hosa OE 'stocking'. ?Hosey Bridge; f.ns. ?Hose Md **2** 98, ?*Hose meade* **4** 250

hospital ModE. Hospital Fm, ~ Island; f.n. Hospital plot **1** 167

hos(s) OE 'shoot, tendril'. ?Hosey Bridge; f.ns. ?Hose Md **2** 98, ?*Hose meade* **4** 250

hostery eModE 'hostelry', f.n. Hostery Gd **4** 299

hovel eModE 'shed', f.n. Hovel pce **1** 167

how(e) ME 'enclosure'. *Beldam Howe* **2** 16, East & West Howe, Heathy How, Howland Coppice. Freq in f.ns., e.g. How Croft **2** 8 [discussion], *le Sheperde Howe* **4** 35 [*cf.* **haga**[1]]

hraca OE 'throat', in AS charters, *budencumbe hracan* **3** 128 [*cf.* **rake, rack**]

hræfn OE 'raven'. ?Rainbarrows, ?*Ramesham*, ?Rampisham

hramsa, **hramse** OE 'wild garlic'. ?*Ramesham*, ?Rampisham, ?Ramsden Coppice, ?Rempstone; f.n. Ransom **4** 88

hremn OE 'raven'. ?*Ramsbury*, ?Ransom Coppice, ?Renscombe Fm; f.n. ?*Remeswelle* **1** 70

hrēod OE 'reed, rush'. ?Mount Radford, Radipole, ?*Radway*, ?Redford Fm, ?Redhole Lane, ?Redlands, ?Redmoor Fm, ?*Redpool*, ?Ridley Pit, Rock Lea Point, Rodford Withy Bed, *Rodlake* (*v.* RNs), Rodlands, Rodmore Coppice, ~ Fm, Rodwell, ?Rodwell **1** 76; f.ns. Radmoor Cl **2** 37, Red Md **4** 464, *Rodland* **4** 19, *Rodmede* **3** 45, Rud Md **4** 69

hrēodig OE 'reedy', ?Radipole

hrēof OE 'rough, rugged'. ?Riveridge Wood; f.n. *Revenhul* **4** 183

hriddel OE 'riddle, coarse-meshed sieve', f.n. ?*Ridelhegh* **3** 214

hring OE 'a ring, a circle, a circular earthwork'. †Badbury Rings, †Buzbury Rings, †Castle Rings, Kennel Rings, †Rawlsbury Camp, †Ringmoor, †The Ring(s), †Rings Hill, †Crawford Castle or Spettisbury Rings; f.ns. *Ryngham* **1** 152, *Rynglynch* **1** 260, *Ryngwell* **4** 205

hringe OE 'a ring'. ?Ringstead

***hring-stede** OE 'place or site with, or in, a ring'. Ringstead

hrīs OE 'brushwood'. ?Riscombe Coppice; f.ns. Hatch Rice **3** 135, *La Rys* **1** 42, *Wodrise* **1** 70

hrīðer, **hrȳðer** OE 'an ox, cattle', f.ns. Rother Close **4** 156, *Rotherford* **2** 11, *Rytheresdene* **2** 251

hrōc OE 'rook'. Rookery Coppice & Fm, Rook Grove, Rookhams, Rooks Hill, ?Ruppen Coppice; f.ns. Rook Ditch **2** 78, Rooke Hayes **2** 82

***hrōst** OE 'rafter, beam', f.n. ?Rostley Md **1** 143

hrung OE 'a pole'. Romford; f.n. ?*Runcley* **4** 465

hrycg OE 'ridge'. Ashridge, *Baggeridge*, Bailey Ridge Fm, ?*Boteridge* **1** 85–6, Boveridge, Bowridge Hill, Broadridge Common, Checkridge Fm, Coppleridge, Cowridge Copse, Crock Ridge, Foxbury/Great Lower Hedge, Hackridge Fm, Lastridge, Marriage Lane, *Maynerigge* **1** 337, ?Pentridge Fm, Ridge (× 2), The Ridge, Ridgeback Lane, Ridge Barn, Ridge Bottom & Hill, Ridge Cliff, Ridge Coppice/Copse (× 3), Ridge Court, Ridge Fm (× 3), Ridge Hill (× 3), Ridge Hill Fm, Rigg Lane, ?Riveridge Wood, Rodden Ridge, Rudge Fm, Standridge, Stonerush Drove, Storridge Fm, ~

Hill (× 2), Waterston Ridge, Windyridge Withy Bed, Wolfridge Fm, Wych Ridge Estate. Freq in f.ns., e.g. *Rudge furlong* **3** 253, *la Rygge* **3** 227, *Wilrige* **3** 45. In AS charters, *brandes hricg* **3** 156, *canu rih* (*v*. Addenda **3** 267), *Hængestes ricg, hengestes rigc* **4** 374, *hricges, rihges* **4** 374, *hricgge* **3** 379, *hruczes* **4** 303, *ricges* **3** 207, *rig'* **3** 253, *smalan hrucg* **4** 303

hrycg-weg OE 'ridgeway'. Ridge Way, Ridgeway (× 2), Ridgeway Hill (× 2), Ridgeway Lane (× 2); f.ns. Ridgeway **2** 63, *Ridgeway Close* **3** 45, Ridgey Grd **3** 29, *Rygeweye* **2** 205. In AS charters, *hricge weges* **3** 207, *hriycsþeg* **3** 111

hūfe OE 'hood-shaped hill, shelter'. Hr & Lr Hove Wood

hulc OE 'shed, hut', f.n. Hulk **3** 314, Hulks **3** 27

hulu OE 'shed, hovel', f.n. *les Hules* **4** 35

Humber pre-English r.n. of uncertain origin and meaning. Heniford Fm & Mill, Humber Wood (× 2), Hummer; f.ns. Hemicombe **4** 382, Humber **3** 127, **4** 209, *la Humber* **3** 375, Humber Mead **2** 220, *Humbermede* **3** 81 (*v*. RNs)

humele OE 'hop plant', f.ns. *Humleclos* **1** 121, *Humley* **4** 304

hund OE 'hound'. *Houndeswellestrate* **3** 148, ?Hound Hill, Hound St 3 361, ?Hounstout Cliff; f.ns. *Houndecrofte* **1** 62, ?Hounslow **3** 27, *Hownystuff* **3** 207, Onyx **1** 193. In AS charters, *houndesete* **3** 257, ?*hunaleze, hunda troh* (*v*. Addenda **4** 193)

hundred OE 'an administrative division of a county, probably consisting originally of 100 hides'. Abbotsbury Hundred, *Albretesberge Hundred*, Badbury ~, Barrow (or Hundredsbarrow) ~, Beaminster ~, Beaminster Forum & Redhone ~, Bere Regis ~, *Blinchesfeld* ~, Brownsall ~, Buckland Newton ~, *Canendone* ~, Cerne ~, Cogdean ~, Combs Ditch ~, Cranborne ~, Culliford Tree ~, Eggardon ~, *Ferendone* ~, Frampton ~, Gillingham ~, Godderthorn ~, Hasler Hundred, Hundred Barrow, *Hundred head* **4** 30, Knowlton Hundred, Little Hundred Coppice, Loosebarrow Hundred, Modbury ~, Monkton Up Wimborne ~, Pimperne ~, Puddletown ~, Redlane ~, Rowbarrow ~, Rushmore ~, St George ~, Sherborne ~, Sixpenny Handley ~, *Stone* ~, Sturminster Newton ~, Tollerford ~, Totcombe ~, Whiteway ~, Wimborne St Giles ~, Winfrith ~, Yetminster Hundred. Also as number 100 (often applied ironically to very small fields) in f.ns. Hundred Acres **1** 249, **3** 287, 314

hūne OE 'hoarhound'. ?Hound Hill; f.n. Shorthound **3** 309

hungor OE 'hunger'. Hunger Hill (× 2), Hungry Hill, Mount Hunger, *North Hungerhill* (in Wareham). Also in f.ns., e.g. Hungar Hill **2** 8, Hunger Hill(s) **1** 28, **2** 42, 143

hungrig OE 'hungry'. Hunger Hill **3** 69, Hungry Ctgs & Down, Hungry Hill; f.ns. Hungry Hill **2** 37, **3** 300

hunig OE 'honey'. Honeybrook Fm, Honeycomb Wood, Honeycombe Fm, Honey Down Plant., Honey Grove, ~ Hill, Honeymead Lane, Honeypits Plant., *Hony Lane* **4** 359; f.ns. Honey Hills Md **3** 213, Honeylands **4** 435, Honey Pinks **4** 329, *Honihulle* **3** 157, *Hony Feld Close* **2** 205

hunta OE 'a hunter'. Hunters Bridge, ?Huntick Fm, Huntingford, Huntley Barn & Lane; f.ns. H(o)untwell **3** 252, *Hunteheche* **2** 140, *Huntemede* **3** 157

hunte ME 'hunt', f.ns. *Burse Hunte Cross* **2** 173, ?Hunt Hay Mdw **1** 324

hunter ModE. Hunters Lodge

hunting eModE 'hunting'. Hunting Bridge; f.n. *Huntynge thornes* **2** 38

huntinghorn ModE, f.n. Huntinghorn Flg **1** 362

hūs OE, **hous** ME 'house'. *Brickhouse* **3** 62, Burnt House Bottom, Canford School, Churchehouse **4** 264, Court House Dairy, Frog House Fm, Great House Fm, Grovehouse Fm, High House Fm, Hole House, Lamb's House Fm, Lane House Fm, Lodge House Fm, Lower House Fm, Newhouse Fm (× 3), New House Fm (× 2), Oxhouse Ctgs, Pigeon House Barn & Fm, Pigeon House Plant., Pitt House Fm, *Pressehous* **4** 485, Red House Fm, Shade House Fm, Sheephouse, Tile House Fm, Valehouse Fm, Water House Fm, White House Fm (× 2), Whitehouse (Rd), Windmill Barn & Knap, Woodhouse Cross, Woodhouse (Fm, Hill), Woolhouse Fm; freq in modern minor names as ~ House, e.g. Beaufort Ho, Black Ho, Country Ho, Court Ho, Fairlie Ho, Parliament House, Round Ho, Thatched Ho, White Ho. In f.ns. *Houshulle* **3** 257, *Oldehouselane* **2** 181. In AS charters, *hus* **3** 102

husk eModE 'rough or worthless plot', ?Husk **4** 282

hūsting lOE 'assembly', f.n. ?*Hustyngesmersshe* **3** 33

hut ModE. *Hut* **4** 107, The Hut, The Hut Dairy, Long Bredy Hut Lane; f.ns. Hut Grd **4** 299, Huts Plat **4** 282

hwǣg OE 'whey'. South Whey; f.ns. Whey & Curds **3** 162, *Wheyden* **2** 254

hwǣte OE 'wheat'. *Caundell Wethill* **3** 320, Watcombe Bottom, ~ Plain & Wood, ?Whatcombe Down, ?Whatcombe Fm & Ho, Whatley Fm, Wheat Close, Whetham Fm, *Whethill* **4** 261, Whetland's Lane,

Whetley (Cross & Fm); f.ns. Wager **3** 30, *Whatcombe* **2** 58, *the Wheat fields* **2** 54, Whetlands **4** 269. In AS charters, *hþete cumb* **2** 58, *whetecombe* **4** 123

hwǣten OE 'wheaten, where wheat is grown'. *Whetenecandel* **3** 320

hwearf OE 'wharf'. Ridge Wharf; f.n. *le Wharfe* **1** 252

hwēol OE 'wheel'. Water Wheel (× 2); f.ns. Wheel Goar **3** 389, Wheel Mdw **4** 269, *La Whele* **3** 259. In AS charters, *hweoul forda* **4** 454

hwet-stān OE 'whetstone', f.n. Whetstone **1** 113

***hwip(pe)** OE 'twig', f.n. Weep hills **2** 106

hwīt OE 'white'. Swyre White Ho, Whitchurch Canonicorum, Whitcombe, ~ Coppice, ~ Fm, ~ Hill (× 2), Whitcombe Hill, White Barn, White Bridge (× 2), Whitecliff Fm, *Whitecliff Mill*, ?Whitecombe, White Copse, ~ Cross, Whitecross, White Down, ~ Fm (× 2), Whitefield (Bottom, Coppice), White Ground Lake, White Hall (× 2), ~ ~ Barn, Whitehall (× 3), Whitehall Ctgs, ~ Fm , *le Whitehaull* **3** 366, White Hill (freq), Whitehill Plant., White Hole, White Horse Hill, White Ho (freq), White House Fm (× 2), Whitehouse (Rd), Whitelands Down, White Lane (× 3), White Mead, White Mill, White Moor, *White Nose* (*Hill*), White Nothe or White Nose, White Pit (× 2), White Post, Whitepost Gate, White Sheet Hill (× 2), Whitesheet Hill, White Way (× 2), Whiteway (Fm, Hill & Plant.), Whiteway Fm, White Way Hill, Whiteway Hill, Whiteway Hundred, Whitey Top, *Whitfield*, Whitfield Barn & Fm, Whitfield Fm, Whitham Ball Pit, ?Whithay Lane & Moor, Whitley Lake, Whitmore, Whitmore Bottom, Whitmore Coppice (& Plant.), ?Whitty Mead, Winterborne Whitchurch; in f.ns. Whitcombe **2** 10, White Hill **2** 215, Whitmoore **3** 58, *Whytestrete* **3** 180, *le Whytlond* **4** 35. In AS charters, *hwitan wyllon* **3** 207 (now **1** 316, *v.* Addenda), *hwitendich* **2** 122, *hwiten welle* **3** 207, *White Wey* **3** 259

***hygel** OE 'hillock'. ?Cocknowle

hyll OE 'hill'. Abbot's Hill (Fm), Allen Hill, Allington Hill, Ambrose Hill, Ash Hills Coppice, Ass Hill, Ayles's Hill, Babylon Hill, Ball Hill, Ball's Hill, †Banbury Hill, Barnes's Hill Fm, Barrow Hill (Fm) (× 5), Bartom's Hill, Barton Hill, Batcombe Hill, Bayard Hill, Bazon Hill, Beacon Hill (× 4), Bell Hill (× 2), Benchy Hill, Bennett's Hill (Fm), Bincombe Hill, †Bindon Hill, Birch Hill (× 3), Birts Hill, Blackdown Hill, Black Hill (× 5), ?Blackhills Plant., Bladeley Hill, Blagdon Hill (× 2), Blake Hill, Blastmoor Hill Barn, Bleax Hill, Boar Hill Lodge, Bonfire Hill, Bonscombe Hill, Bourne Hill Plant., Bowcroft Hill, Bowden Hill, *Bowdens Hill*, Bower Hill Plant., Bowridge Hill, Bow's Hill, Boys Hill, Brake Hill, Break

Heart Hill, Bredy North Hill, Breeze Hill, Brick Hill (Plant.), Brierhill Wood, Brimley Hill, Brock Hill, Bronkham Hill, Broom Hill, Broomhill, Broomhill Coppice, ~ Copse, ~ Drove, ~ Fm (× 2), Broom Hill Lane, Broomhills, Brownsall Hundred/Brunsell's Fm & Knap, Brown's Hill, Broxhill, Bubb Down Hill, Buck Hill, Buckhill Stall, Buckland Hill (× 2), Buckshill Coppice, Bulbarrow Hill, Bullhill Copse & Fm, Bull's Hill, Bunker's Hill Plant., Burchill, Burton Hill Plant., Burt's Hill, Canon Hill, Castle Hill (freq), *Cathille* **4** 289, *Caundell Wethill* **3** 320, Chaffeymoor Hill, Chalbury Hill, Chammen's Hill, Chapel Hill (× 3), Champs Hill Plant., Chardown Hill, Charlock Hill, Charlton Hill Coppice, Charlton on the Hill, Chelborough Hill, Chesters Hill, Chicks Hill, †Chilcombe Hill, Chilfrome Hill Barn, Chine Hill, Chitten Hill, Church Hill (× 3), Churchill (× 3), Church Hill Fm, Church Hill Lane, *Churchulle* **4** 158, Clanden Hill, Clandon Hill, *Clay Hill*, Clay Hill (Ho), Cleevehill Lane, Clifton Hill, Clouds Hill, Clown Hill, Clump Hill, Cobbett Hill, Cockett Hill, Cockhill Fm, ?Cocknowle, Colehill, Cole Hill Wood, Colmer's Hill, Coltleigh Hill, Combe Hill, Common Hill, Conegar Hill (× 2), Coneygore Hill, Constitution Hill, Con(e)ygar Hill, Coombe Hill, Coombe Down Hill, *coppehulla* **3** 143, Coppet Hill, Corfe Hill/Corfehill Fm, Cornford Hill, Cornhill Ctg, Corton Hill, *Courtehilles* **4** 333, Court Hill, Cowdon Hill, Cowdown Hill, Cox Hill, Crabb's Hill, Crane Hill Plant., Creech Hill, Crete Hill, Long & More Crichel, Crooked Oak Hill (× 2), Crook Hill (× 3), Cross Hill, Crow Hill, Cuckoo Hill, Cuckoo's Hill, ?Cuddle, ?Cudnell, Custard Hill, Damer Hill, Dancing Hill, Danes Hill, Dark Hill Plant., Darknoll Fm, Daw's Hill, Dean Hill, Denhay Hill, Dewlands Hill, Dickley Hill, Dimstone Hill, Ditton's Still, Dodpen Hill, Dogbury Hill, Doghill Coppice, Dole's Hill Plant., Dolman's Hill, Dragon's Hill, Dry Hill Moor, Dugberry Hill, Duke's Hill, Duncliffe Hill, †Dungeon Hill, Dungrove Hill, Dunyeat's Hill, ?Durdle Door, Earl's Hill, Eastfield Hill, East Hill (freq), East or Ice Hill, Easton's Hill Coppice, †Eggardon Hill, Eldon Hill, Ellston Hill, Emmetts Hill, Everley Hill Fm, Evershot Hill Plant., Farm Hill (Bottom & Copse), *Fearne Hill* **4** 324, Fern Hill (× 3), *Fern Hill*, Ferny Barrows, Fifehead Hill, Fir Hill, Folly Hill, Fontmell Hill Ho, Fore Hill (Plant.), West Fossil Dairy & Fm, Foul Hill, Four Barrow Hill, Fox Hill (× 2), Foxhill, Frenchman's Hill Barn, Frizzel's Hill, Frome Hill, Furze Hill (freq), Furzehill Fm, Furzhill, Gainsborough Hill, Gales Hill, Gallows Hill (× 3), Garvey Hill, Gate's Hill, Gaul Hill, Gerrard's Fm, ~ Hill, Giant Hill, Glue

Hill, Goathill, Gold Hill (× 2), Goose Hill, Gore Hill Plant., Gore's Hill, Gould(')s Hill, Grain's Hill, Grammars Hill, Grange Hill, Gravel Hill (× 2), Great Hill (× 2), Greenhill, Green Hill (× 6), Gussage Hill, Hack Hill, Hackthorn Hill, *Hakerygehulle* **4** 403, *Halfway Hill*, Hall's Hill, Halstock Hill, †Hambledon Hill, Ham Hill, Hammiton Hill, Hampton Hill, Handley Hill, Hanover Hill, Harbour Hill, Harper's Hill Copse, Hart Hill Plant., Hartland Moor, *Hasel hyll* **4** 42, Hawker's Hill, Hawkhill Ditch & Mill, Hawkmoor Hill, Haydon Hill, Haydon Hill Wood, Hewish Hill (Buildings), Highcomb Hill, Higher Hill (Barn & Bottom), High Hill, High Lea Hill, *Hill*, Hill, Hillamsland, Hill Barn (freq), Hillberry, Hill Bottom (freq), Hill Buildings (× 2), Hillbutts, Hill Close (Coppice), Hill Coppice (× 5), Hill Copse (× 3), Hill Ctg(s) (freq), Hill Crest, Hill Cross, Hill Fm (freq), Hillfield (Hill), Hill Flower Coppice, Hill Ho (× 2), Hill Pit, Hill Plant. (× 3), Hillside (Fm), Hill Street, Hilltop Bungalows, Hill View (× 2), Hill Wood (× 2), Hilton Hill Buildings, Hincknoll Hill, Hine's Hill, *Hockerill*, Hodder's Hill, †Hod Hill, Hog Cliff Hill, *Hogenhull* **3** 223, Hog Hill (× 4), Hollis Hill, Hollow Hill (Plant.), Holt Hill, Holway Hill, Honey Hill, Horn Hill, Horner Hill, Horsehill Fm, Horse Hill Lane, Hound Hill, *Hull*, *La Hulle* **3** 260, **4** 379, Hunger Hill (× 2), Hungry Hill, Hunt's Hill, Hyde Hill (× 2), Hyde Hill Plant., Ibberton Hill, Ivory Hill Copse, Iwerne Hill, Jack's Hill, Jan's Hill, Jock's Hill, Jordan Hill, ?The Kendalls, Kington Hill, Kite Hill (Coppice), Kitehill Plant., Kit Hill (× 2), Kithill Plant., ?Kittle Barrow, Knacker's Hill Lane, Knap Hill, Knaps Hill Fm, Knowl(e) Hill (× 6), *Lakehyll* **4** 220, Lambert's Castle Hill, Lambert's Hill, Langdon Hill, Laurel Hill, Legge's Hill Barn, Lewesdon Hill, Lime Hill, Limekiln Hill, Linton Hill, Little Minterne Hill, Little Puddle Hill, Livett's Hill Coppice, Loader's Hill, Lillington Hill, Lockyer's Hill, Loders Hill, Lodge Hill (× 2), Longbury Hill, Lophill Fm, Loscombe Hill, Lotmoor Hill, Low's Hill Lane, Luccombe Hill, *Luters Hill*, *Lut*(*t*)*shill*(*s*) **1** 358, Luxmore Hill Plant., Lyme Hill, Magiston Hill, Mangerton Hill (× 2), Manor Hill, Mansion Hill, Mappledore Hill, Marnhull, Martins Hill, Melbury Hill, Merry Field Hill, Merry Hill, Middle Hill (Plant.), *Middle Hill*, Midney Lane, Mill Hill, Mintern's Hill, Moathills Coppice, Monkwood Hill Fm, Monmouth's Hill, Mounter's Hill, la *Mulehulle* **4** 397, Muse Hill, Mythe Hill, Noon Hill, Nordon Hill, North Field Hill, North Hill (× 6), North Hills Plant, *North Hungerhill* (in Wareham), Norton Hill, Notting Hill, Notton Hill Barn, Oak Hill, Oakhills Coppice, Oborne Hill, Ochill Wood,

Okeford Hill, Old Hill, *Old Hill*, Orchard Hill, Osehill Green, Osmington Hill, Packers Hill, Pamphill, Paper Hill Copse, Park Hill, Parsonage Hill, Patson Hill (Fm), Peak End Hill, Peat Hill, Peat's Hill, Pen Hill, Penney's Hill Coppice, Penn Hill (× 2), Penny Hill Fm, *Pennys Hill*, Pentridge Hill, Pen Well Hill, Pheasant Hill, Philip's Hill, ?Pidney, Piper's Hill, Pistle Hill, Pitcher's Hill, Pleck Hill, Plush Hill, Poorton Hill, Porter's Hill, Porthill Plant., Portisham Hill, Potterne Hill, Pottle's Hill Plant., Pound Hill (× 2), Povington Hill, Quarr Hill (× 2), Rake Hill, Rakehill Fm, Rampisham Hill, Ram's Hill, Ramshorn Hill, Ranston Hill, Redpost Hill, Revels Hill, Rew Hill, Ridge Hill (× 6), Ridgeway Hill, Rimpton Hill, Rings Hill, Rock Lea Hill, Rod Hill Lane, Roger's Hill, *Rood Hill*, Rooks Hill, Round Hill (× 4), Row Hill, Rull Fm, Rushton Hill, Ryall, Ryalls Cross, Ryall's Fm & Lane, Ryal's Plant., Rye Hill, St John's Hill, St Martin's Hill, Salt Hill Plant., Sandhill Clump, Sandhills (× 2), Sandhills Coppice, Sandley, Sandy Hill Fm & Lane, Sandy's Hill, Scotchey Hill, Seaborough Hill, Sea Hill, Seldon Hill, Sharnhill Green, Shave Hill, Shearplace Hill, Shedrick Hill, Shepherd's Hill, Sherrin's Hill, Shillingstone Hill, Shipton Hill, Shorn Hill, Silly Hill, Slape Hill, Sliding Hill, Sloes Hill, Smedmore Hill, Snow Hill, Soapers Hill Plant., Southern Hill, South Field Hill, South Hill Estate, Spring Bottom Hill, Spring Hill Ctgs, Stapehill, Sticklehill Barn, Stinsford Hill, Stoat's Hill, Stock Hill (× 2), Stockland Hill, Stonebarrow Hill, Stone Hill Gate & Wood, Stone Hills Plant., Storridge Hill (× 2), Stour Hill, Sugar Hill, Summer Hill Ho, Summerlug Hill, Sutton Hill, Sweet Hill, *Sweet Hill*, Hr Sweethills, *Swire Hill*, Swyre Hill, Sydling Hill, Symes's Hill (Copse), Tabbit's Hill Fm, Taddle Fm, Tan-hill Copse, Tark's Hill, Telegraph Hill, Tenant(')s Hill (× 2), Thistle Hill, Thomas's Hill Plant., Thornford Hill Plant., Thorn Hill Fm, Thornhill (Fm), Thornhill's Coppice, Tilly's Hill, Timber Hill, Tokenhills Coppice, Tophill or Uphill, Totnell, Tout Hill, Tower Hill, Town Hill, Tulk's Hill, Underhill (Fm), *Uphill*, Vanity Hill Wood, Venn Hill, Vesey's Hole Hill, Wabmill Fm, Waddon Hill (× 2), ?Wadhill Coppice, Walton Elm Hill, Wancombe Hill, Wardon Hill, Warne Hill, (Old) Warren Hill (× 3), Wash Hill Coppice, Waterpot Hill, Watton Hill, Watts Hill, Weam Common Hill, Wears Hill, Wedge Hill, Welcome Hill, Wennet Hill, West Hill (freq), Weston Hill, Wether Hill, Whistle Barn, Whitchurch Hill Barn, Whitcombe Hill (× 3), White Hill (freq), Whitehill Plant., White Horse Hill, White Sheet Hill (× 2), White Way Hill, Whiteway Hill, Windmill Hill (× 2), Winyards

Hill Coppice, Woodbury Hill, Wood Fold Hill, Wood Hill, Woolland Hill, Wootton Hill, Wynman Hill Copse, Yaffle Hill. Freq in f.ns., e.g. *Bromhull* **4** 34, *Doddehulle* **4** 411, Furze Hill **4** 22, *Thornehill* **3** 35. In AS charters, *ac hylle* **3** 119, *bromhulle* **3** 255, *cing hille* **3** 128, *ciric hylle* **3** 281, *cuceles hylle* **4** 28, *deoh holes hyllæ* (*v.* Addenda **3** 267), *fern helle* **3** 102, *fyres hylleford* **4** 303 (*v.* Addenda **4** 193), *grenenhille* **2** 57, *hean wifeless hylle* **3** 288, *hul* **3** 257, *morhelle* **3** 207, *Rumanhelle* **3** 171, *sandhellesled* **3** 111, *stocchylle* **4** 301

***hylu** OE 'hollow'. ?Cocknowle

hymele OE 'hop plant', f.n. ?*Hymercombe* **1** 194

hynt Welsh 'way, path'. ?*Hyntwey* **1** 90

***hyppels** OE 'stepping stones', f.n. Hipple Md **3** 293, *magnum huppyll* **3** 207

hyrdel OE 'hurdle'. Hurdle Coppice; f.ns. Hurdle Cl **3** 266, **4** 282, Hurdle lands **4** 329, Hurdle Md **1** 249

hyrne OE 'angle, corner, land in a river-bend'. Heron Coppice, ?Heron Grove, Northernhay, *Parkeshurne* **4** 379, Rhymehorn, *Trehurne* **4** 154; f.ns. Ditch Herons **2** 199, Gt Hern **4** 423, Heron **3** 27, *Kingesheigeshurne* **2** 162, Woodhorn **2** 70. In AS charters, *dice hyrnan* **4** 34, 454, *diches hirne* **3** 206, *heges hirne* **3** 206, *hirnen* **1** 31, *horngetes hirne* **2** 82, *hyrn* **2** 161, *mæde hyrnan* **4** 35, *tunes hirne* **2** 82, *wulfrices dice hyrnan* **4** 454

hyrst OE 'copse or wooded hill'. Bedchester, *brokhurste* **2** 196, Hogchester, Holnest, ?Horchester, Horse Pit, Hurst (Fm), ?Hurst Mill; f.ns. *Dunhurst* **3** 32, *Podelhurste* **4** 411, *Pykyd Hurst* **2** 52. In AS charters, *achirste* **2** 162, *drian hyrstes lace* **4** 454, *readan hyrstes* **4** 304

hys(s)e OE 'shoot, tendril', f.n. *Hussle(y)wode* **1** 333

hȳð OE 'harbour, landing place'. *Huthe* **1** 255, Langton Hive Point, Rodden Hive; f.ns. *Hythefeldes* **3** 73, Hythe 20 Acres **4** 17

-ic OE noun-forming suffix. ?Gussage Brook (*v.* RNs), Gussage All Saints/St Michael/St Andrew

ice-house ModE. Icehouse; f.ns. Ice House Orchd **4** 166, Icehouse Pce **1** 127

īdel OE 'empty, deserted', f.n. *Ydelwerk* **4** 160

īfig OE 'ivy'. ?*Iveley* **1** 115, Ivy Ctg (× 5), Ivy Cross, Ivy Green Ctgs, Ivy Ho; f.ns. ?*Ivescripple* **3** 6, *Yuymede* **2** 231. In AS charters, *eficlif* **1** 31

-ig[3] OE adj. suffix. Bramble Coppice; f.ns. Bennety Coppice **2** 204, *Bremelifurlang* **2** 173, *le levery acre* **1** 238, *Meisimor* **4** 133,

Myntylond **1** 327, Ruffetty **4** 330, *Wedylond* **1** 216, *Wexihamme* **2** 284

igil, **īl** 'hedgehog', possibly also 'leech'. ?Ilcombe Fm

ile ME 'an island'. Brownsea Island, Eye Bridge & Mead, Furzey Island, Green Island, Isle of Purbeck; f.ns. Isle **1** 227, *La Yle* **2** 54

impa, **impe** OE 'sapling', f.ns. Hamslade Flg **1** 362, *North-*, *Suthymstal* **1** 216

in OE 'in, inner'. ?Eastington, Incombe (Wood), *le Inne Hundred* **1** 6, Inpark Fm, Instow, Ryme Intrinseca, Uppingtown, ?Weston; f.ns. Inn Md **4** 25, *Inslade* **2** 14

independent ModE. Independent Quarries

inferior Lat 'lower'. Netherbury **4** 230

-ing- OE medial connective particle indicating 'called after' or 'associated with'. ?Allington, Baltington, Beaminster, Bovington, (Hr) Chalmington, ?Chedington, Didlington Fm, Drimpton, Ibberton, ?Ilsington, ?Ilsington Wood, Leeson Ho, Lillington, ?Linnington, Mannington, Nottington, Osmington, Pallington, Povington, Poyntington, Putton, Rollington Fm, White Lackington, Woonton Fm, *Wulfheardig stoke* **4** 338. In AS charters, *bacging berghe* **2** 122, *beorreding mæd* **3** 286, ?*boddingc weg* **2** 136, *Winteintune* **3** 113

-ing[1] OE noun suffix. ?Pop Mallers Coppice

-ing[2] OE place-name forming suffix. ?Armswell Fm, Dodding's Brick Yard & Fm, ?*dydelingdune* **2** 135, ?Evenings Ford, Fordington, ?Ilsington, ?Ilsington Wood, ?Linnings Copse, ?Nottington (*v.* Addenda), ?Pop Mallers Coppice, ?Snelling Fm, ?Stottingway, Uddens Ho, *Wanetinge* (*v.* RNs), ?Wavering Lane. In AS charters, ?*birninge*, *Byrnynge* **3** 254, ?*boddingc weg* **2** 136

-ing[3] OE patronymic suffix. ?Armswell Fm

-ingas (gen. pl. **-inga**) 'followers, people'. *Boretonyngehamme* **3** 170, Gillingham, *hore tuning gemære* **2** 159, *Pentringtone* **2** 235, *Purbicinga* **1** 1–2; f.n. ?Coreans **1** 79

-ingtūn OE, *v.* **-ing-**, **tūn**

inheche, **inheching** ME 'land enclosed for cultivation', f.ns. *le Inhechyn* **4** 385, *Inichin* **1** 345

inhoke ME 'land temporarily enclosed', f.ns. Enoch 4 328, Innock **2** 171, Innox **3** 27, Inox **2** 105, In Ox **2** 25

inland ME 'land near a residence', f.ns. England **4** 423, Englands Md **4** 267, In Lands **3** 27, *la Inlonde* **2** 250

inmost OE 'furthest in(land)'. Inmosthay Quarries

inn ME, ModE 'dwelling, lodging, inn'. Chequers Inn, Revels Inn Fm. Freq in modern inn names, e.g. Nags Head Inn, New Inn, Picnic Inn, Plough Inn, Pure Drop Inn, Ship Inn, Stone Inn, Trooper Inn; f.ns. *the Newe Inne* **2** 53, New Inn **2** 172, *Platellesynne* **3** 158

inner eModE 'closer'. Inner Bracket's Coppice, Inward Lardenhall; f.n. *the Inner Batelease* **2** 175

***innōm** OE 'land taken in', f.ns. Inhams **3** 279, 387

insula Lat 'island'. Furzey Island, Green Island, Isle of Purbeck

intake ME 'land taken in', f.n. Durdalls Intake **2** 25

intrinseca Lat 'inner, lying within the bounds'. Beaminster Hd, Ryme Intrinseca

***-iol** PrWelsh adjectival suffix. Cerne Abbas

iron ModE, f.n. Iron Box **1** 41, Iron Grd **2** 114

iron foundry ModE. *Waterloo Iron Foundry* **2** 7

irre OE 'wandering, winding', f.n. ?*Irepool* **4** 364

-isc OE noun and adj. suffix, f.n. *Cnollisschefurlang* **3** 255

īse(r)n OE 'iron'. ?Icen Barrow, ?Icen Dairy & Lane; f.n. Ison Hill **4** 68, *la Ysenebeme* **4** 93

island ModE. Gigger's Island, Hospital Island, Martins River Island, Otter Island, Pergin's Island, Stone Island, Weston's Island

īw OE 'yew-tree'. Nine Yews, The Yews, Yew Ho, Yew Tree Ctg, Yew Tree Fm (× 3), Yewtree Ctg, ~ Fm (× 2), ~ Pit; f.ns. *le hiis* **2** 215, Yew Tree Fd **2** 215

ïw Brittonic 'yew-tree'. ?R. Iwerne (v. RNs), giving name to: Iwerne Courtney, ~ Minster, Steepleton Iwerne

***Īwernā** Brit goddess-name 'she who moves in a customary course'. ?R. Iwerne (*v.* RNs), giving name to: Iwerne Courtney, ~ Minster, Steepleton Iwerne

jack ModE dial. 'small', f.ns. Jack Close **4** 204, Jackhams **1** 79, Jackleasehays **2** 100, Jack Wd **4** 463

Jack-a-Lent eModE 'figure set up in lent to be pelted', f.n. Jackalints **1** 60

jackass ModE, f.n. Jackass Grd **3** 97

jasmine ModE, f.n. Jessamin **4** 409

jaw ModE, f.n. Picked Jaws **2** 26

jetty, jutty ModE 'jetty'. Cobb Gate Jetty, *le Jety* **1** 256, Lucy's Jetty, Rock Lea Jetty

Jew ModE, f.ns. Jews Fd, Md & Plot **2** 41

jolif ME 'pleasant'. ?*Jolyfmer* **1** 118

jubilee ModE. Jubilee Column, ~ Coppice, ~ Hall, ~ Oak (× 2), ~ Plant., ~ Pond, ~ Walk, ~ Wood; f.n. Jubilee **3** 27

Judas eModE, f.n. *Judas garden* **1** 169
jug ModE slang 'prison, jail', f.n. Gt & Lt Jug **4** 424
junction ModE. Bokerly Junction, Hamworthy Junction, Holes Bay Junction, Yeovil Junction Lodge
juniper ModE. Juniper
justing ME 'jousting', f.n. *Joustyng croft* **3** 157
keep ModE 'dungeon', f.n. ?*Keepe crosse* **1** 362
keeper ModE 'gamekeeper'. Keeper's Ctg Plant., Keepers House, Keeper's Lodge (freq); f.n. *Keepers ham* **2** 10
keeve, **kive** Do dial. 'large tub for wort when brewing', f.n. Keeve Md **1** 202 [OE **cȳf**]
kenel ME 'kennel'. Dog Kennel Copse, Kennel Copse (× 2), Kennel Dairy Ho, ?Kennel Fm/Hill Plant./Rings/Wood, Kennel Pond, Kennel Quarter, The Kennels (freq), Kennels Lane; f.n. *Kenele* **4** 254
kex, **kecks** eModE 'hollow-stemmed plant', f.n. Keck **4** 172, Kick Acre **3** 78
key, **kay** ME 'quay'. Hyde's Quay, *la Keye* **1** 252, Ower Passage, Quay **1** 156, The Quay **2** 42, Quay Ctgs, Redhorn Quay, Russel Quay, Wytch Passage; f.ns. *Hope Quay* **1** 252, Quay 1 325
kiddle eModE 'dam. weir', f.n. Kiddle Nap Well **4** 452
kide ME 'young goat', f.n. *Kedeham* **3** 111
kidney ModE, f.ns. Kidney **4** 25, Kidney Cl **4** 435
kike(n) ME 'to watch, peep'. Kick Hill Coppice
kiln, *v.* **cyln**
kirk ME 'church', f.n. Kirck barton **4** 181
kit-cat eModE 'a game', f.n. ? Kitcats Cl **1** 80
kitchen, *v.* **cycene**
knacker eModE 'horse-slaughterer'. Knacker's Hill Lane, *Knacker's Hole* **1** 241, Knacker(')s Hole (× 4). Also in f.ns, e.g. Nackers Hole **1** 41
knappede ME 'provided with a top, or with a roof ornament'. ?Knap Barrow; f.ns. *Knappydehall* **1** 184, Napperty Hall **3** 373 [*cf.* **cnæpp**]
knappy ModE 'hilly'. *Nappy Colcombe* **4** 262; f.ns. Knappy Grd **1** 29, Nappy Cl **3** 253, **4** 151, 424 [*cf.* **cnæpp**]
knob ME 'a knoll'. Knobcrook, Knob's Crook, Nicodemus Knob, Worbarrow Tout; f.n. Nob Crook Orchd **3** 197
knorre ME 'gnarled tree', f.n. *Knorreknol* **4** 133
knot ME 'hillock', f.n. Black Knot Down **4** 498
kyn(e) ME, **kine** ModE dial. 'cows, cattle'. ?The Kendalls, ?Kine Bush Lane

labourer ModE, f.ns. Labourers Gdn **1** 293, **4** 79

laburnum ModE. Laburnum Villa

lacer ME 'maker or seller of cord, etc'. Lazerton (*v.* Addenda **2** 117)

lacu OE 'stream'. *Alrecumeslake* **4** 377, *Attwell lake* **2** 21, Baglake Barn & Fm, Balls Lake, Bidlake Fm, Bride Head Lake, Castle Lake Spring, Chafey's Lake, ?Charlock Hill, Clay Lake (*v.* RNs), Claylake, Coombe Lake, Coward's Lake, Creekmoor Lake, *Culmelake* (*v.* RNs), Culverlake Lane, East & West Lake (Side), Ebblake, Featherlake, Graston Lake Copse, Green Island (Lake), *Kokenhule Lake* (*v.* RNs), Lake, Lake Copse, ~ Fm (× 2), Lakegate Lane, Luckford Lake (*v.* RNs), Middlebere Lake (ib), Morcombelake, Morkhams Lake, North Haven Lake, Ower Lake, Parkstone Lake, *Parkwellelake* **4** 165, Pink Lake Coppice, Pulham Lake, Putlake Fm, Radipole Lake, Ramshorn Lake, Redhorn Lake, *Rodlake* (*v.* RNs), Rope Lake Head, Shortlake, Shot Lake Wood, Silklake Quarries, Silverlake Fm, South Lake, Spring Close Lake, Stone Island Lake, Upton Lake, Vale Acre Copse & Fm, Water Lake, West Spring Lake, White Ground Lake, Whitley Lake, Woodlake, Wych Lake (*v.* RNs). Freq in f.ns., e.g. *Conditeslake* **4** 10, Dock Lake **3** 26, *Holelake* **4** 10, *Huckelake* **2** 221, *Oselake* **2** 11, Woodlakes Cl **2** 10. In AS charters, *bridian beara lace* **4** 453, *drian hyrstes lace* **4** 454, *fulan lake* **2** 158, *halgan weies lake* **3** 110, *holan lace* **4** 454, *lace* **3** 298, 380, *lake* **3** 102, 111, *lacmere*, *late mere* **3** 41, 128, *litlen lake* **3** 102

(ge)lād OE, ModE dial. **lode** 'water-course, river-crossing'. Load Lane, ?Lode Wood, ?Winslow; f.ns. ?Load **4** 159, *Westlade* **1** 230

ladder ModE dial., f.n. Ladder Stile **4** 108 (*v.* NDEFN 239)

lady, *v.* **hlǣfdige**

***læc(c)**, ***lecc**, ***lece** OE, **leche** ME 'stream, bog'. Latchmore Pond, ?Lecher Water (*v.* RNs), Lytchet Bridge & Lane; f.ns. *the Harry Leach* **1** 62, Latch **2** 200, Latch Acre **2** 25

***lǣcere** OE (WSax) 'a leech gatherer or leech dealer'. ?Lazerton Fm [but now *v.* Addenda **2** 117]

***lǣd** OE, **lead** ModE 'water-course'. Broadley Wood; f.n. *le lede* **1** 121

lǣfer OE 'rush, reed', f.ns. *Lavermeade* **2** 230, *Leverbeddes* **3** 273, *le levery acre* **1** 238, *Liver close* **1** 194

***lǣge**, ***lēge** OE 'fallow'. *Layfield*; f.ns. Bayley Ley **3** 43, Layhill **4** 55, Ley Cl **2** 28, 270, *la Leyecroft* **4** 19, *Leyforlang* **1** 250

lǣs OE 'pasture, meadow-land'. Belleaze Fm, *Boles Leas* **4** 195, Broadleaze, Cowleaze, Cowleaze Bottom, ~ Copse, ~ Knap, Cow Leaze Stall, Cowleaze Withy Bed, Crossleaze Coppice, Culeaze

Coppice/Fm/Ho, Deans Leaze Fm, Easterly Copse, Eweleaze Barn, ~ Spinneys, *Ewe-lees*, Higher/Home/Lower Eweleaze, *Fordelese* **4** 275, Foxleaze Withy Bed, Green Leaze, *hoggen lease* **3** 118, Hog Leaze, Hogleaze Fm, Horse Leys Pit, James Lears, ?Lastridge, Lease Acre, The Leaze, Leaze Lane, Marley's Row, Mogers Leaze, New Leaze Fm, *North End lease* **4** 215, Overleaze Coppice, Oxenleaze Coppice, ~ Plant., Oxleaze Coppice/Copse, ~ Wood, Pitman's Leaze, Popes Leaze Aldermore, Potter's Leaze Plant., Somerleigh Court, Steeple Leaze Fm & Wd, *Studland Castle*, West Lears Fm, Westleaze, Woodleaze, Yewleaze Barn. Freq in f.ns., e.g. Hogleaze **4** 25, Lambs Leaze **2** 22, *la Oxenelese* **4** 19, Weather Leys **4** 26, Wildlese **2** 217

(ge)lǣt(e), *v.* **weg-gelǣt(e)**

***lagge** OE 'marsh', f.ns. *the Bridge Lagg* **2** 90, *Le/The Lagge* **3** 56, 79, Lt Lag **2** 25, *Lagdon* **4** 476, Lagland St. **2** 42

lagger eMod dial. 'long narrow strip', f.n. Laggers **4** 337 (*v.* NDEFN 240)

lake ME 'lake, pool'. Crichel Lake, The Lake (× 2, *v.* RNs), Lake Fm, Lake Hill Plant., *Lake Mill Stream* (*v.* RNs), Lake Plant. (× 2), Lucerne Lake, Melbury Lake, Parnham Lake, Sherborne Lake [*cf.* **lacu**]

lām OE 'loam'. ?Lambrook

lamb OE 'lamb'. ?Lambrook, Lamb House Fm, ~ ~ Woods, Lamblands Copse, ?Lamb's House Fm

lambing ModE. Huish Lambing Yard, Lambing Yard (× 2), Lambing Yard Plant.

Lammas ModE 'church festival 1st August', f.n. Lammas **3** 121

lampe ME 'lamp', f.ns. *Lamp Close* **3** 33, *Lampeacre* **3** 226, *Lampeland* **2** 53, *the Lamp-House* **4** 68, *the Lamp Light Close* **1** 81, **1** 121

lanch, **lanchet**, *v.* **land-sc(e)aru**

land OE 'cultivated land, tract of land, estate, strip in a common field'. Ashlands Coppice, Barrowland Fm, Beerlands Fm, Befferlands Fm, Blackland Plant., Blacklands, Blackland's Close, Blacklands Lane, *Blakelond*, Bracklands, Brooklands (Pit), *Cauzsuslond* **4** 256, Chalklands Quarries, Clayland Coppice, Claylands Lane, Dewlands Hill, ?Early's Fm, Emlands Bridge, Fanny's Land Plant., Flaxland Lane, Footlands Coppice, ~ Lane, Fox Ground, Frog Lane **3** 52, Furzeland Barn, Glebeland Lane, Godlandfoot, Gooseland Lane, Grazeland, Greenland, Hatchlands, Heatherlands, Heavland Lane, Hillamsland, Holland's Hill Coppice, Howland Coppice, Kingsland,

Lamblands Copse, Langlands Cross, Loders Land, Longlands (Copse), Long Lands Barn, ?Marland Pit, Moorland (Lodge), Morgan's Land, Newland (× 2), Newland(')s Ctgs, ~ Fm, Newlands (Gate & Plant.), New Lane **3** 53, Oatlands Fm, Old Lawn Coppice & Fm, Pitlands Coppice, Portland, Priorslawn Dairy, *Rammeshameslond* **4** 443, Redland Plant., Redland(')s (Fm), Red Lane **2** 18, Rodlands, Rowlands, Ryeland(s) Coppice, Ryeland's Barn, Ryland Plant., *St Johns Lande* **1** 357, Scotland (× 2), Sheeplands Copse, Smellands Lane, *Stockland*, Stockland, Hr Stockland, Stony Lawn, Swalland Fm, Studland, Wayland, Whetland's Lane, Whitelands Down, Woodlands (freq), *Wooland*, Woolland, Woolland Grove, *Wulsieslande* **4** 261, Yarlands, ?Yearlings' Bottom & Drove, *La Yerdelonde* **4** 276. Freq in f.ns., e.g. *Crabbelond* **4** 10, *Everlond* **4** 10, *Flexlond* **1** 137, *Rodland* **4** 19. In AS charters, *Ierðlonde* **3** 207

landmark ModE 'boundary mark'. Land Mark; f.n. The Land Mark Bank **2** 41

land-sc(e)aru, **land-scoru** OE 'boundary, share of land'. Lanch Lane, Landchard Lane, Landshare Coppice, Landshire Bridge, *Landsore* **2** 240 (*v*. RNs), *le Lanschore* **3** 20, The Launch. Also in f.ns, e.g. Lanches **3** 132, Landshard (Down) **3** 216, **4** 25, *le Landshare* **1** 267, *lansshoredych* **1** 31, *Lawnse* **4** 19, *threlansshorys* **3** 227. In AS charters, *land sceare* **4** 454, *landscher' ford* **3** 199, *land scorforð* **3** 136, *land scorhlinc* **3** 207, *land share* **2** 82, *land share hegen* **3** 138, *thre lanschere* **3** 258, *þrem land sharen* **3** 139

landshut eModE dial. 'a flood', f.n. ?Landshut Fd **1** 286

landslip ModE. Southwell Landslip

lane, **lanu** OE 'lane'. *Boyeslane* **3** 336, Bramdon Lane, *Chapyllan'* **4** 158, Dean Lane, Lane Coppice, Lane Cottages, Lane End, ~ ~ Fm, Lane House Fm, Long Lane **2** 154, **3** 4, Love Lane (freq), Marriage Lane, Pilford Lane, Redlane Hundred, ?Stickland Coppice, Street Lane, *Terrieslane* **3** 74, Thickthorn Lane, Tibble's Lane, Wavering Lane, Winterborne Stickland. Freq in modern minor names and street names as ~ Lane, e.g. Ansty Lane, Back Lane, Birch Lane, Blind Lane, Bockhampton Lane, Bovington Lane, Bugs' Lane, Butt Lane Hollow, Carraway Lane, Chalk Pit Lane, Chapel Lane, Charminster Lane, Chine Hill Lane, Church Lane, Claypits Lane, Cow Lane, Cuckoo Lane, Dark Lane (freq), Dirty Lane Coppice, Druce Lane, Duck Lane, Easton Lane, Froom's Lane, Gaddy's Lane, Gascoyne Lane, Green Lane, Highgate Lane, Icen Lane, Knighton Lane, Louse Lane, Lytchet Lane, Miles's Lane, Milom Lane, Muston Lane, Nottington Lane, Putton Lane, Radipole Lane,

Rod Hill Lane, School Lane, Slyer's Lane, Snag Lane, Snow Hill Lane, Spear's Lane, Sturt Lane, Three Lanes End, Watergates Lane, Water Lane, Waterston Lane, Watery Lane (freq), West Lane, White Lane. Freq in f.ns., e.g. *le Blyndelane* **2** 70, *Champflourslane* **4** 10, *Lane end* **2** 205, *Roddokeslane* **2** 216. In AS charters, *fildene lane uppende* **2** 122

lang[1] OE 'long'. Cheselbourne, Chettle Long Barrow, Hyde Ctg & Wd, Ibberton Long Down, Lamplands Copse, ?Lancombes Cross, Langbeer Fm, Langbourne, Langcombe Bottom (× 2), Langdon (Fm, ~ Hill), *Langeberge Hundred*, *Langehyde* **1** 38, Langford Fm, Langham, *Langhgeley* **2** 147, Langlands Cross, *Langley*, Langmoor Fm, ~ Lane, Langton Herring, Langton Long Blandford, Langton Matravers, Lankham Bottom, ~ Eweleaze, Long Acre Quarries, Long Aldermore, Long Ash Fm, Long Ash Lane, Long Barrow (× 3), Long Barrow Hill, ~ Lane, Long Bottom (× 4), Long Bredy, Long Bridge (× 2), Long Burton, Longbury, Long Close, Long Close Fm (× 2), Longcombe Bottom, Long Coppice (freq), Long Copse (freq), Long Crichel, Longcroft Coppice, Long Cross, Long Cross Pit, Longcutts Fm, Long Ebb, Longelm Fm, Longfleet, Longford Bridge, Long Ground Plant., Long Gully, Longham, Long Hay Coppice, Long Hayes, Long Ho, Long Island, Longlands (Copse), Long Lands Barn, Long Lane (× 4), Long Ledge, Longley Leg, Long Lynch, Long Mead Coppice (× 2), Longmeadow Coppices, Longmead Plant., Longmoor Coppice, ~Fm, ~ Manor, ~ Plant., Long Park, Long Plant. (freq), Long Row Coppice, Longstone Ope Quarries, Long St. **3** 361, Long Thorns, Longthorns, Long Walk (Plant.), Long Wood. Freq in f.ns., e.g. *langacre* **2** 14, *Langedon* **3** 33, *Langforlang* **4** 19, Lankeys **3** 28. In AS charters, *langan hlinc* **2** 136, *langen cumbes hauede* **3** 104, *langengroue* **3** 257, *langhecgan* **3** 138, *langhelee* **3** 207

lang[2] OE 'long strip of land', f.n. *Haverlang(e)* **4** 270

***lann** PrWelsh 'church'. *Lanprobi* **3** 358

larch ModE 'larch-tree'. Langdon Larches, Larch Plant. (× 2)

larder ME 'larder', f.n. *Lardir* **4** 133

large ModE. Large Bar Barn & Hill

larking-glass ModE 'a machine with mirrors, used to attract larks to the net'. Larkinglass Fm

launde OFr, ME 'an open space in woodland, a forest glade, a woodland pasture'. Black Lawns Coppice, Broad Lawn, Cleverlawns, Damas Lawn Bridge, Ilsington Lawn, Land Coppice,

Lawn Bridge & Fm, *The Lawns*, Lawn Fm, Organ Lawn, Rabbit Lands Copse, Rose Lawn Coppice; f.n. The Lawns **4** 435

launder ME 'person who washes linen', *La(u)nder Lane* **3** 147

laundry eModE 'laundry'. Clyffe Laundry, *Landry Lane*, *the laundry-house* **3** 147, White House Laundry

laurel ModE. Laurel Avenue, ~ Bottom, ~ Ctg(s), ~ Covert, ~ Lodge, The Laurels (× 3)

lavender ModE, f.n. Lavender Acres **2** 191

lawe, **laghe** ME 'law'. ?Foxbury/Great Lower Hedge; f.ns. *La Laghemede* **3** 178, *the lawe gutter* **3** 372, *Lawthorne* **4** 182

lāwerce, **lǣwerce**, **lāferce** OE 'lark'. Laverstock Fm; f.ns. *Larkedole* **1** 238, Larks Leaze **3** 353, **4** 452, *Lerkefurlong* **1** 345

laye ME 'a pool'. ?Laymoor Copse & Lane, ?Laymore

le, **la** OFr def. art. 'the'. The Hyde, La Lee Fm, *Lopegarston* **3** 174, Lynch Fm; f.n. *Lyle* **1** 227

lēac OE 'leek, garlic', f.ns. *Lecledde* **1** 345, *lez lekebeddys* **1** 128, Lickhays **2** 19

lēac-tūn OE 'leek enclosure, herb garden'. Layton Ho, ?Letton, ?*Wynterburne Leton* **2** 83; f.n. *Medeheyesleighton* **3** 56

lēah OE 'wood, woodland clearing or glade', later 'pasture, meadow'. Ashley, Ashley Barn, ~ Ctg, ~ Fm, ~ Wood, Bainly Bottom, Bartleys Row & Wood, Barnsley Fm, Batterley Drove, Beaulieu Wood, Bewley Down & Fm, ?Bingledon Wood, Bingley Fm, Bladeley Hill, ?Bokerley Ditch, Bradle Barn & Fm, Brimbley Fm, Brimley, Broadley Wood, Bugley, ?Bulbury Camp, ?Burl Fm & Moor, Catley Copse, Chalkley Coppice, Cobley Fm, Coltleigh Fm, Cotley, Delcombe Botton & Fm, Downley Ctg & Coppice, *Earnley*, ?Emley Lane, ?Everley (Hill) Fm, Filford, ?Fillymead, Hail, Flaxley Coppice, Furleigh Fm, Halstock Leigh, Harely Fm, Harley Down & Wood, Hartley Fm, Hatherly Fm, Hawley Fm, Henley, High Lea Fm, High Lea Hill, Hookley Copse, Huntley Barn & Lane, *Iveley* **1** 115, Kimerley Wood, *Langley*, Hr & Lr Lea, La Lee Fm, Lay Wood, Lee Barton, Lee Ctgs, Lee Wood, Leigh (× 3), Leigh Fm, Leigh Lane, Leigh Wood, Lily Fm, Longley Leg, Lousley Wood, Loverley Fm, Lower Lye, Marley, Matterley Ctgs, Merley Hall Fm, Nutley Clump, ~ Way, *Oak Hill* **3** 119, Oakley (Fm), Oakley Down, ?Oakley Lane, Oakley Wood, Patley Wood, Philleyholme Coppice & Mill, Pleckley, Plumley Wood, *Primesley*, ?Ratleigh Wood, Ridley Pit, *Rudgeley* **3** 22, Shitley, Sixpenny Handley, Southley Fm, Spratley Wood, ?Stanley Court, Hr & Lr Stockley Fm, *Studley*, Tiley, Tytherleigh, Urless Fm, ?Wadhill Coppice, Wantsley Fm,

Waterley Wood, Weekley Coppice & Lane, Westley Fm & Wood, West Parley, Whatley Fm, Whetley (Cross & Fm), Whistley Fm, Whitley Lake, *Wysselay*, Yardley. Freq in f.ns., e.g. *Blanchardisley* 2 13, ?*beorhlem* **2** 122, *Eldelee* **1** 260, *Northele* **4** 425, Pimperne Leaze **3** 29. In AS charters, *æsc leage* **3** 288, *aescleȝe* (*v*. Addenda **4** 193), *beorhleage* **3** 281, *bradeleighe* **3** 255, *cliveleghe* **3** 255, *cludesleghe* **3** 110, *cristemeleighe* **3** 255, *earna lea weȝe* **4** 290, *eastlea* **2** 161, *filithleighe* **3** 199, *foxlee* **3** 128, *ȝyrdlea* **4** 292, *hunaleȝe* (*v*. Addenda **4** 193), *ketan lege* **4** 454, *langhelee* **3** 207, *leas ende* **4** 304, *limben lee* **2** 58, *linleaga mor* **2** 162, *litlen ac lee* **2** 266, *Mapoldorlea* **2** 286, *preosta leȝe* (*v*. Addenda **4** 193), *ruwan leges gete* **2** 162, *Schepleghʼe* **3** 258, *stiþeleighe* **3** 258, *tryndlea weȝe* **4** 305, *tudesleghe* **3** 178, *wadeleighe* **3** 200

leat eModE 'conduit for water', f.n. *le letes* **4** 385

leather ModE, f.n. Leather Close **2** 139

leche, *v*. ***læc(c)**

***lẹ̄d** PrWelsh 'grey'. Lytchett Matravers & Minster

ledge ModE, **legge** ME ledge, shelf'. Bar Ledges, Black Head Ledges, Bran Ledge, Broad Ledge, Canary Ledges, Dancing Ledge, Dover Ledge, Frenchman's Ledge, Grey Ledge, Hannah's Ledge, Howcombe Ledge, Kimmeridge Ledges, Long Ledge, Lucy's Ledge, Maple Ledge, Mupe Ledge, Perry Ledge, Pool Ledge, Raffey's Ledge, Ringstead Ledge, St Gabriel's Ledge, Tanville Ledge, Washing Ledge, Western Ledges, Yellow Ledge; f.ns. *Leggeforlang*, *Leggeham* **4** 182

leeward ModE 'sheltered', f.n. Leaward Hays **2** 22

leg ME 'leg, long narrow meadow'. Dunbury Leg, Humber's Leg, The Leg, Legg Lane **2** 185, Leg Row, Longley Leg, Sammel's Leg; f.ns. *Barleg* **1** 119, Coombs leg **2** 104, Leg Plot **2** 171, Lr & Lt Leg **3** 40

leman ME 'lover, sweetheart'. Landscombe Lane

lencten OE 'Spring, Lent'. Lenthay Common; f.n. *Leteneheie* **1** 203

leper ModE. Leper's Well

***leuc-** Brit 'bright'. ?Luckford Lake (*v*. RNs)

leye, **laye** ME 'pasture land, grass-land'. ?Lays Coppice

leyne, **lain** ME 'layer, tract of arable land, arable strip'. Ashy Drove, Broad Lines Plant., *Bromlines* **4** 118, Laines Plants., Shroton Lines, Water Lane Fm; f.ns. Horse Langs **3** 262, Lains Md **3** 28, *Layndone* **2** 85, *le layns* **2** 216, Sparrowline **2** 66

lias ModE. Lias Rocks

liberty ModE 'district with privileges'. Alcester Liberty, Alton Pancras ~, Bindon ~, Broadwindsor ~, Dewlish ~, Fordington ~, Gillingham ~, Frampton ~, Halstock ~, ?Liberty Goyle & Plant., Loders & Bothenhampton Liberty, Piddlehinton ~, Piddletrenthide ~, Portland ~, Powerstock ~, Ryme Intrinseca ~, Stour Provost ~, Sutton Poyntz ~, Sydling St Nicholas ~, Wabyhouse ~, Wyke Regis & Elwell Liberty

līc OE 'corpse', f.ns. Leachway **4** 348, *Lichforlang* **1** 372, *Lychewey* **4** 227. In AS charter, ?*lyc hagen* **4** 304 (*v.* Addenda)

lich-gate ModE. Lych Gate (freq)

licgende OE 'fallen', in AS charters, *ligcgendum æc* **4** 454

līc-tūn OE, **litten** ModE dial. 'burial ground', f.ns. *the Abbie Litton* **3** 359, *le Lytton* **3** 366, *la Neuelitton* **3** 158, New Lidden **2** 214

lightning ModE, f.ns. Lightening Plot **4** 234, Lightning Close **4** 151

lily ModE. ?Lilymoor Copse

***līm-** Brit 'water'. R. Lim (*v.* RNs), Lyme Regis

līm OE 'lime'. *Limost* **3** 307, *lym hous* **4** 360; f.ns. Lime Croft **4** 452, Lymeste **4** 132

limekiln eModE 'lime kiln'. Limekiln Beacon, ~ Bldgs, ~ Cave, ~ Coppice (× 3), ~ Copse, ~ Ctgs (× 2), ~ Fm, ~ Hill, ~ Plant., (Old) Limekiln(s) (freq); f.ns. Lyme Keel Cl 2 45, *lyme kyll* **2** 95

limepit eModE 'lime pit'. Lime Pit Coppice. Also in f.ns, e.g. Lime Pits **4** 132, 166, *la lymepytt* **1** 52, *Madelymepittehill* **3** 80

lime-tree ModE. Limetree Ho

līn OE 'flax'. ?Limbury Fm, ?Linen Hill Fm, ?Linnings Copse, Linton Barn & Hill, ?Lymburgh's Fm; f.ns. *Lincumbe* **4** 14, ?Linhays **3** 56, Linings **4** 382, *Linlega* **4** 374. In AS charters, *lindune* **2** 151, *linleaga mor* **2** 162, *lyngærstun* **3** 380

linch, **linchet**, *v.* **hlinc**

lind, **linde** OE 'lime-tree'. ?Limbury Fm, ?Lymburgh's Fm; f.n. *Lyndhull* **2** 216. In OE charters, *gemearcodan lindan* **2** 162

linden OE 'growing with lime-trees', in AS charters, ?*limbenlee* **2** 58

linhay Do dial. 'lean-to shed'. The Bushes Linhay, The Cliff Linhay, Old Park Linhay, Sheep Linhay; f.ns. *Bove smithelyngheye* **4** 181, Linhay **4** 116, 424

***lisc** OE 'reed'. Lyscombe Bottom; f.ns. Liscomb **1** 362, Liswell **2** 19. In AS charters, *liscbroc*, *lisebroke* **3** 203 (*v.* RNs)

***litano-** Brit 'broad'. R. Lodden (*v.* RNs), R. Lydden (ib)

lið-wyrt OE 'dwarf elder', f.n. *Lithewortforlang* **3** 200

living ModE 'a property'. Croomes Living, Martin's Living; f.ns. Higher Living **3** 339, *West Living* **2** 205

***lobb** OE 'something heavy, steep slope'. Lob Gate, Lophill Fm; f.ns. Little Lob **4** 132, *Lobbe Thorne* **1** 375, Lobbs **4** 337, Lob Plot **4** 382

loc OE 'lock, fold', **loca** OE 'enclosure', f.ns. *Kingeslocke* **1** 229, *Lockeforlange* **2** 53, *Locklondes* **4** 142, Loke Acre **3** 252, *Loklond* **1** 313

***loch** Corn 'pool, inlet'. ?Broad Looe, ?Loders, ?Shag Looe, ?Wood Bar Looe

locust ModE. Locust Clump

lode, *v.* **(ge)lād**

***lo-dre** Brittonic 'habitation site, homestead'. ?Loders

lofte ME 'loft', f.n. *Le Ottlofte* **3** 150

log(g)e ME 'lodge'. Bussey Stool Lodge, *Cobley Lodge* **2** 236, Cowper's Lodge, East Lodge, Gamekeeper's Lodge, Holt Lodge Fm, Iwerne Lodge, Keeper's Lodge (freq), Little Lodge, Lodge Copse, Lodge Down, Lodge Fm (× 3), Lodge Hill, Lodge House Fm, Lodge Wood. Freq in modern minor names and house names as ~ Lodge(s), e.g. Audlem Lodge, Beckford Lodges, Came Lodge, Creech Barrow, Denby Lodge, Elwell Lodge, Lewell Lodge, (The) Lodge (freq), Lodge Plant., Lodge Wood, Morn Lodge, North Lodge (freq), North Lodges, Osmington Lodge, Park Lodge, Poxwell Lodge, Rendezvous Lodge, South Lodge (freq), Syward Lodge, Wareham Lodge (× 2), Warmwell New Lodge, West Holme Lodge, West Lodge (freq); f.n. *Le Lodge* **2** 216

lome ME 'implement, machine, loom'. ?Loomcroft

lookout ModE 'place to observe from'. Ladies' Lookout, Look Out, (The) Lookout (freq), Lookout & Rocket Post

lopped ME 'cut', f.n. *Lopethorn* **3** 294

***lort(e)** OE 'dirt'. Lodbourn, Lodburn, *Lorteborn* (*v.* RNs), Lorton Ctgs/Fm/Ho, Ludbourne Rd **3** 361; f.ns. *Lorteborn* **1** 184, *Lortemerforlange* **2** 72

lousi ME 'infested with lice', f.ns. Lousy Cmn **4** 40, Lousy Grd **4** 435, Lowsy Bank **2** 45

lover ModE. Lover's Grove

lowe ME 'low-lying', f.ns. *Lowelande* **2** 174, *Lowewode* **3** 111

lower ME 'lower down'. Kingston Lower Dairy, Lower Barton, Lower Common, Lower Covert, Lower Down, Lower Fm (freq), Lower Grounds, Lower Hamworthy, Lower Lodge, *Lower Okeford*, Lower Plant., Lower Rookery, Lower Street (× 3), Lower Waterston, Lower Wood, Lower Wraxall

lowest ME 'most low-lying', f.n. ?*Loweste* **1** 313

***lūce** OE '?a kind of enclosure'. ?Look(e) Fm

lucerne ModE 'purple medick, a fodder crop', f.ns. Lucerne **2** 200, ?Luesome Clover Ld **3** 109

***luf-** OE 'river valley, low wet place'. ?*Lovard*

lufu OE 'love'. ?*Lovard*, Love Lane (freq); f.n. ?*louecotesforlong* **1** 70, Love Hill **4** 132, *Loveland* **3** 188

lug ModE dial. 'a square pole or perch'. ?Summerlug Hill; f.ns. Hundred Lug **2** 157, Lugbarrow **4** 391, Lugland **4** 424

lumber ME 'timber' or 'odds and ends'. Lumber Lane; f.ns. Lumber Leaze **4** 399, 463

lūs OE 'a louse', perhaps sometimes describing something small or insignificant. Loosebarrow Hundred, Louse Lane, ?Lowsome Lane, Lush Barrows; f.ns. Lousebarrow **2** 19, *le Louseslappe* **2** 289, Lucehills **2** 105, *Lusarne* **1** 121 [*cf.* **lousi**]

***lutā** Brit 'mud'. ?Lodmoor

***lycce** OE 'lock, enclosure', f.n. ?Lytch Croft **4** 499, in AS charters, ?*lyc hagan* **4** 304

lyn(n)et ME 'flax, lint'. ?Lintern Lane

lȳtel OE 'little'. Frome St Quintin, Hinton Parva, Little Ball, Little Bredy, Little Canford, *Little Cheselbourne*, Little Cliff, Littlecombe Bottom, Little Common, ~ Coppice (freq), ~ Copse, ~ Court, ~ Covert, *Little Crawford* **2** 181, *Little Crichel*, Little Down (freq), Little Francis, Little Hanford, Little Kimmeridge, Little Kington Fm, Little Knoll Copse, Little Lane (× 2), Little London, Little Marsh, Little Mayne Fm, Little Mill, Littlemoor, *Little Moors*, Little Moors Fm, Little Nutford, Little Park, Little Puddle Fm, Littlestone Coppice, Littleton, Little Waddon, Littlewindsor, Littlewood, Little Wood (freq), *Lytelpryourreswode* **4** 225, *Lytilparrok* **4** 404; f.n. *Lytellonde* **2** 53. In AS charters, *litlan graf* **4** 374, *litlen ac lee* **2** 266, *litlen lake* **3** 102, *littlen wde* **2** 291, *lutlen seaxpennes suðeke* **3** 104

mackerel ModE. ?Mackerel Rock

mad ModE 'foolish'. ?Madgrove; f.n. Madmans Pce **2** 288

madam ModE, f.n. Madams md **1** 119 (*v.* NDEFN 264)

made ME (p.p. of *make*) 'produced', f.n. *Madelymepittehill* **3** 80

mǣd OE 'meadow'. Alder Mead, Allington Mead, Alton Mead Lane, Bolter's Mead, Boynmead Fm, Broadmead, ~ Ctgs, ~ Lane, *Bucke mead* **2** 257, Charmouth Mead Plant., Chaw Meadow, Common Mead, *Earl's Mead*, East Meade Ctgs, Eye Mead, Fair Meadow, Fillymead, Fishy Mead Copse, *Fordemede* **4** 275, Frome Mead, Grange Mead, Hammond's Mead, High Mead Lane, Hine's Mead Lane, *Hollish*, *Holowroudemede* **3** 154, Homemeadow Coppice, Honeymead Lane, *King's Mead*, Lady Mead, Long Mead Coppice

(× 2), Longmeadow Coppices, Longmead Plant., ?Madgrove, Mead Coppice, Mead Lane (× 2), The Meadows, ?Metford Mill, *le Milmede* **4** 277, *Mylmede* **4** 379, Mythe Mead Willow Bed, Netherwood Mead, New Mead Copse, New Meadow Coppice, North Mead Copse, Nurserymead Coppice, Palemead Coppice, Park Mead Bottom, Pond Mead Stall, Red Mead Coppice, Rushmead Fm, Splitmead Coppice, Tucker's Meadow Copse, Westmead Ho, White Mead, Whitty Mead. Freq in f.ns., e.g. *Clyvemede* **4** 411, Inn Md **2** 19, Meadow Marsh **2** 66, *Rownde Meadowe* **3** 34. In AS charters, *babbanmede* **3** 254, *beorreding mæd* **3** 286, *breþling made* **3** 178, *deuelisc made* (*v.* RNs, Addenda **1** 304), *mæd* (*æcras*)...*on ige* **2** 136, *mæde hyrnan* **4** 35, *med ham* **2** 122, *sucgimade hauede*, *sugging made* **3** 136, 138, *west mæd* **3** 289

mægden OE 'maiden'. †Maiden Castle, Maidenford Weir, Maiden Newton, Maiden Plant., Maiden's Grave Gate, *Maidshole* **3** 174, *Maydenecroft* **3** 174; f.ns. Maiden Land **4** 435, *Maidenwelleforlong* **1** 70, *Maydenbrigge* **4** 11 [*cf.* **maiden**]

mǣle OE 'variegated, multicoloured'. Melbury Abbas, ~ Bubb, ~ Osmond, ~ Sampford, Melplash. In AS charters, *mealeburg* **3** 116

(ge)mǣne OE 'held in common, communal'. Main Down, Mancroft, ?Mandeville, ?Mannington (*v.* Addenda **2** 151), Manor Coppice & Hill, ?Maumbury, *Le Menehyde* **1** 38, ?*Milborne Mamford*; f.ns. Mandling **1** 223, Manwood **4** 268, *Menelande* **4** 411, *Menewode* **3** 257

(ge)mænnes OE 'community'. ?Chapman's Pool

(ge)mǣre OE 'boundary'. Five Marys, ?Gatemerston, Haremere Wood, ?Wood Mire Copse. In AS charters, *adwines imare* **3** 163, *bissopes imare* **3** 136, *hina gemære* **2** 146, *hina imares* **3** 136, *hore tuning gemære* **2** 159, *imeren hole weg* **3** 102, *kinghes imare* **3** 136, *leouen imare* **3** 136, *meairweie* **2** 82, *mearhhilde mere* **4** 173, *merehale* **3** 181, 200, *merehawe* **3** 200, *mere sled* **4** 173, *merewege* **3** 129, *merhawin* **3** 257, *Meleburge imare* **3** 129, *Wlgares imaren* **3** 136, *Wulfrices gemære* **2** 161

mǣw OE 'seagull'. ?Mupe Bay, Ledge & Rocks

mag, **madge** ModE dial. 'magpie', f.ns. Madge Acre **4** 282, ?*mogacre* **3** 257

magdalen eModE 'home for the refuge of paupers or prostitutes'. Magdalen Almshouses **1** 355, *the Maudelyn* **3** 150

magging ModE 'chattering, talking'. Magging Stone

maggot ModE dial. 'magpie'. Maggot Clump; f.ns. ?*Magattehayes* **2** 70, ?Maggot Md **1** 324

magotte ME 'worm or grub', f.ns. ?*Magattehayes* **2** 70, ?Maggot Md **1** 324, Maggotty Plot **4** 486

magna Lat 'great'. Broadwindsor, Fontmell Magna, Kimmeridge, Kington Magna

magnolia ModE. Magnolia Fm

maiden ModE 'tree grown from seed, unlopped tree', f.n. ?Marsh maidens **2** 105

main ModE 'chief in size'. Main Channel, ?Main Covert, Main Cross

main[1] ME 'demesne land', f.ns. *Mayne pale* **4** 284, ?*Mayngore* **2** 114, ?*Mayne Place* **2** 99

***main**[2] PrWelsh 'a rock, a stone'. Broadmayne

māl lOE 'law-suit, rent'. ?Mallams **1** 217

malemit OFr meaning uncertain, a prison (1280) at Corfe C. (*v.* discussion **1** 11)

malkin ME 'wench', f.n. ?Makenhays **3** 272

mallard ModE, f.ns. Mallards Close **2** 62, Mallards Green **1** 378

malm OE (WSax **mealm**) 'sandy or chalky soil, soft stone'. Mampitts Fm, ?Maumbury; f.ns. Malm Grd **3** 28, 186, Malm Pit(ts) **1** 151, Mampitts Lane **3** 146

malthouse ModE. Malthouse, Malt House, Malthouse Ctgs (× 2), ~ Fm, ~ Lane

***malu** OE 'gravel ridge'. ?Mallams **1** 217; f.n. ?*Maldich* **1** 365

mamelles Fr 'breasts'. ?The Mumbles

***mamm** PrWelsh 'breast, breast-like hill'. ?East & West Man

manere ME 'manor'. Branksome Manor, Catherston Manor Fm, Charlton Manor, Cheselbourne Manor, Hallings or Hallam Manor, Hilfield Manor, Manor Ctg, ~ Fm (freq), ~ Ho (freq), ~ Park, Manor Yard, Moorcourt Fm, Pilsdown Manor, Radipole Manor, Rew Manor, South Lytchett Manor, Steepleton Manor, Wolfeton Manor; f.n. *Beauchampesmaner* **2** 51

mange ModE 'a cutaneous disease of animals caused by a parasite'. ?Manswood; f.n. Manswood Coppice **2** 237

mangere OE 'trader, merchant'. Mangerton

mann OE 'man'. Dead Man, Deadman's Coppice, Deadman's Covert & Stile, ?East & West Man, *Westmyngton*; f.n. *Southmanworth* **3** 274

manse ModE. The Manse

mansion ModE. Mansion Hill, Mansion House (× 2)

mapel ME 'maple-tree', f.ns. *Mapeles* **4** 284, *Mapilfurlonge* **1** 327

mapuldor OE 'maple-tree'. Mapperton (× 2), Mappledore Hill, Mappowder, Matterley Ctgs; f.n. Maplecrate **4** 202. In AS charters,

Mapelder **3** 163, *Mapoldorlea* **2** 286, *mapulder* **3** 289, *mycclan mapeldran* **4** 454
māra OE 'bigger', f.n. *Moresuth'lee* **3** 198
marche OFr, ME 'boundary', *La Marche* **1** 71
mare Lat 'the sea'. Longfleet
mareis OFr, ME 'marsh'. Caundle Marsh, *Le Marays* **3** 164
maresc(h)al OFr, ME 'one who tends horses'. ?Marshalsea
market ONFr, **merket** ME 'market'. Castleton Markett **3** 301, *le Cornemarkett* **3** 364, *la market* **2** 185, Market Place, Market St. **1** 253, **3** 364
marle OFr, ME 'marl'. ?Marland Pit, Marl Close, Marl Lane, The Marl Pit, Marlpit Cross, Marlpits Fm, Marl Pits Wood, Old Marl Plant.; f.ns. (*La*) *Marl*(*e*) **1** 313, **2** 72, *Marle Close* **4** 11, Marl Leaze **3** 28, *þe Schortmarle* **2** 81
marlede ME 'spread with marl', f.n. (*la*) *Marledelond*(*e*) **1** 327, **3** 129
marliere ME 'marl pit', f.n. *le Marler* **2** 90
marling ME 'marl pit'. ?Marland Pit, The Marl Pit, Marl Pits Wood; f.ns. Marling Pits Fd **2** 214, *Marlyngputte* **2** 132, *Marlyng Pyttes* **3** 237
marrow, *v.* **mearh**
marrow-bone ModE. Marrowbone Barn; f.ns. Marabone Park **4** 486, Marrow Bone(s) **4** 329, 491
mase ME 'maze', f.n. *le Mase* **1** 128
mash ModE 'malt mixed with hot water to form wort'. ?Mash Water (*v.* RNs)
maslin ModE 'mixed grain', ?Muslins **3** 127
matte ME 'mat', f.n. ?*Matcomb* **4** 344
***māwett** OE 'mowing', f.n. ?*Mowetecombe* **1** 258
may ModE 'May', f.n. May Md **4** 452
maypole ModE, f.n. *le Maypole* **1** 258
mearc OE 'boundary mark'. Chilmark, Mackey's Copse, ?Markham Ho, Sea Mark Clump. Also in f.ns., e.g. *Merkewey* **1** 31. In AS charters, *mearc broces heafde* **2** 136, *mearchagan* **2** 162, *richte merke* **1** 238
gemearcod OE 'marked', in AS charters, *gemearcodan lindan* **2** 162
mearh, **mearg** OE 'marrow', fig. 'fertile ground'. ?Marley's Row; f.n. Marrow Md **3** 314. In AS charters, *mearhhilde mere* **4** 173
***m(e)arn** OE 'soft stone'. ?Marnhull
mearð OE 'marten, weasel', f.n. *Martleg'* **3** 33
meeting-house ModE, f.n. Meeting House Fd **3** 236

***mę̄l** PrWelsh 'bare hill'. Fontmell Brook (*v*. RNs) giving name to Fontmell Magna & Parva

melancholy ModE. Melancholy Lane

***melce, milce** OE 'milch, yielding milk'. ?Melcombe Regis; f.n. Milch Plot **4** 348

melde OE 'orach', f.n. *Milde Close* **4** 465

menagerie ModE. The Menagerie

meoluc, milc OE 'milk'. Melcombe Horsey, Melcombe Regis; f.ns. *Melk*(*e*)*well* **2** 242, **3** 81, **4** 19, *Milkcrofte* **3** 33, Milk Wells **2** 98

mēos OE 'moss, lichen' or 'bog, swamp', f.n. ?Moose **1** 105, in AS charters, *mes delle* **3** 117

***mēosig** OE, **mesy** ME 'mossy', f.n. *Meisimor* **4** 133

mere[1] 'a pool, a pond'. Almer, Ashmore, Burton Mere, Cashmoor, Colmer Copse, ~ Fm, ?Dymor, Enmore Green, *Fordmere* **3** 146, Frogmore Fm, ?Frogmore Fm & Hill, ?Gatemerston, Holton Mere, *Jolyfmer* **1** 118, Latchmore Pond, Mear's Pond, Meerhay (Fm), The Mere, *le Mere* **1** 118, Merley Pond, Penmore Rd, Ringmoor, Whitmore Coppice, *widemere* **2** 212, ?Wood Mire Copse; f.ns. *Chalmere* **3** 226, *Frankemere* **2** 174, *Fysschemere* **3** 256, *Meracr'* **4** 11. In AS charters, *coccelmæres dene* **4** 302, *cranemere* **3** 41, *cranmere* **3** 128, *elmere* **2** 55, *lacmere*, *late mere* **3** 41, 128, *mere* **2** 162, **3** 377, *riscemeres* **2** 67

mere[2] 'a mare'. Mare Pond, Marriage Lane, ?Meriden Wood

merkin eModE 'female pudendum', f.n. *Merkynhill* **1** 184

mermaid ModE. Mermaid Pond; f.ns. Mermaid **4** 424, Mermaid Orchd **4** 330

merry Do dial 'wild cherry'. ?Merry Hill; f.n. Merry Fd **4** 391, *Merrylands* **4** 235, Merry Md **2** 237 [*cf.* **myrge**]

mersc OE 'marsh'. Bere Marsh Fm, Bincombe Marsh Dairy, Blackmarsh Fm, Bodsmarsh Lane, Caundle Marsh, Champernhayes Marsh, Chapel Marsh, Dickens's Marsh, *Filcumysmersche* **4** 376, Font Le Roi, Gibbs Marsh Fm, Guy's Marsh, Hawkmoor Marsh, Little Marsh, Margaret Marsh, The Marsh (× 2), Marsh Barn (× 2), Marsh Bridge (× 2), Marsh Common, *Marshcoombe* **4** 246, Marsh Copse (× 4), Marsh Court, Marsh Fm (freq), ~ Lane (× 3), ~ Wood, Marshwood, ?Mash Water, Middlemarsh, Peacemarsh, *Peals Marsh*, Salterne, Wyke Marsh, Wytherstone Marsh. Freq in f.ns., e.g. *Holemers* **3** 33, Marsh Cl **2** 9, Marsh's Md **3** 28. In AS charters, *merse* **1** 31

merveille OFr 'marvel', f.ns. ?*Merveile* **2** 10, ?*newe Marvyll* **2** 53

mete ME meaning uncertain (*cf.* OE *mete-* 'measure, allowance', ME *mete* 'boundary). *Met(e)lond* **1** 184 (*v.* discussion), ?Metland's Wood; f.n. *Metforlands* **2** 105

micel, **mycel** OE 'big, great'. Kington Magna, Mistleberry Wood, *Mochelpryouryswode* **4** 225. Also in f.ns, e.g. Michelmoor **1** 151, Mitchells Md **2** 172, *Mochecroft* **1** 31, *Mochylmede* **2** 53. In AS charters, *miclan corf* **4** 302, *miclan cruc* **4** 290, *mycclan mapeldran* **4** 454

Michaelmas ModE, f.n. Michaelmas **4** 152

***midd** OE 'middle'. Midney Lane; f.ns. *Middechave* **3** 33, *Mideforlong* **2** 53, Midlakes **1** 69, *Midnam* **3** 6

middel OE 'middle'. ?Metherhills, Middlebere Fm, Middle Bottom, Middle Bridge, Middlebrook Ctg, Middle Bldgs, Middle Channel (*v.* RNs), Middle Coppice/Copse (× 4), Middle Fm (freq), Middle Field Lane, Middle Heath, Middle Hill (× 3), *Middle Hill*, Middle Ho, Middle Lodge (× 2), Middlemarsh, Middle Mill Fm, Middle Plant., Middle St, *Middlestreet*, Middle Wood, Milton Abbas, Milton on Stour, West Milton; f.ns. *Middeldone* **2** 216, *Middelforlang* **1** 260, Middle Fd **2** 214. In AS charters, *midelgete* **2** 82

middelman ME 'maker of iron wire', f.ns. *Middelmannedon, -londe* **3** 227 (*v.* discussion)

middeweard OE 'middle', *on cruc middeweardne* **4** 290 (AS charter)

midding ME 'dung-hill'. *Midenhull* **4** 21; f.ns. Midden Hills **4** 481, *Myddenhull* **3** 227

midsummer ModE, f.n. Midsummer Md **3** 297

mīl OE 'mile'. Fair Mile Rd, Miles Cross, Two Mile Coppice

milc, *v.* **meoluc**

milce, *v.* ***melce**

milchy ModE 'producing good milk', f.n. Milchy Plot **4** 471

***mīl-stān** OE 'mile-stone', f.ns. Milestone Fd **1** 151, **2** 229, Millstone Flg **4** 34, Millstone Pce **2** 148

milk-house ModE, Milkhouse (Md) **4** 22, 249

milking ModE 'act of milking cows'. Milking Barton Coppice; f.ns. Milking Cl **2** 51, Milking Plot **4** 160

milleward ME 'miller, f.n. *Milwardlond* **1** 238

mine eModE 'mine'. *Alum* **1** 51, *Mines heath*, *The Mynes* (both **2** 40)

minnow ModE. Minnows Hole

minte OE 'mint'. Minterne; f.ns. ?Mynty Cl **1** 172, Myntylond **1** 327

mire ME 'a mire, a bog'. Oakhill Mire Pond; f.ns. *Attaremire* **3** 33, *Miremedoo* **3** 273, *la Myre* **1** 272, **2** 250

misery ModE. Mount Misery
mist OE 'mist'. *Mystecombe*
mite ModE. *Napper's Mite* **1** 355
mixen OE 'dung-hill'. *Fowley Mixen* **3** 248, The Mixen, Mixen Plant.; f.ns. *the Misone* **2** 45, *Mixon Acre* **4** 43
mizmaze eModE 'a labyrinth or maze'. Maze, Miz Maze; f.n. Mizmaze **1** 94
moel ModWelsh 'hill', f.n. Cock Moile **4** 399
moitie eModE 'moiety', f.n. *le Moitie* **4** 465
***moke** ME, **mock** ModE dial. 'stump, log', f.n. Mock Pits **4** 116, cf. ?*hlosmoc* **4** 374 (AS charters)
molde OE 'earth, soil', f.n. ?*Southmalde* **4** 235
mole eModE 'pier'. *Molehead* **1** 256
molin OFr 'a mill'. Corfe Mullen
molle ME 'mole', f.n. ?*Molewudecombe* **3** 375
monachus Lat 'monk'. Cerne Abbas
mōnandæg OE 'Monday', f.n. Long Monday **1** 286
monastery ModE. Monastery Fm & Lane
money ME 'money'. ?Money Pit Lane; f.n. Money Plot **4** 152
mont OFr, ME 'a mount, a hill'. Belmont Ho, Boswell's Mount, Fir Mount, Holme Mount, (The) Mount (freq), Mount Ararat, Mount Coppice/Copse, Mount Hunger, Mount Misery, Mount Pleasant (Fm) (freq), Mount Poynings, Mount Radford, Mount Skippet, Mount Sloe, Mount St **2** 42, Mount Sylva, Rydal Mount Poultry Fm, St Michael's Mount. Also in f.ns., e.g. Mount **1** 172, Mount Folly **3** 67, Mount Misery **4** 55, Mount Pleasant **4** 34
monteyne ME 'lofty hill'. ?Mountain Clump
monument ModE. Monument Coppice
moorhen ModE, f.n. Moorhen Hole **1** 242
mooring eModE 'place for a boat to be moored', f.n. *Moringe* **4** 465
mōr OE 'marshy ground'. Alder Moor (× 6), Aldermore (× 5), Alley-Moor, Alleymoor Wood, Bagmore Fm, Baker's Moor, Banger's Moor, Birdsmoor Gate, *Bittlesmore*, Black Moor, Blackmoor Forest & Vale, Blackmore Ctgs, ~ Fm, Blastmoor Hill Barn, Bluntsmoor Bridge & Fm, Bourne Moor, Broadmoor Coppice, ~ Fm, Broomhill Coppice, Burl Moor, Catmoor, Chalmington Moors, Chantmarle Moor, Chickerell Moor, Chilmore, Chitmoor, Church Moor Copse, Cocks' Moor Pond, Coles Moor, Cook's Moor, Cosmore, Cranes Moor, Creekmoor Bridge, Crocker/Crockham Moor, Deadmoor Common, Donkey Moor, Dry Hill Moor, Duckett's Mor, *Dudley Moor*, ?Dymor, *Eastmore*, Egg Moor, Elder Moor, Elmore Fm,

Flower's Moor, Foxmoor Coppice, Foul Moor Lane, *Frogmoor*, Frogmoor, Frogmore Cottages, ?Frogmore Fm & Hill, Frogmore (Dairy) Ho, ?Gatemerston, Gatmore Fm, Goodsmore Coppice, Great Moor, Hankmoor Coppice, Hare Moor, Hartmoor (Fm), Hawkmoor Fm, Haymoor Bottom, *Haymore*, Highmoor, Hollymoor Common Lane, Horse Moor (Coppice), Hurdley Moor Fm, King's Moor Coppice, Lr & Upr Kitemoor, *Kittemor* **4** 278, Langmoor Coppice & Manor, ~ Fm, ~ Lane, Latmore Fm, Laymoor Copse & Lane, Laymore, Lilymoor Copse, Littlemoor, *Little Moors*, Little Moors Fm, Lodmoor, Longmoor Fm, ~ Plant., Lotmoor Hill, Luxmore Hill Plant., Mark's Moor, ?Meriden Wood, Millmoor, The Moor (× 2), Moorbath, Moor Coppice (× 2), Moor Court Fm, ?Moore's Fm, Moorfield Ho, Moor Ho, Moorland, Moorland Lodge, Moor Lane (× 3), Moormill Withy Bed, Moor Plant., The Moors (× 2), Moorside, Moortown Fm, Moorway Lane, Morden, More Crichel, Moreton, Moryatt Fm/Marriotts Fm, Nettlemore Fm, Northmoor Fm (× 2), *Nyaysshe More* **4** 407, Pains Moor Copse, Parlour Moor, Peat(y) Moor, Pennymoor Pit Fm, Pigs Moor Coppice (× 2), Pillmoor Bottom, Pinnocks Moor, *Portmanmore*, Portymoor Coppice, Prestwood Moor, Pymore, Ramsmoor, Redmoor Fm, Rimsmoor Pond, Rodmore Coppice, ~ Fm, Rossmore, Roughmoor, Rushmoor pond, Rushmore Fm, Rushmore Hundred, Sackmore Lane, *Sacmore* **1** 169, Sandbury Moor, *Seckeymoor*, Shipmoor Point, Shortmoor, Short Moor, Smedmore Ho, Snipe Moor, Southmoor, Springmoor Coppice, Stanley's Moor, Starmoor Plant., Stroud's Moor Withy Bed, Swaine's Moor, Sweed Moor, Tidmoor Point, West Moors, White Moor, Whithay Moor, Whitmore, Whitmore Bottom, Whitmore Coppice & Plant., Wigmore Coppices, Winham's Moor, Woodmoor Copse. Freq in f.ns., e.g. *Brademore* **4** 10, *Cossemoore* **3** 294, *Craynsemore* **3** 32, *Dychemore* **4** 10, Long Moorelay **4** 40. In AS charters, *bytelesmor* **3** 153, *cinerget mor*, *cuterget mores* (*v*. RNs.), *cran mor* **2** 158, *ears mores heaued* 3 128, *linleaga mor* **2** 162, *lodomor* **1** 232, *mor* **1** 31, *morhelle* **3** 207, *rupemor* **1** 32, *sand mor* **3** 198

morass ModE 'boggy ground', f.ns. (the) Morass **1** 337, **3** 373

morgen OE 'morning'. Morning Well; f.ns. *Morrow Leaze* **3** 33, *le Morwelese* **3** 81, *Mourwelese* **3** 273

morgen-gifu OE 'morning gift', f.n. *Mo*(*u*)*rweyefehey* **3** 273

mōrig OE 'marshy', f.ns. *Moorye Close* **2** 201, Morey Md **4** 190, *Mowry Close* **1** 203

morrow-mass ME 'the first mass of the day', f.n. *the Morrowe Masse Landes* **3** 33

***mort** OE 'young salmon or similar fish'. ?Morcombe Fm, ?Morcombelake, ?Mosterton; f.n. *Martlake* **4** 223 [*cf.* ***murt**]. For the possibility that ***mort** may have had a topographical sense 'stump, hill', *v.* Addenda **4** 308.

morte eModE 'harlot', f.n. ?*Mortcleaus* **1** 271

morter ME 'mortar used for building', f.ns. Mortar Holes **2** 200, Morterhay **4** 363

(ge)mōt OE 'meeting'. ?Moathills Coppice, *Modbury*, Modbury (Fm), Modbury Hd, Motcombe; f.ns. Moot Hill Pce **3** 5, *Motborow* **1** 327, *Motcomb* **4** 304, *Motweie* **4** 284. In AS charters, *motbeorh* **4** 35

mote ME 'moat, ditch'. *Le Mote park* **2** 193; f.ns. Moat **4** 268, Mote Md **3** 40

mound ModE, f.n. Mound Knowle **1** 227

mow Do dial. 'rick' (**mūga** OE), f.ns. Moo Crate **4** 25, Mowbarton **4** 22, Mowcrate **4** 471, Mow Md **3** 136, Mow Plot **2** 51

mowere ME 'one who mows'. ?Mower Lane; f.n. Mower Md **1** 29

mudde, **mode** ME 'mud', *the Mudlands* **2** 44, Mudoak Wood; f.n. *Modewell* **4** 11

***muddig** OE 'muddy'. Muddicombe Coppice, Muddyford Lane

***mukel** ME 'dirty, muddy place' (a derivative of ME **muk** 'dung, muck, dirt', *cf.* ModE S dial. **muckle** '(a heap of) manure'). ?Muckleford; f.ns. Muckle Close **2** 200, Muckle Croft **4** 330, Muckles **2** 172, ?*Muggle paroc* **3** 111

mūl OE 'a mule'. ?Moulham; f.n. ?*Mowlesmede* **3** 237

mulberry ModE 'mulberry'. Mulberry Gardens; f.n. Mulberry Md **1** 29

munuc OE 'monk'. Cerne Abbas, Friar Waddon, Moncton Bushes, Monks' Pond, Monk's Wood, Monkton Up Wimborne, Monkton Wyld, Monkwood Hill Fm, *Myleburn Munketon* **1** 282, Tarrant Monkton, Winterborne Monkton, Hr & Lr Monkwood; f.ns. *Monketon* **3** 227, Monks How **2** 229, *Monkynwell* **4** 374

murrain ModE 'disease of cattle', f.n. ?Murren Plot **1** 142

***murt** OE 'small fish', in AS charters, ?*murter* [?*murtes*] *wyll* **4** 304 [*cf.* ***mort**]

mūs OE 'mouse'. Musberry Lane; f.ns. *Mousehill* **3** 375, *Musebergh* **2** 250

music ModE, 'the Music Barrow' **1** 259 Addenda

must ME 'must, new wine', *le Musthous* **4** 3, *la Mustmull* **4** 277

mustard ModE, f.n. Mustard Pound **1** 203

mutton ModE. ?Mutton Cove

mūða OE 'mouth'. Charmouth, Dunch Mouth, Eype Mouth, Green Island, Mouth Rocks, St Gabriel's Mouth, Small Mouth, Weymouth; f.ns. *Chidiokes Mowthe* **4** 384, *Dichemouth* **4** 10

myln OE 'mill'. *the Abbey Mill* **3** 367, Abbotsbury Mill, Affpuddle Mill, Alderholt Mill, Alton Mill, Arfleet Mill, Arish Mell, Baker(')s Mill, *Barnabesmull* **3** 11, Beaminster Mill, Bedmill, Bere Marsh Mill, Bidcombe Mill, Binnegar Mill, Bradford Abbas Mill, Brimley Mill, Buckham Mill, Burton Mill, Bushell Mill Fm, Cann Mill, Cards Mill Fm, Castle Mill Fm, Cattistock Mill, Chamberlayne's Mill, Champernhayes Mill, Cheselbourne Mill, Chetnole Mill, ClenhamMill, Cobb Mill Ctgs, Corn Mill (freq), Corscombe Mill, Creekmoor Mill, Crichel Mill, Cut Mill, Dewlish Mill, Didlington Mill, *Drakes Mill*, Duntish Mill, Durweston Mill, *Eccliff Mill* **3** 12, Fiddleford Mill, Fifehead Mill (× 2), Flax Mill, Folly Mill, French Mill, Gear's Mill, Goathill Mill, Halstock Mill, Haselbury Mill, Hawkhill Mill, Heifer Mill Ctgs, Hembury Mill, Heniford Mill, High Bridge Mill, Higher Mill, *Higher Mill*, Hinton Mill, Holwell Mill, Hurst Mill, Keynston Mill, King's Mill, *Langgedonesmuill* **4** 243, Lewell Mill, Little Mill, Litton Mill, Loders Mill, Lodge Mill, Loud's Mill, Lower Mill, *Meath Mill* **4** 306, Melplash, Middle Mill Fm, Milborne St Andrew, Milborne Stileham, *Milbridge* **4** 277, Milburn Lane, Mildron Copse, *Milford Ctg*, Mill (freq), Mill Bog, Millbrook Ho, Mill Bush, Mill Coppice, Mill Copse, Mill Ctgs (× 3), Mill Dam, Mill Down, Mill Fm (× 2), Millhayes, Mill Hill, Mill Hill Close, Mill Ho (× 2), Mill Lane (freq), Millmore, Mill Pond(s) (freq), Mill Race (freq), Mills (Fm), Mill Stream (freq, *v.* RNs), Millum Head, Millwater Fm, Millway, Millway's Lane, *le Milmede* **4** 277, Milom Lane, ?Milton **2** 17, Milton Mill (× 2), Morden Mill, Mosterton Mill, *Mulnehale* **4** 277, *la Mustmull* **4** 277, *the myle ryme close* **4** 277, Nether Cerne Mill, New Mills, Norris Mill, North Mills (× 2), Nottington Mill, Okeford Mill, (The) Old Mill, Osmington Mills, *Oushullesmulle* **3** 269, Over Compton Mill, Overmoigne (sic) Mill, Oyle's Mill, Peggs Mill Bridge, Philleyholme Mill, *Picket Mill* (× 2) **4** 244, **4** 334, Piper's Mill, Powerstock Mill, *Prutelesmille* **2** 124, Puddle Mill, Puncknowle Mill, Purns Mills, *Pyd*(*d*)*emylle* **3** 368, Pymore Mills, Lr Radipole Mill, Rolls Mill, Romford Mill, Rowden Mill Fm, Rushton Mill, Sadborow Mill, Sandford Corn Mill, Shermel Gate, Slape Mill, South Mill, *Sprag*(*g*)*s Mill*, Stanbridge Mill, Stander's Mill Plant., Steepleton Mill, Stockham Mill, Stoke Mill (× 3), Stour Provost Mill, Stratton Mill, Sutton Mill,

Tadnoll Mill, Toller Mill, Tolpuddle Mill, Town Mills, Trent Mill, *Tucking Mill*, Tucking Mill, Tuck Mill, Upper Mill, Wareham Mill, Warmwell Mill, Waterloo Mill, *la Weremull* **3** 368, Westbury Silk Mills, Westford Mill, West Mill(s) (× 5), Whatley Mill, Whetham Mill Cross, *Whitecliff Mill*, White Mill, Witchampton Mill, Withyhook Mill, Woodbridge Mill, *Woodmill* **4** 405, Yetminster Mill. Freq in f.ns., e.g. *Hodyesmull* **3** 33, Mill Droves **4** 19, Mill Ham(s) **2** 9, 139, *Releishe Mill* **3** 34. In AS charters, *eadelmes melne* **3** 138, *melenbroc* **3** 236, *miliere* **3** 207, *molenhame* **2** 58, *muledich* **1** 272, *myle broce* **4** 32, *mylenburnan* **3** 298, *mylenburnna* **3** 380, *mylen stede* **3** 117

myncen OE 'nun'. Minchington; f.n. *Mynchinweyslow* **4** 183

mynet OE 'coin'. ?Minterne

mynster OE '(the church of) a monastery, a (large) church'. Beaminster, Charminster, Iwerne Minster, Lytchett Minster, St Peter's Abbey, Sturminster Marshall, Sturminster Newton, Wimborne Minster, Yetminster

myrge OE 'pleasant, merry'. Marley Bottom/Pond/Wood, Merryfield (× 2), Merry Field Hill, ?Merry Hill. Also in f.ns., e.g. *Merghedene* **2** 216, Merry Fd **1** 29, **1** 80, **3** 40 [*cf.* **merry**]

myrtle ModE. Myrtle Ctg (× 2); f.n. Myrtle Grove **4** 391

(ge)mȳðe OE 'a confluence of streams'. Mythe Cottages; f.n. Mythe **3** 379. In AS charters, *ge middu'* (*v.* Addenda **3** 267)

nacod OE 'naked, bare'. Naked Cross

nǣddre OE 'adder'. Adder's Copse; f.ns. Adder Hill **4** 470, *Addreputte* **2** 81, *Nadderham* **1** 184

næss OE 'promontory, headland'. Cotness (Wd)

nattok ME '?marsh plant', f.n. Noddick **3** 213

navvy ModE 'labourer'. Navvy Shovel

nearu OE 'narrow'; Narrow Bridge Coppice, Narrow Plant.; f.n. Melcombe Narrows **1** 257, The Narrows **1** 54 [here sb.]. In AS charters, *nearn þan anstigan* **2** 82

nēat OE 'cattle'. ?Netmead Lane, Notton; f.n. Neat Md **3** 40

neathowse ME 'cattle shed', f.n. Nethouse Md **1** 167

nebb ME 'point', f.n. *Nebb therne* **3** 294

nēd OE 'need', in AS charters, ?*neodbournen* (*v.* RNs)

needle ModE, f.n. Needle **3** 70 (*v.* NDEFN 294)

neoðerra, **niðerra** OE 'lower'. Bingham's Melcombe, Lower Kingcombe, Lower Sturthill, *Neather Yepe* **4** 427, Netherbury, *Netherbury*, Nether Cerne, Nether Compton, Nether Coombe, *Netherdownes* **4** 220, Nether Grove, Netherhay, *Netherlym* **4** 392,

Netherstoke, Netherton, Netherton Fm, Netherwood Mead, *Nether Wothe* **4** 322; f.ns. *the Netherlynche* **3** 227, Netherwood Mdw **2** 9, *le Nethircrofte* **2** 174. In AS charters, *neþan pol* **3** 207, *niþerstoce* (*v*. Addenda **4** 193)

nest OE 'nest'. Kite's Nest Lane, The Nest; f.ns. *Crowenest* **4** 10, Duck's Nest **3** 251, *le Swannesneste* **1** 329

***nesu** OE 'nose, headland'. Nath Point

netel(e) OE 'nettle'. Nettlebed, Nettlecombe (× 3), Nettlemore Fm; f.ns. *Netelbedde* **3** 273, *neteldene* **2** 255, Nettlecombe **3** 253

nethermost ModE 'lowest', f.n. *Nethemustetwyste* **3** 238

new-found ME 'newly discovered', f.n. *la Nywefoundelond* **1** 314

newman ME 'newcomer', f.n. *Newmanfeld* **2** 140

nightingale ModE. Nightingale Copse, Nightingale Lane **2** 42

nigon OE 'nine'. Nine Acre Coppice/Copse, ~ ~ Pit, ~ ~ Plant., ~ ~ Withy Bed, Nine Barrow Down, Nine Bottle Ctgs, Nine Cornered Copse, Nine Corners Close, †Nine Stones, Nine Yews; f.n. The 9 Acres **2** 51

ninepins ModE 'a game', f.n. Nine Pins **4** 471

nineteen ModE. Nineteen Acre Hanging

nithe, **nythe** ME 'envy, malice', f.ns. ?*Nithewode*, *Nytheclose* **3** 99

nīwe OE 'new'. Maiden Newton, New Barn (freq), New Barn Plant., New Barton, *New Bridge*, New Bldgs (freq), New Bridge, Newbury Ho, New Close (Coppice, ~ Fm), New Coppice (freq), New Covert (× 2), Newcross Ctgs, New Cross Gate, Newell **3** 362, New Entrance, New Field Coppice & Lane, Newfield Fm, New Gate (Ctgs), New Gorse, *New Ground*, New Hall, New House (× 2), Newhouse Fm (× 4), New House Fm (× 4), New Inn (freq), Newland (× 2), Newlands (× 2), Newlands Ctgs, ~ Fm, Newland's Fm, Newlands Gate & Plant., New Lane (freq), New Leaze Fm, New Lodge (× 2), New Mead Copse, New Meadow Coppice, New Mills, New Moreton, *Newnham*, Newnham Fm, New Plant. (freq), New Planting, Newport, New Rd (freq), New Street Ctgs & Lane (× 2), Newton (× 2), Newton Bay, Newton Fm (× 3), Newton (Fm & Manor), *Newton Montacute*, Newton Peveril, New Town (freq), Newtown (× 3), Newtown Fm, Newtown (Hill), New Withy Bed, Sturminster Newton, Warmwell New Lodge. Freq in f.ns., e.g. *la Newecroft* **2** 216, Newlands Md **4** 410, Newnham's **1** 127, *la Nywhegge* **3** 227

nok ME 'nook of land', f.n. the Nook **4** 34

noman ME, **nān mann** OE 'no man'. No Man's Land; f.ns. *Nomandesplott* **4** 254, *Nonemannes lond* **2** 214, **3** 232

norð OE 'north'. Norden Fm, ~ Hill, ?Nordon, Nordon Hill, Nordons Estate, Northay Fm, North Barn (freq), North Bowood, North Bridge, Northbrook Fm & Ho, *Northbrook or Norbrook Fm*, *Northbury* **4** 123, North Causeway, North Channel, ~ Chideock, ~ Coppice, Northcote, North Dairy Ho, North Dibberford, North Down, *North Down*, Northdown Barn & Fm, North Eggardon Fm, North End (Dairy Ho, ~ Fm), Northernhay, North Ewe Leaze, Northfield Fm, North Fm (freq), Northfield Barn, ~ Fm, ~ Ho, ~ Plant., North Field Hill, Northgate, Northground Dairy, North Haven Point, Northay, North Hayes Fm, North Heath, North Hill (× 6), North Hill(s) Plant., North Holt Fm, North Lane, North Lodge (freq), North Lodges, Northmead Copse, North Mills, Northmoor Fm (× 2), *Northover* **4** 91, Northover Ho, *la Northpark* **4** 225, North Plain Plant., North Plant. (× 2), North Poorton, North Slip, North Wood (× 2), Northwood (Coppice & Fm), North Wootton, Norton Coppice, ~ Ctgs, ~ Hill, ~ Lane (× 2), ~ Wood, Norway Lane, Norwood; f.ns. *Nordene* **2** 53, *Northdon* **3** 227, *Northebury* **3** 33. In AS charters, *norðeage* **2** 161, *north ecge* **3** 106, *norðheal* **2** 162, *norðræwe*, *norð rewe* **4** 374, *sledes Northecge* **3** 123

norðan OE 'north, (lying) north of'. Northbrook, Northport, Norwood Coppice & Park

norðerne OE 'northern', in AS charters, *northene stod dich* **1** 30

norðerra OE 'more northerly', f.n. *the Norther Lake* **2** 174

norðmest OE 'northernmost'. North Hayes Fm

nose OE 'promontory, headland'. *Puncknoll Nose* **4** 39, *White Nose (Hill)*, White Nothe or White Nose

nowhere ModE. Nowhere

nugget ModE 'lump'. Noggets

nunne OE 'nun', f.n. ?Nunbarrow **1** 69. In AS charters, *minnanlinche*, *nunnen linc* **3** 163, 200

nunnery ModE. Nunnery Grove

nursery ModE. (The) Nursery (freq), Corfedale Nurseries, Dorset Nursery, Hermitage Nursery, Nursery Copse, ~ Lane, Nurserymead Coppice, Old Nursery, Roundham Nursery, South View Nursery

nutting ModE 'action of gathering nuts'. ?Nut Coppice

obit ModE 'commemoration of death', f.n. Obits Fd **4** 80

odde ME 'odd', f.ns. *Oddelinche* **4** 70, *Odfurlonge* **3** 163

ofen OE 'oven, furnace'. ?Gt Ovens Hill; f.n. *atte Ouene* **2** 289

ōfer[1] OE 'bank, shore'. East Over, East Stour, ?*Northover* **4** 91, Northover Ho, ?Shotover Moor, ?Southover, Southover (Fm & Ho),

Westover Bridge & Fm, West Stour, Yondover (Fm); f.ns. *Est overe* **4** 218, Northover **4** 435

***ofer**[2] OE 'hill, ridge'. Bentover Coppice, Hanover Hill, Hendover Coppice, ?*Northover* **4** 91, ?Southover

ofer[3] 'over, above, across', f.ns. *Hoverþeweye* **2** 52, Overgang **3** 353, *la Overgange* **3** 200

***oir** Brittonic 'cold'. ?Owermoigne (*v.* Addenda)

ōlde OE 'steep slope', f.n. ?*Vernolde* **2** 66

old harry ModE 'the devil'. Old Harry, Old Harry's Wife

old nick ModE 'the devil'. Old Nick's Grd

ope ME 'open, unenclosed', f.n. Ope Md **4** 310

ope Do dial. 'opening in the cliffs'. Church Hope/Ope Cove, Clay Hope/Ope, Longstone Ope Quarries

ōra[1] OE 'bank, shore, slope, ridge'. Blacknor, Broadwindsor, Fitzworth, Goathorn, God Nore, Hasler, Hasler Hundred, Horner Hill, Littlewindsor, Ower (Fm), *Owre Common*, *Portland Nore*, ?White Nothe or White Nose; f.ns. Nore Hill **1** 206, ?Woor Md **2** 127

orceard OE 'garden, orchard'. Broad Orchard, Carter's Orchard Plant., Cherry Orchard Fm, Dodge's Orchard, East & West Orchard, High Orchard, Morey's Orchard, Norchard, Old Orchard, Orchard Coppice (× 2), ~ Ctg(s), ~ Fm, ~ Hill, ~ Plant., ~ Villa, Orchards, Pit Orchard. Freq in f.ns., e.g. Orchard **2** 69, Orchard Grd **3** 28

ord OE 'point, promontory'. Jerry's Point, Redhorn Lake & Quay, ?White Nothe or White Nose; f.ns. Ord Cl **1** 80, *Richemans Orde* **1** 52, *Shullourd* **2** 45. In AS charters, *Saggelorð* **1** 272

or(i)gan ModE dial. 'wild marjoram, pennyroyal'. ?Organ Lawn

osier ME 'species of willow'. Drinkwater Osier Beds, Pottle Oysters Coppice; f.ns. Hoziers or Oziers **2** 62, Ozier Bed **3** 287, Ozier Grd **3** 8

otemel ME 'oatmeal', f.n. *Oatemell Row* **3** 149

oter OE 'otter'. Otter Island

outhouse eModE, f.n. *la Outhouse* **3** 150

outside ModE 'outer', f.ns. Outside Six Acres **1** 94, Outside Strap **1** 100

oval ModE. Oval Plant.

overmost ModE 'highest', f.n. *Ovemisttwyste* **3** 238

oxa OE 'ox'. Oxbridge, Oxencombe Bottom, Oxenleaze Coppice, ~ Plant., Oxhouse Ctgs, Oxleaze Coppice/Copse, ~ Wood; f.ns. Fat Ox **1** 305, 312, *Ocslad* **1** 260, Oxenleys **2** 19, *Oxunfold* **4** 11. In AS charters, *oxene bricge* **3** 163

oyster ModE, f.n. Oyster Bank **2** 45
***padde** OE 'toad', f.ns. *Padell Close* **4** 223, Padmoor **2** 9, *Padner* **2** 201
***padduc** OE 'frog', f.n. *Paddecdelf* **1** 101
pæð OE 'path'. *Addle Path*, Bridle Path, Coombe Bottom **1** 10, Dernepath **1** 126, *Gatpathe* **4** 378, Glide Path Hill & Rd, Loose Path Coppice; f.n. *Hull'path* **3** 207. In AS charters, *Alfstanes paþ* **1** 30, *ealde paþe* **3** 206, *fugelan pæþ* **2** 136, *hegen paðe* **3** 110, *hippe pad* **3** 206, *Smalan pæþ* **4** 35, *stigla pæþ* **2** 136, *strut heardes paþe* **1** 10
pageant ModE. Pageant Gardens
pāl OE 'pole, stake'. ?Pole Coppice (× 2); f.n. ?*Palmede* **1** 211
palatable eModE 'agreeable to the palate'. ?*Paltable Coppice* **2** 202
pale ME 'fence, boundary'. Palemead Coppice; f.n. *Mayne pale* **4** 284, Pail Md **4** 217, Gt Pale Md **3** 28
palefrei ME 'palfrey', f.n. *Palfrayhyll* **1** 298
paleis OFr, ME 'a palace'. ?King's Court Palace
palis, paleis OFr, ME 'a fence of pales, a palisade'. ?King's Court Palace
palmer ME 'pilgrim', f.n. *Palmecruch* **2** 143
***pamp** OE 'hill'. ?Pamphill
pancake ModE, f.ns. Pancake **2** 9, Pancake Cl **3** 78
panne OE 'a pan', f.n. *Les Pannes* **3** 257
paper-mill ModE, f.ns. Paper Mill Fd **2** 191, Paper Mill Md **2** 136
***papol** OE 'pebble', f.ns. ?*Papilhill* **2** 78, *Papyllane* **4** 167, Popple Street **3** 78, *Popwellake* **3** 375 [*cf.* ***popel**]
paradis ME 'paradise'. *paradice* **4** 360, Paradise (× 2), Paradise Ctgs, ~ Fm, ~ Lane, ~ Pond, ~ Road, ~ Wood; f.ns. Hither Paradise **3** 279, Paradice Md **2** 82, Paradise **2** 14, 139, 157, **3** 28, 57
park OFr, ME 'enclosed tract of land for beasts of the chase'. The Abbey Park, Alderholt Park, Alexandra Park, *Benville Park* **4** 292, Blagdon Fm & Wood, Branksome Park, Bryanston Park, Bushy Park, Bussey Stool Park, Came Park, Canford Park, Cerne Park, Charborough Park, Chelborough Park, Cold Park Wood, Compton Park, Crichel Park, Deer Park (freq), Deer Park Fm, Dewlish Park, *Duntish Parke* **3** 243, Eastbury Park, East Park, Frampton Park, Frome Park Coppice, Gunville Park, Hamworthy Park, Harbin's Park, Henbury Park, Holmwood Park, Holnest Park, *Holte Parke* **2** 151, Hooke Park, Horn Park, Horse Park Plant., InparkFm, Iwerne Minster Park, Kingston Lacy Park, Kingston Park, Leigh Park, Leweston Park, Little Park, Long Park, Lulworth Park, Manor Park, Melbury Park, Melcombe Park, Merley Park, Milton Park Fm &

Wood, Minterne Park, Morden Park, Moreton Park, Motcombe Park, *Northpark* **4** 225, Old Park (freq), The Park (freq), Park Belt, ~ Bottom (Road), ~ Chapel, ~ Coppice (freq), ~ Copse (freq), ~ Corner, ~ Ctg, Park Fm (freq), *Park Fm*, Park Gate, Park Gully, Park Hill, Park Ho (× 2), Park Ho Fm, Park Lane (× 3), Park Lodge, Park Mead Bottom, Park Plant. (× 2), Park Pond, Park Quarries, Park Row, The Parks **4** 279, Parkstone (Park), Parks Withy Bed, Parks Wood Plant., Park View Point, Park Water Ctgs, Park Wood (× 3), Parnham Park, Poole Park, Radipole Park Gdns, St Giles's Park, Sheep Park Coppice, Sherborne Park, *Southeparke* **4** 225, Stafford Park Copse/Fm, Stalbridge Park, Stock Gaylard Park, Stone Park, Thornhill Park, Uddens Park, Victoria Park, Westford Park, Whatcombe Park, Woodlands Park. Freq in f.ns. e.g. Gt Park **3** 231

parlur ME 'small piece of ground'. Parlour Moor, Pugs Parlour. Also in f.ns., e.g Cuckolds Parlor **1** 27, *the Parlore* **3** 34, Parlour Cl **2** 66

parsley ModE, f.n. Parsley Bed **3** 300

parsonage ModE. (The) Parsonage (freq), Old Parsonage, Parsonage Barn (× 3), ~ Coppice (× 4), ~ Copse, ~ Fm (× 3), ~ Hill, ~ Ho, ~ Lane

parsonatus Lat 'ecclesiastical benefice, parsonage'. *Beminster Parsonatus* **4** 311

part(ing), **party** ModE 'shared or divided land', f.ns. *Parte Acres* **2** 106, Parting Mdw **3** 177, Party Md **4** 352

partridge ModE, f.n. Partridge Hedge **3** 309

parva Lat 'small'. Blandford St Mary, *Charlton Parva*, Fontmell Parva, Frome St Quintin, Hinton Parva, *Little Crichel*, Little Kington Fm, Littlewindsor, Minterne Parva

pas OFr, ME 'passage, crossing place', f.n. Pass **4** 330

passage OFr, ME 'passage, crossing place'. Ower Passage, Passage **4** 21, *Passage* **2** 44, *Passage Houses* **1** 269, Wytch Passage; f.n. *the passadge* **2** 21

pasture OFr, ME 'pasture', f.n. Boggy Pasture **2** 62, Pasture Fd **2** 200

patche ME 'small piece of ground'. *Jordan Hill Patch* **2** 212; f.ns. Alder Patches **3** 132, Patch **2** 200, Patch Cl **4** 282, Patch Croft **1** 168

paternoster ME 'the Lord's Prayer', f.n. *Pater Noster Crosse* **2** 174

patron ME 'holder of an advowson', f.n. *Patrone hayes* **4** 364

***pat(t)e** OE 'mud, marsh'. ?Patley Wood; f.ns. *Patapuldor* **4** 386, ?Pat Close **4** 268, *Patcroft* **4** 465, Pat Plot **4** 335, 343

***pēac** OE 'knoll, peak, pointed hill'. The Peak, Peak Coppice, Peak End Hill, Peakey Coppice; f.ns. Peak **4** 410, Gt & Lt Peak **2** 57

peacock ModE. Peacock Lodge (× 2)

peaky ModE 'with a peak, pointed', f.ns. Peaky Fd **2** 85, Peaky Leigh **4** 452

***p(e)all** OE 'ledge'. *Palbroke*

pearroc OE 'small enclosure, paddock'. Paddock's Fm, Park Fm **4** 404; f.ns. *Cokerellesparrok* **3** 32, Paddock Orchd **3** 5, *the parock* **4** 19, *Topps Perocke* **1** 129

peaty ModE. Peaty Moor [*cf.* **pete**]

pece ME 'piece of land', f.n. Piece **3** 314

peck-mill ModE 'a pecker-mill'. Peck Mill

peere, *v.* **pier**

peked lME 'with a peak, pointed', f.ns. Peaked Close **3** 57, 227, Peaked Higdens **2** 199 (*cf.* **pīcede**)

***pempe** OE 'hill'. ?Pamphill

pening OE 'penny'. Penny Farthing Wood, Penny Hill Fm, ?Pennymoor Pit Fm, Penny Plot Rd **4** 393. Also in f.ns, e.g. *Pennie Meade* **3** 6, Pennys Plot **1** 134, Penny Wells **4** 486, Pennyworth **3** 8, *Penyham* **1** 328

***penn**[1] PrWelsh, PrCorn 'a head, a top, a hill'. Dodpen Fm & Hill, Pen Barn & Ctgs, †Penbury-Knoll, ?Pen Coppice, Pen Hill, ?Penn (Fm & Hill), Penn Hill, Penn Wood, Pentridge, ?Pentridge Fm, ?†Pilsdon Pen, Sixpenny Fm, Sixpenny Handley (Hundred); f.n. Penhill **2** 241. In AS charters, *lutlen seaxpennes suðeke* **3** 104

penn[2], **pænn** 'a pen, a fold, an enclosure'. Cuckoo Pen, ?*Lobepenn* **2** 108, ?Pen Coppice, ?Penn (Fm & Hill), The Pens, ?Pentridge Fm, ?Pen Well Hill, ?Pilsdon Pen, ?Ruppen Coppice; f.n. *Gospannes* **2** 158

penning eModE 'pen, enclosure', f.ns. *Mancelles pennyng* **4** 223, Penning **3** 29, Penning's **3** 67, *Penynge closes* **4** 35

***pennuc** OE 'small pen', f.n. Pinock Hill **3** 129

pentis ME, **penthouse** ModE 'an outhouse or shed with sloping roof, an annex'. Paynthouse Fm, *Pentice house* **1** 350; f.n. *la Pentys* **1** 349

pēo, **pīe** OE 'gnat or other insect'. ?Penmore Rd, ?Pye Lane, Pymore. Also in f.ns., e.g. ?Pyehill **1** 257, ?Pylands Acre **1** 151

pepper ModE. ?Pepperclose Trees. Also in f.ns., e.g. Pepper Alley **1** 286, ~ Md **1** 29

peppercorn ModE. Peppercorn Lane; f.n. Pepper Alley **1** 286

perche ME 'pole, stake', f.n. *le Perche* **1** 42

persone OFr, ME 'parson'. Parson's Bridge, ~ Copse, ~ Piece; f.ns. ?Parsons **3** 29, Parson's Grove **3** 29

***perth** PrWelsh, PrCorn 'a hedge, a bush, a brake'. Pert Copse

peru, **pere** OE 'pear'. Parnham, The Pear, Pear Close Copse, Pear Tree Fm, West Parley; f.n. Pear Tree Mdw **2** 200

pete ME 'peat'. Peat Hill, Peat Moor; f.ns. Peat Moor **1** 151, Peat Mower **2** 200

petit, **pety** ME 'small'. *Little Crichel*, Littlewindsor, Pettycrate Lane; f.ns. Petty Cl **2** 78, Petty Croft **4** 436, Pitty Cl **4** 330

***petru-** Brit 'four'. ?R. Parrett (*v*. RNs)

pheasant ModE. Pheasant Hill; f.ns. Pheasant Coppice/Copse **1** 69, **2** 234

pheasantry ModE 'enclosure for rearing pheasants'. Pheasantry (× 2)

pīc[1] OE 'point, pointed hill'. Peaked Corner, ?Pickaxe Cross, Picket Fm; f.n. *Pykeacre* **3** 227, ?*Pykemull* **2** 87

***pīcede** OE 'pointed'. Peaked Corner, ~ Post, Peakey Coppice; f.ns. Picked Acre **2** 205, Picked Cl **2** 200, **3** 57, Picked Jaws **2** 26, Picket Plains **4** 300 [*cf.* **peked**]

pickaxe ModE, f.n. Pickaxe **2** 119

piddle ModE dial. 'small enclosure', f.ns. Gt & Lt Piddle **2** 273, Parsons Piddle **2** 249

***pide**, ***pidu** OE 'marsh, fen'. ?Pitcombe Dairy; f.n. ?*Pyd*(*de*)*mylle* **3** 368

***pidele** OE 'marsh, fen'. R. Piddle (*v*. RNs), giving name to: Affpuddle, Piddlehinton, Piddle Lane & Wood **4** 137, Piddletrenthide, Puddletown, Tolpuddle, Turner's Puddle; f.ns. Piddle Hays **4** 181, Piddle Knap **4** 471, Piddle Moor **4** 452

pie ME 'magpie'. ?Pye Lane. Also in f.ns, e.g. ?Pyehill **1** 257, ?Pylands Acre **1** 151

piebakere ME 'baker of pies', *Pybakers house* **2** 186

piece ModE, **pece** ME 'piece of land'. Elsworth Piece, Hanging Piece Belt, Heathypiece Bungalow, Parson's Piece, ?Piece Copse

pier ModE, **peere** ME 'pier'. *Chine Pier* **1** 222, The Cobb **4** 394, Durdle Pier, Folly Pier, King's Pier, Lilliput Pier, *New Pier* **1** 222, Pier (at Castletown, Swanage), Pile Pier, Pottery Pier, Salterns Pier, South Pier, *Underbank Pier* **1** 222, Victoria Pier

pig ModE 'fragment'. ?Pig & Whistle

pigeon ModE. Pigeon Clump, Pigeon House Barn & Copse, Pigeon House Cottages, Pigeon House Fm, ~ ~ Plant.

pigge ME 'young pig'. Pig Oak, ?Pigs Moor Coppice, ?*Piggeshey* **1** 314, *the Pigg market*t **1** 155, *Pyggeslowse* **4** 3; f.ns. Pighays **3** 98, *Pig Hill* **3** 364, Pigs Hams **3** 340, *Pyȝedon* **4** 35

pightel ME 'small enclosure', f.n. Pightle **2** 9

pīl OE 'stake, shaft, pile', also 'peak'. Barbers Piles **2** 42, Pile Lane, Pile Pier, ?Pilford, ?Pill Ash, Pillwell, †Pilsdon, *la Pylelane* **3** 149, Pyle Well; f.ns. Piles Hill **3** 98, *Pilewen* **2** 90

pilede ME 'made with, or marked by, stakes or piles', f.ns. *le piledlond* **2** 58, *the piled weare* **1** 138

piler ME 'a pillar', *la Pyl(l)er* **3** 366

pill ModE dial. 'small stream', *v.* **pyll**

pillori ME 'wooden frame for punishment'. *Pillory St.* **2** 185; f.ns. *la Pillori* **3** 146, *le pillory* **4** 374, *le Pullery* **4** 11

***pimp** OE 'small hill'. ?*Pimpern* (*v.* RNs), ?Pimperne (× 2)

***pïmp** PrWelsh, ***pimp** PrCorn 'five'. ?*Pimpern* (*v.* RNs), ?Pimperne (× 2)

pīn OE 'pine-tree'. The Pine Clump, The Pines, ?Pinford; f.ns. ?*Penwell* **3** 187, *le Pynehayes* **4** 386

***pinc** OE 'minnow' or ***pinca** 'finch'. Pin Bridge, ?*pinguedine* **4** 54, ?Pink Knoll Hollow, ?Pink Lake Coppice; f.ns. *Peinke Pitt* **2** 205, Pinkmore **3** 57, *Pynckemede* **4** 364

pincushion ModE, f.n. Pincushion (Head) **2** 205

pin(e)appel ME 'pineapple'. Pineapple Ctg & Fm

pinion ModE 'a wing'. Pinion Rock

pinn OE 'pin, peg'. ?Pinford; f.ns. ?Penwell **3** 187, ?*Pynehey* **1** 152

pinnacle ModE 'a natural peaked formation, pointed rock'. Old Harry, The Pinnacles (× 2)

pintel OE 'penis', f.ns. *Pintle Mead* **1** 230, *Pynteldirne* **3** 375

pioneer ModE. Pioneer Rd

pīpe OE 'pipe, conduit', f.n. ?*Pipe Hilles* **2** 175

pirige, **pyrige** OE 'pear-tree'. Perhay Fm, Perry Coppice, ~ Copse (× 2), ~ Fm (× 2), Perryfield, Purcombe Fm; f.ns. *Perriefeild* **4** 411, Perryhay **2** 98, *la Purihegge* **3** 34

piscary ME 'fishery', f.n. *la Paskarie* **3** 375

pise, **peosu** OE 'pease'. Peacemarsh, Peascombe Copse, ?Piece Copse, *Piselane* **1** 350; f.ns. Pease Cl **4** 43, Pease Lands **4** 22, Piscombe **1** 198, Pissmoor **2** 9, Pistol **4** 300, *Pushyll* **3** 111

piss ME 'piss', f.n. Piss Lake **4** 268

pitch ModE 'pitch'. *Martin's Pitch* **1** 169

pixie, **pixy** eModE 'pixy'. *Pexy'shole* **2** 94

place OFr, ME 'open space, plot of ground, residence'. Alington Place, *Anketil's Place*, Bennett's Place, *Biddles Place* **3** 51, Brickplace Copse, *Brytys place* **2** 265, *Comptonesplace* **3** 19, Court Place, *Damory Court*, Denzil's Place, Hibernia Place, *Hullettesplace* **3** 366, Payne's Place, *Pease Leaze*, Place Fm, *Ponsonby Place*,

Portland Place, Prince's Place, Prospect Place (× 2), Rural Place, Shearplace Hill, Strode Manor Place, Tunnel Head Place, Twelvetrees Place, *Wagges Place* **4** 280. Freq in f.ns., e.g. *Cleyardesplase* **2** 63, *Filthamesplace* **3** 32, *Pyttes place* **4** 411, *Stretesplase* **4** 22

plack, *v.* **plek**

***plæsc** OE 'pool', surviving as dial. **plash** 'marshy pool'. Melplash; f.ns. Place Lane Cl **2** 229, *The Plashmore* **4** 133

plain OFr, ME 'open tract, flat piece of land'. Abbotsbury Plains, Battle Plain, Binnegar Plain, Job's Plain, Milton Plain, North Plain Plant., The Plain, Plain Bottom, Plain Copse, Planefield Ho, Watcombe Plain. Also in f.ns., e.g. Broad Plains **4** 300, Great Plain Cl **4** 436, *Pellies plaine* **2** 99, Plain fd **1** 168

plaissiet OFr 'enclosure made with wattles', f.ns. ?Plashett **1** 203, ?*Plasschett* (*Meade*) **3** 341, 382

plantation ModE 'a wood of planted trees'. The Plantation(s) (freq), Plantation Coppice. Freq in modern minor names as ~ Plantation, e.g. Beacon Plant., Boswell's ~, Bound Stone ~, Bourne Hill Plant., Bowley's Plant., Brick Hill Plant., Brick Kiln Plant., Bridport Rd Plant., Brown Down Plant., Brownjohn's Plant., Bunker's Hill Plant., Burleston Plant., Coffin Plant., Common Plant., Cook's Plant., Coombe Plant., Coronation Plant., Coux Plant., Dick o' th' Banks Plant., Dole's Hill Plant., Dover's Plant., Drive Plant., Duddle Plant., East ~, Eight-Hatch Plant., Fanny's Land Plant., Fir Plant., Ford Plants., Forty Acre Plant., Goldsmith's Plant., Gorse Plant., Great War Plant., Grimstone Down & Fds Plant., Half Moon Plant. (× 2), Hamar's Plant., Hart Hill Plant., Horse Lynch Plant., Horse Plot Plant., Howdes Barrow Plant., Ivo's Plant., Jackman's Plant., Jubilee Plant., Kenneth's Plant., Kidney Plant., Langford Plant., Larch ~, Lawyer's ~, Limekiln Plant., Long Walk Plant., Maiden Plant., Millicent's ~, Moreton ~, Muckleford ~ , Newlands Plant., New Plant. (freq), Nigel's Plant., North Hills Plant., North Plant. (× 2), Oval Plant., Penns Plant., Prisoner's (sic) of War Plant., Salt Hill Plant., Seven Barrow Plant., Shepherd's Plant., Skippet Plant., South Plant., Starmoor Plant., Stratton (Down) Plant., Table Plant., Three Cornered Plant., Warren Plant., Watcombe Plant., Wellclose Plant., Well Plant., West Field Plant., Wreden Plant.

planting ModE 'plantation'. New Planting

plaschiet OFr 'marshy pool', f.ns. ?Plashett **1** 203, *le Plasshet* **4** 228, ?*Plasshett* (*Meade*) **3** 341, 382

plat[2] ME 'plot', f.ns. Great Plat **3** 26, *Naysshe Plattes* **4** 407, The Plat **4** 282, Plat Mdw **2** 205, Tithing Plat **3** 5

playing ModE 'used for games', f.n. *Playing closse* **3** 188

pleasant ModE. Mount Pleasant (Fm) (freq), Pleasant Spot

pleasure ModE. Pleasure Ground; f.n. Pleasure House Plant. **2** 139

plega OE 'play, sport'. Plaisters Lane, (*The*) *Play Close* **2** 90, **4** 92, Play Cross **2** 105, **3** 187, Playford Cl **2** 229, *Pleistret* **3** 34, *la Pleycrosse* **3** 200, *Pleymersch* **1** 287, *Pleystret*(*e*) **1** 328, **3** 34

plek ME 'small plot of ground'. Boyne's Lane Pleck, Pleak House Fm, Pleck (× 5), Pleck Ctgs (× 2), Pleckley. Also in f.ns., e.g. *Greneplekke* **1** 313, Plack **2** 200, The Plecks **2** 114

plente OFr 'abundance'. *Plentey* (a tower) 1280–5 (at Corfe C. **1** 12)

plim ModE dial. 'filled out', f.n. ?Plympthorn Coppice **2** 98 [but rather **plȳme**, *v.* Addenda **2** 98]

plock ModE dial. 'small plot of ground', f.ns. Deer Plock **4** 156, Plocks **2** 89 [*cf.* **plek**]

plot lOE, ME 'small piece of ground'. Carriage Plot, Curfew Plot, Dinkey's Plot, Dumb Maid's Plot, Gallows Plot, Horseplot Ctg, Horse Plot Plant., ~ ~ Withy Bed, *Minsomes Plat* **4** 279, *Nasheplott* **4** 407, (The) Plot, Plot Plant., Wagg's Plot

ploughed ModE, f.n. Ploughed Close **2** 9

ploughman ModE, f.n. Plowmans Grd **2** 37

plover ModE 'plover'. Plovers Stays **1** 206

pludde, **plodde** ME 'pool, puddle', f.ns. Pliddy Moor **4** 391, Plod Md **4** 464, *Pluddie-lane* **2** 42, Pluddy Marsh **1** 180

plum(b) ME 'lead', *the plumbe-house* **3** 366

plumb ModE 'sheer', f.n. ?Plunb Pits **2** 72

plūme OE 'plum'. Plumber Fm, Plumley Wood; f.n. *Plumetenement* **4** 386, Plumley **3** 300

plump ModE dial. 'clump of trees', f.n. ?Plumpland **3** 136 (*v.* NDEFN 336)

ply ModE 'bend', f.n. ?Ply Cl **4** 436

plȳme OE 'plum', f.ns. *Plimcroft* **4** 332, *Plymhous* **4** 386, Plympthorn Coppice **2** 98

***plysc**, ***plisc** OE 'pool'. Plush (*v.* also *Plusshebroke* in RNs)

po ME 'pea-fowl'. ?Povert Bridge [but now *v.* Addenda **2** 103]

***poc(c)e** OE 'a frog'. ?Poxwell

pode ME 'toad', f.ns. *Podehegh* **3** 257, *Pudemore* **4** 271

***pohha** OE 'a pouch, a bag'. Bowridge Hill (*v.* Addenda **1** 143, **3** 11)

point ModE 'promontory'. Anvil Point, Ballard Point, Bower Point, Brand's Point, Bran Point, Bum Point, Clay Hard Point, Cleavel

Point, *Copcourt Poynt* **1** 125, Countess Point, Crown Point, East Point, Egmont Point, The Foreland or Handfast Point, Goathorn Point, Gold Point, Green Point, *Harcuits Point* **2** 21, Holton Point, Jerry's Point, Keysworth Point, Langton Hive Point, Mead Point, Nath Point, North Haven Point, Nothe Point, Park View Point, Patchins Point, ?Payne Coppice, Peveril Point, Redcliff Point, Redend Point, Rock Lea Point, Rodden Hive Point, St Alban's or St Aldhelm's Head, Seventeen Acre Point, Shipmoor Point, Shipstal Point, South Haven Point, Stanford Point, Swineham Point, Tidmoor Point, Water Point, Weston's Point, West Point

poison ModE, f.ns. Poison Close **4** 417, 464

poke ME 'bag', f.ns. *Flaxepoke* **1** 229, *ij Pokes* **2** 175

pol[1] OE 'pool, pond, creek'. Blue Pool, Bradpole, Caggypole Fm, Chapman's Pool, Cowpool Fm, *Deuelepole* **3** 317, Duckpool, Empool Bottom & Heath, *Frog Pool*, Gutch Pool Hole, Horsepool Lane, Kimmeridge Bay, Parson's Pool **3** 146, Poker's Pool, ?Pole Coppice (× 2), Poll Bridge, Poole, Pool Ledge, Pool Pond, *Poulewey* **4** 3, ?Povert Bridge (*v.* Addenda **2** 103), Pulham, Radipole, *Redpool*, Washingpool Fm; f.ns. *Horse Poole Close* **3** 6, Pool Md **3** 29. In AS charters, *ennan pol* **3** 298, *hlyda*(*n*) *pol* **4** 299, *neþan pol* **3** 207, *roddan pol* **2** 136, *schort mannes pol*, *seortmannes pol* **1** 64

pollard ModE 'pollarded tree'. ?Pollards Wood; f.ns. Pollard Grd **4** 348, Pollard Oak Pce **3** 340

pols ME 'pulse, plant yielding pulse', f.ns. *Pulselynche* **4** 22, *Pulshull* **2** 251

ponde ME 'pond, pool'. Abbey Ponds, Bathing Pond, Bowldish Pond, Brewer's Pond, Brickells Pond Fm, Butt's Pond, Cocks' Moor Pond, Common Pond, Coy Pond, Decoy Pond (× 2), Deer Park Ponds, Firs Pond, Fish Pond, Forty Pond, Great Pond (× 2), Green Pond, Hatch Pond Rd, Heedless William's Pond, Higher Pond, High Woods Pond, Hilly Pond, Hook Pond, Jubilee Pond, Kennel Pond, Latchmore Pond, Lower Pond, Mare Pond, Mayers Pond, Mear's Pond, Merley Pond, Mermaid Pond, Mill Pond(s) (freq), Monks' Pond, Oakhill Mire Pond, Paradise Pond, Park Pond, Pond Bottom (× 2), Pond Close (× 3), ~ Coppice (freq), ~ Ctg, ~ Eweleaze, ~ Fm (× 2), ~ Ground, ~ Head, ~ Ho, Pond Mead Stall, Pond Plant., *Ponds*, Pool Pond, Revel's Pond, Rimsmoor Pond, Rushmoor Pond, Rushpond, Skating Pond, Spittle Pond, Spring Pond, Stag's Pond, Swan's Pond, Sycamore Pond, Veiny Cheese Pond, West Bexington Pond, Whatley Pond; f.ns. Pond Cl **4** 222, *Ponde Close* **2** 193 [*cf.* **duckpond**, **fishpond**]

pony ModE. Pony Drive
poor ModE. Poor Allotment, Poor Common (× 2), Poor Delf, Poor Ho (× 2), Poorhouse Lane, Poor Lot(s), Winterborne Poor Lot. Also in f.ns., e.g. Poor Altmt **1** 249, *the Poor Land* **1** 29, Poors Grd **1** 29
***popel** OE 'a pebble'. ?Pop Mallers Coppice; f.ns. *Popelhous* **4** 386, Popple Cl **4** 203 [*cf.* ***papol**]
poplar(-tree) ModE. Poplar Copse, Poplar Fm, ~ Plant., The Poplars, Poplartree Fm; f.n. Poplar Md **3** 29
porch eModE 'porch', f.ns. *at Porche* **3** 188, *Baggetts Portch* **4** 363, *la Porche* **1** 328, *Porch Haie* **1** 230
porcus Lat 'pig'. Toller Porcorum
port[1] OE 'harbour'. ?Bridport, Portland, *Waymouthe Porte* **1** 255
port[2] OE 'market town'. Bimport **3** 145, ?Bridport, Northport, Porthill Plant., Portisham, ?Westport; f.ns. Partland Fd **4** 475, *Portbrigge* **3** 178, Port Lane Cl **2** 180
port[3] OE 'gate'. ?Newport, ?Westport; f.n. *Newporte* **4** 167
port-mann OE 'townsman'. *Portmanmore*; f.ns. *Portmanfelde* **4** 364, *Portmanneslond* **3** 158
port-weg OE 'road leading to market town'. Barnston Fm, Partway (Ctgs & Lane), Portway, Portymoor Coppice. Also in f.ns, e.g. *Eld(e)portwey* **3** 148, Partway **3** 216, *Portewey* **1** 52, **1** 199, **2** 78, 175, **4** 3, *Portway* **1** 121, *la Portweye* **1** 199
post OE 'post, pillar'. Lookout & Rocket Post, Peaked Post, Post (Fm), Post Green, Post Row, Red Post (× 3), Redpost Hill, Stony Post Ctgs, White Post, Whitepost Gate; f.ns. Post (Cl) **3** 297, *Post Hedge* **2** 53
post-house eModE 'stable for post-horses'. *le posthowsse* **1** 314
potsherd ModE, f.n. Potsherd Green **2** 200
pot(t) lOE 'pot'. Potterne Fm; f.ns. Potten Hill **2** 214, ?*Pothull* **1** 199, ?*Pottmoore* **2** 175
potte ME 'deep hole', f.ns. *Potescombe* **3** 200, ?*Pothull* **1** 199, ?*Pottmoore* **2** 175
pottery ModE. Kinson Pottery, Old Pottery Kiln, Pottery Fm, ~ Pier
pouche ME 'pouch', f.ns. *Harepoche* **4** 385, *Pucchelake* **4** 386
poultrie eModE 'poultry', f.n. *le Pultrey Cross* **3** 148
premises ModE. Glove Premises
***prenn** PrWelsh, PrCorn 'tree'. ?*Pimpern* (*v.* RNs), ?Pimperne (× 2)
prēost OE 'priest'. Bryants Puddle, Hampreston, ?Pressham Wood, Presswood Copse, *Presteswlle* (*v.* RNs), Preston, Preston Fm, Preston Hill, Prestwood Moor, Priestlands **3** 363, Priest's Chamber & Chapel, Priest's Way, Priests Way, ?Prystock Fm; f.ns. Press

Mdw **2** 191, Press Street **3** 177, Prestill Mead **3** 236, Priest('s) **3** 29. In AS charters, *preosta leȝe* (*v.* Addenda **4** 193)

preserve ModE 'place where hunting or fishing is preserved'. The Preserve

pric(c)a OE 'prickle', f.ns. ?Prigwell **1** 344, *Prycketherne* **3** 257

prickled eModE 'prickly', f.n. *Pryclet Closse* **4** 167

prima Lat 'first'. Upbury Fm **4** 230

primrose ModE. Primrose Ctg(s), ~ Fm, ~ Hill

prince ME 'prince'. Prince's Bridge, Prince's Lane & Wood, Prince's Place

prine ME, eModE 'the holm or evergreen oak'. ?Prime Fm

prior ME 'a prior'. ?Briar's Wood, *Canford Prior*, *Melebir Priur* **4** 214, Priors Copse, Prior's Down, Prior's Lane, Priorslawn Dairy; f.ns. Priors Grove **2** 175, *Pryoryshavyn* **4** 11

priory ModE. *Canford Prior*, *Cranborne Priory*, (The) Priory, Priory Fm, St Monica's Priory

prisoner ModE, Prisoner's of War Plant. (sic) **1** 374

propriation ModE 'action of making one's own'. Propriation Mdw **1** 181

prospect ModE 'a wide view'. Prospect Ctg (× 2), ~Fm, ~ Place (× 2); Prospect **4** 348

providence ModE. Providence Ctg

pryfet OE 'privet'. Pribdean Wood; f.n. *Prevet* **4** 218

pūca OE, **puke** ME 'goblin'. Pucksey Brook (*v.* RNs), Puckstone, Pug's Hole, Putlake Fm, Pug Pit. Also in f.ns., e.g. Packmoor **3** 57, *Pokelane* **3** 149, *Pokeput* **1** 328, *Pokershe* **3** 207, *pook acre* **4** 144, *Poukepitte* **1** 214, Puck (Marsh) **1** 80, Puckhay **3** 101, *Pucklake* **4** 205, Puckmoor **3** 98, Puckway, Puckwell **4** 249, *Puckysbarry*, *Puckysway* **1** 185, Pug Pit **3** 78, *Pukefurland* 1 243, *Puxeye Wood* **4** 157, Puxy Gdn **1** 203, Puxy Md **1** 203

pūcel OE 'goblin', f.ns. *Pokelmour* **4** 35, *Pokulchurche* **1** 272

***puddel** OE, **podel** ME 'muddy pool, pond'. Holmes & Puddleway Coppice, Pottle Oysters Coppice, Puddle Coppice, ?Puddle Dock Dairy Ho, Puddle Mill, *Puddle St.* **2** 186. Also in f.ns., e.g. *Podelhurste* **4** 411, Puddle Hole **1** 203, Puddleton(s) **4** 383

pudding ME 'bowels, entrails', figuratively 'soft ground', f.ns. *Podingstede* **4** 139, Pudding Cl **3** 198, Pudding Fd **2** 51, Pudding Lane **2** 186

pudge ModE dial. 'puddle', f.n. Podge Lake Orchd **4** 424

pulpit ModE. Pulpit Rock; f.ns. Pulpit **1** 236, Pulpit Md **3** 353

***pull** OE 'pool, creek', f.n. Poulter Cl **3** 177

pulse, *v.* **pols**

pump ModE 'pump'. Pump Ho (freq), f.ns. Pump Close **4** 283, Pump Holes **1** 113

punchbowl ModE. Punchbowl Coppice

pund ME 'pound, enclosure for cattle'. Court Pound, Cowpound Wood, Cuckoo Pound (× 5), Fir Pound, Fox Pound (× 4), Pound (freq), Pound Close, Pound Ctg, Pound Covert, Pound Drove Copse, Pound Fm (× 3), Pound Hill (× 2), Pound Ho, Pound Lane (× 4), Pound Wood, Round Pound, Sadborow Pound, Wytch Fir Pound; f.ns. Cuckoo Pound **3** 287, *le Mullepounde* **2** 14, Parish Pound **2** 205

***pund-fald**, ***pynd-fald** OE 'a pinfold, a pound'. Penfold Belt, Pondfield, Pound (freq), Pound Lane **3** 22, *punfalde* **4** 171. Also in f.ns., e.g. *pinfald* **2** 140, Ponfill fd **1** 61, *Ponfolde* **2** 146, *Poundesfald* **2** 181. In AS charters, *punfald* **4** 171

pūr OE 'a bittern (or snipe)'. Isle of Purbeck. Also in f.ns, e.g. ?*Porcombe* **4** 344, ?*Purfurlonge* **1** 294, Purley **3** 373

purchase ModE, f.n. Purchase **3** 128

purgatory ModE. Purgatory; f.n. Purgatory **4** 453

purlewe ME 'outskirts of a forest'. *Purley Wood*; f.n. Purlieu **2** 9

purse eModE 'money bag', f.ns. Cut Purse **3** 321, Pick Purse **2** 26, Purse **4** 80, ?Purselands **1** 172

pyll OE 'pool in a river, small stream, tidal creek', surviving as SW dial. **pill** 'small stream'. Pellwell Fm, ?Pilford, ?Pill Ash, Pillmoor Bottom, Pill Well. Also in f.ns, e.g. Pill Acre **3** 44, Pill Cl **2** 41, *Pillecrofte* **4** 465, Pills Moor **2** 172, *Whitepill Lake* **1** 81

pylon ModE, f.n. Pylons **2** 38

pyrige OE 'a pear-tree'. Perry Copse (× 2), ~ Fm, Purcombe (Fm)

pytt OE 'pit'. Alderholt Gravel Pit, Alderney Gravel Pit, Ash Pit, Bateham's Pit, Bragg's Pit, Brimpit Fm, Bull Pit, Bullpitt Beacon, Bloody Pits, Brooklands Pit, Bullpits, Burwood Pit, Bushes Pit, Chalk Pit (× 2), Lt Charborough Pit, Chill Pits, Clay Pit(s) (freq), Clay Pit Plant. (× 2), Coal Pit Copse, Cobblers Pit, Combehays Pit, Coombe Pit Plant., Crumpet's Fm, Cuckolds's Pit, Dr Palmer's Pit, Dungeon Pit (× 2), Dunspit Lane, Eady Pit, Earthpit, Earthpits Lane, Endless Pit, Flint Pit, Gallpits Gorse, Gipsies' Pit, Glastonbury Pit, Glydes's Pit, Gore Pit, (Old) Gravel Pit(s) (freq), Gravel Pit(s) Plant., Gravel Pits Wood, Great Pit Lane, Green Pit Bridge, Griffiths' Pit, Harpitts (Fm), Hell Pit, Hill Pit (× 2), Holcombe Head Pit, Honeypits Plant., Horse Leys Pit, Horse Pit, Huish Pit, James's Close Pit, Johnny Vizer's Pits, Kimbert's End, Langham's Pit, Lewell Pits, Long Cross Pit, Mampitts Fm, Manfield Shale Pit,

Marland Pit, The Marl Pit, Marl Pits Wood, Money Pit Lane, Nine Acre Pit, Pennymoor Pit Fm, Pit Bottom, Pit Close Copses, Pit Coppice (× 4), Pit Copse (× 2), Pit Croft Copse, Pitfield Fm, Pit Ho, Pitlands Coppice, Pitman's Lane, Pit Orchard, Pit Plant., Pitt (House) Fm, Pitt's Ctg, *Poukepitte* **1** 214, Pug Pit, Purbeck Stone Quarries, *Puthus* **3** 150, *Puttes staple* **4** 94, Redhills Pit, Red Pit, Ridley Pit, Rock Pit, Hr & Lr Rough Pits, Sandpit (Fm), Sandpits, Slough Pit, Spearpit Coppice, Spinney Pits Coppice, Square Pit, Standpitts Lane, Stone Pit(s), Washers Pit, Wherry Pit, White Pit (× 2), ?Winspit, Yewtree Pit. Freq in f.ns, e.g. *Addreputte* **2** 81, *Ladewynyspytte* **3** 226, *Leunespitt* **2** 58, *Peinke Pitt* **2** 205, *Rydeputte* **1** 129. In AS charters, *cealc pyt* (*v.* Addenda **3** 267), *col pytte* **4** 302, *ludmannes putte* **3** 111, *tweie pettes* **3** 208, *wæter pytt* **4** 305, *water pet* **2** 69

pyttel OE 'a hawk, a mousehawk'. ?Piddles Wood

quaggy ModE dial. 'boggy', f.n. Quaggy Brakes **1** 41

quagmire eModE 'boggy ground', f.n. Quag Mire Orchd **3** 198

quaker ModE, f.ns. Quaker's Grove **1** 257, Quakers Ld **4** 453 (*v.* NDEFN 346)

quarentene eModE 'quarantine', f.ns. *Quarentaine* **2** 125, Quarrentine Orchd **4** 330

quarrere, **quarre** ME 'quarry'. *Admiralty Quarries*, Bill Quarries, Bottom Coombe Quarries, Bowers Quarries, Breston Quarries, Broadcroft Quarries, Butts Quarries, Chalklands Quarries, Coombefield Quarries, Cottonfields Quarries, Duncecroft Quarries, France Quarries, Freshwater Quarries, Gannetts Quarry, Great Down Quarry, Independent Quarries, Inmosthay Quarries, King Barrow Quarries, Long Acre Quarries, Longstone Quarries, Old Quarr(y) (× 2), Park Quarries, Purbeck Stone Quarries, (The) Quarr (freq), Quarr Fm (× 2), Quarr Hill (× 2), Quarr Lane **3** 363, Quarr Shrub, Quarry (freq), Quarry Coppice, Quarry Corner, *Quarry Head*, Quarry Hill, Quarry Lane (× 2), Quarry Wd, Sheat Quarries, Shepherd's Dinner Quarry, Silklake Quarries, Tout Quarries, Trade Quarries, Wakeham Quarries, Waycroft Quarries, Weston Quarries, Wide Street Quarries, Withies Croft Quarries; f.ns. Quar **3** 29, 198

quart, **quarter** eModE 'one of four parts', also 'district, residence'. Kennel Quarter, School House Quarter, Solomon's Quarter; f.ns. *Brechquart* **1** 326, *Hoggesquart* **1** 327

quick ModE 'quickset or hawthorn hedge', f.n. ?Quicks **4** 453

quintaine ME 'post set up to be tilted at'. ?Quince Hill Wd; f.n. *Quyntine* **3** 382

rā OE 'roe-buck'. *Rocombe*

rabbit ModE. Rabbit's ~, Rabbits' Copse, Rabbit Warren; f.ns. Rabbit Brake **1** 344, Rabbit Grd **2** 172, Rabbit Holes **1** 84

race eModE 'a rush of water, a water-channel'. Mill Race (freq), *Race of Portland* **1** 220, ?Race Fm

race ModE 'horse race'. Blandford Race Course, Race Down, Racedown Fm; Race Post Fd **2** 238, Racing Grd **1** 61

rack, *v.* **rake**

rād OE 'suitable for riding on', later 'road'. Broad Road Fm, Old Road Wood, ?Rhodehorn Plant., Road Plant., Roads Lane, ?Rodway Fm. Freq in modern minor names and street names as ~ Road, e.g. Arrowsmith Road, Fair Mile Road, Harvey Road, Newton Road, Palmerston Road, Whitehouse Road [*v.* also **road** in sense 'channel']

radar ModE. Radar Hill (*v.* Addenda **1** 133)

rād-weg OE 'road suitable for riding'. ?*Radway*, Rod Hill Lane; f.ns. *Radweye* **4** 304, *Rodways* **2** 53

***rǣc** OE 'a reach, a strip of land or stretch of water', f.ns. *Long Reach*(*e*) **1** 345, Lt Reach **2** 98, Reach **1** 143, *reche voc' la Wade* **2** 181

rǣge OE 'roe', f.n. *Reigheham* **4** 368

ræt OE 'a rat'. ?Ratcombe Wood, ?Ratleigh Wood; f.n. Ratsley **4** 268

rǣw, **rāw** OE 'row (of house or trees, etc)'. Bartleys Row, Blackrow Cmn & Fm, Boxbury Row, Bushy Row, Captain's Row, *Chepemanrewe* **3** 365, Chesters Coppice & Hill, ?Cockrow Fm, Combe Row, Cooks Row **2** 184, Dungrove Rows, Elderen Row, Fancy's Row, Fives Court Row, Furzey Close Row, Gatehouse Row, Harvey's Wood Rows, Hr & Lr Row, Leg Row, Long Row Coppice, Lynch Fm, Marley's Row, Parkham Rows, Park Row, Percy's Row, Perham's Row, Philpott's Row, Post Row, Rew (× 2), The Row (× 3), Row Copse & Hill, Rue Fm, Sheepcroft Row, Sheephouse Row, Stafford Row, Stour Row, Ten Acre Row, Westrow Ctgs, *Widehokesrewe* **4** 206, *le Woderewe* **3** 76, Woodrow (Fm) (× 3); f.ns. *Rewe ground* **2** 216, *le Rewes* **2** 11. In AS charters, *firesrewe* **3** 256, (*h*)*linc reawe* **1** 15, **3** 128, 207, *norð rewe* **4** 374, *rœw apoldran* **4** 454, *reowe* **3** 238, *þorn rewe* **3** 200

raff lME 'rubbish' or 'a kind of timber', f.n. *Raffehowse* **3** 111

ragge ME 'rough stone', also probably 'moss, lichen'. Rag Copse (× 2); f.ns. *Ragg* **3** 334, *the Ragge Copse* **2** 158, Rag Md **4** 352, Rag Moor **3** 213

ragged eModE 'ragged, rough'. Ragged Firs, Ragged Rocks; f.ns. Ragged hedge **2** 119, Ragged Row (Grd) **2** 158, 249
raile ME 'wooden bar', f.ns. The Rails **3** 129, *Rayles* **3** 145, *Wulton rayles* **1** 346
rainbow ModE, f.n. Rainbow Flg **1** 363
rake, **rack** ME 'narrow hill-path, rough path'. ?*Outrack Pryme* **4** 197, Rake Bottom & Hill, Rakehill Fm, Rakespit Coppice & Plant., Shire Rack; f.ns. Rack Cl **4** 192, *Le Rakehay* **1** 365 [*cf.* **rakke**]
rakke ME 'tenter-frame', f.ns. *le Racke stare* **2** 297, Rackhay **3** 8, *Rackheis* **2** 183, *Rakcloose* **2** 175 [*cf.* **rake**, **rack**]
ramm OE 'ram'. ?Old Ram, Ramesham, Ram Lane, ?Rampisham, ?Ramsden Coppice, Ramshorn Lake
range ModE 'row (of trees)', f.ns. The Range **2** 81, Range Fd **3** 205
rān OE 'boundary strip', in AS charters, *ran wylle* **3** 380
ranke ME 'vigorous in growth', f.n. Rankley **4** 410
rāp OE 'rope'. Rope Fm, ?Rope Lake Head; f.ns. *Ropemelne* **4** 70, *ropes* **2** 230
rape ModE 'rape (plant)', f.n. Rape Plot **4** 424
raspberry OE. Raspberry Copse, Raspberry Lane **3** 147
raw ME 'raw', f.n. ?*Westmostrawland* **2** 11
rēad OE 'red'. ?Mount Radford, *Radeslow*, ?*Radway*, Red Barn, Red Barn Fm, Red Bridge (× 2), Redcliff Fm, Redcliff Point, Red Cow Dairy, Red Cross, *Red Cross*, Redd quarre **4** 8, Redend Point, ?Redford Fm, Red Hill, ?Redhole Lane, Redhone Hd, Redhorn Lake & Quay, Red House Fm, Redland Plant., ?Redlands, Redland's Fm, Red Lane (× 4), Redlane Hundred, ?Redmoor Fm, ?*Redpool*, Red Post (× 3), Redpost Hill, ?Ridley Pit, Rodden. Freq in f.ns., e.g. *Radegare* **4** 40, *Redd Hill* **4** 11, *Redewode* **2** 216, *la Rydeforhurde* **3** 34. In AS charters, *radenwey* **3** 257, *readan hyrstes* **4** 304, *readan weg* **2** 135, *readan weȝ* (*v.* Addenda **4** 193), *rede hane* **2** 122
rectory OE. Old Rectory (× 3), Rectory (freq), Rectory Fm
redcap ModE 'sprite or goblin', also 'goldfinch', f.n. Red Caps Md **4** 192
reef eModE 'narrow ridge or chain of rocks, shingle or sand'. ?Reforne
(ge)rēfa OE 'reeve', f.ns. Reeve Md **1** 271, Reeve Plot **3** 78, *Riveham* **3** 111
remedy ModE. Remedy Coppice, ~ Ctgs, ~ Gate
reservoir ModE. Waterloo Reservoir
restharrow ModE plant-name. Rest Harrow
gerēþru OE 'a bank of oars', in AS charters, ?*gereþrenc* **3** 110
retreat ModE. The Retreat

rex (gen. sg. **regis**) Lat 'king'. Bere Regis, Lyme Regis, Wyke Regis
ribbe OE 'ribwort, hound's tongue', f.ns. *Ribbecroft* **4** 59, Riblands **4** 436
***ric** OE 'strip of land, stream, narrow road'. ?*Boteridge* **1** 85–6, Kimmeridge
ricke ME 'heap, stack'. Shadrach (Fm), Shadrach St. **4** 240, Shedrick (Bridge & Hill); f.ns. *Reckeham* **3** 298, Rick Hay **3** 198, Rick Plot **4** 424, Rickyard Cl **3** 109
rickebarton eModE 'rick yard', f.ns. *le Reeke Barton* **1** 281, **2** 78
***rïd** PrWelsh, ***rid** PrCorn 'a ford'. ?Chetterwood, ?R. Parrett (*v.* RNs)
ride ME 'bridle-path'. Bridmore Ride, The Old Ride; f.n. ?*Ridesherde* **1** 346
ridel ME 'riddle', f.n. ?*Ridelhegh* **3** 214
riding eModE 'bridle-path'. Ridingway
riff eModE 'a rift, a chink'. ?Reforne
rifle ModE, f.n. Rifle Pce **3** 225
riht[2] OE 'straight', in AS charters, *richte herepath* **1** 31, *richte merke* **1** 238, (*i*)*richt wege*, *-wei* **1** 31, *riȝtwey* **3** 257
rima, ***rioma**, ***reoma** OE 'rim, edge, border, boundary'. ?Rempstone, ?Rempstone Gate, Rhymehorn, ?Rhymes Copse, ?Rimbrow Coppice, ?Rimsmoor Pond, *Ryme Extrinseca*, Ryme Intrinseca
***rip(p)** OE 'strip, edge', f.n. ?*Ryp*(*p*)*ewey*(*e*) **2** 14
risc, ***rysc** OE 'rush'. Ruscombe Lane, Rushcombe Bottom, Rushmead Fm, Rushmoor Pond, Rushmore Fm, Rushmore Hundred, Rushpond, Rushton, *Rushton* **2** 232, Rushy Ctgs, Russon; f.n. *Rushley* **2** 140. In AS charters, *riscemeres* **2** 67, *risshdene* **3** 257
risc-bedd OE 'rush-bed', f.ns. Ricksbeds **3** 392, *Ryxbedd* **3** 162. In OE charters, *ealden reshbed* **3** 206, *rissbed* **3** 257
riscen OE 'rushy', f.ns. Rixen Md **4** 424, *Ruyschenam* **1** 169
riscig OE 'rushy'. Rushy Bottom; f.ns. *Risschiaker* **1** 260, *le Risshiham* **1** 264, Rixey Moor **3** 162, Rushy Plot **4** 368
rīð OE 'a stream'. ?Chetterwood; f.n. *La Rytha* **2** 230. In AS charters, *alor riðe* **3** 233, *riþe* **3** 111
rivere ME 'river' also 'river bank'. Blackwater River (*v.* RNs), Corfe River (ib), Mangerton River (ib), Martins River Island, *Moones Bridge River* (*v.* RNs), Moors River (ib), River Ho, River's Edge, Riverside Works, Rock Lea River (*v.* RNs), Sherford River (ib), Wriggle River (*v.* RNs), Wriggle River Lane; f.ns. *the North River* **1** 194, *le Ryver* **1** 204

road eModE 'channel, sheltered water where ships may ride'. Brownsea Road, *the North Rhoade* **1** 256, *Portland Road* **1** 223, *Weymouth Road* **1** 257 [*v.* **rād**]

roadstead ModE 'place where ships anchor'. Roadstead Fm

roche[1] OFr, ME 'rock', f.n. *Rochelon* **2** 72

rocket ModE. Lookout & Rocket Post

rocky ModE. Rocky Common, Rocky Knap

***rod**[1], ***rodu** OE 'a clearing'. Road Lane Fm, Rhode Barton, ?Spruce Road; f.ns. ?Bushrods **1** 119, ?Rod **2** 26, *la/the Rode* **3** 109, *Rode Meade* **4** 391. In AS charters, *ealde rode* **1** 31

rōd[2] OE 'rood, cross'. Holyrood Fm, *Rood Hill*; f.ns. *Hollirodehill* **1** 313, *Rood close* **4** 304

rodde OE 'rod, shoot', f.ns. *Roddele* **1** 328. In AS charters, ?*roddan pol* **2** 136

roguery ModE, f.n. Roguery **4** 348

rokke ME 'rock'. Black Head, Black Rock (× 3), *Black Rock*, Black Rocks, Black Venn Rocks, Boiling Rock, Broadrock, Butter Rock, Christmas Rock, Church Rock, Cove Rocks, King Rock, Lane House Rocks Rd, Lias Rocks, Mackerel Rock, Mouth Rocks, Mupe Rocks, Norman Rock, Pinion Rock, Pulpit Rock, Ragged Rocks, ?Rocket, Rock Ho, Rock Pit, Shag Looe, Split Rock, Tar Rocks, Virtle Rock, Watch Rock; f.ns. Black Rock **1** 257, Rock Cl **4** 84, Rocks **1** 257

Roman ModE. Roman Bridge, ~ Road

roofless eModE 'roofless (tenement)', f.ns. *Roveles* **4** 35, *Stoffordes Roveles* **3** 112

rookery ModE. Denhay Rookery, Heave Rookery, Lower Rookery, N Bowood Rookery, North End Rookery, Rookery (Ctgs), The Rookeries, The Rookery (freq), Rookery Coppice & Fm, ~ Copse, ~ Lane, ~ Wood

rooted ModE 'full of roots', or 'grubbed up, rooted and dug up by swine', f.ns. Rooted Copse **4** 55, Rooted Wood **1** 297

ropere ME 'rope-maker'. Roper's Lane **1** 156

ropewalk eModE 'shed or alley for rope-making'. Rope Walk, Ropewalk, *Rope Walks*

ropeway eModE 'ropewalk'. *Ropeways* **4** 358

rose ModE. Rose Ctg (freq); f.n. Rose Eddish **4** 283

ross ModE dial. 'marsh, moor', ?Rossmore

rōt OE 'cheerful'. ?Ratcombe Wood

roten ME 'rotten', f.ns. Rotten Md **4** 25, *rotten okes* **4** 284

round OFr, ME 'round'. Round Barrow, Roundbush Wood, Round Chimneys Fm, Round Clump (× 2), ~ Coppice/Copse (× 6), ~ Down (× 2), ?Roundham Nursery, ?Roundhayes, Round Hill (× 4), ~ Ho (freq), ~ Island, ~ Knoll, ~ Plant., ~ Pound, ~ Wood; f.ns. *Rounde Meadowe* **3** 34, *Round Hamm Furlonge* **2** 175

roundabout ModE 'piece of ground surrounded by trees, streams or roads'. Roundabout (Plant.); f.n. Rounabout Cl **4** 283, Roundabout **4** 414, 453

round robin ModE 'a game', f.n. Round Robbin **4** 109

rubble ModE, f.n. Rubble Batch **1** 228

ruffet ModE 'rough ground' (OE ***rūhet**), f.ns. Ruffett **4** 424, Hr Ruffetts **4** 410, Ruffits **4** 335, White Roughit **1** 258

ruffety ModE 'rough', f.ns. Ruffety **4** 330, Ruffity **3** 57

rug ModE dial. 'stony ground', f.n. Rug **4** 343, Rug Md **4** 352

rūh OE 'rough'. Rough Bulbury, Rough Close (Barn), Rough Copse (× 2), Rough Corner Copse, Rough Down Coppice, ~ Lane, Rough Ground Coppice, Roughmoor, Hr & Lr Rough Pits, Rough Plant., The Roughs (× 2), Roundham Nursery, ?Roundhayes, *Row Barrow*, Rowbarrow (× 2), Rowbarrow Barn, Rowbarrow Hundred, Rowbarrow Lane, Row Close Copse, Rowden Bridge, ~ Coppice, ~ Hill, ~ Mill Fm, Row Down Coppice, Row Hill, Rowlands. Freq in f.ns., e.g. *Roudoun* **1** 129, Rough, Row **1** 94, Row Borrow **3** 213, *Rowefurlong* **2** 14, *Rowemeade* **1** 264. In AS charters, *Ruanberghe* **3** 257, *rugen bergh* **3** 207, *rupemor* **1** 32, *Ruþanþorn* **2** 58, *ruwan beorȝ* **4** 304, *ruwan leges gete* **2** 162, *ruwe dich* **3** 257

ruins ModE. Ruins Lane, Ruins Plant.; f.n. Ruins **4** 331

rūm OE 'roomy, spacious'. ?Ram's Hill; f.n. Rum Pits **1** 84. In AS charters, *Rumanhelle* **3** 171

rum-hous ME 'a privy'. ?Room Fm

running eModE 'pasturage, stream', f.ns. *le running acre* **2** 20, *le Running Ham* **2** 158, Running Md **3** 29

rust OE 'rust, rust colour', f.n. Rust Bridge **2** 32

(ge)rydd OE 'cleared', f.ns. *Ridelond* **4** 35, *la Rydelond* **2** 283

***ryde**, ***rȳd(e)**, ***rēod** OE 'clearing'. ?Chetterwood (*v.* Addenda); f.ns. *Medryde* **3** 257, *Northrede* **4** 167

***ryden** OE 'clearing', f.n. *Ridencroft* **4** 304

***ryding** OE 'clearing'. ?Rydens; f.n. *la Ridyngge* **4** 411

ryegrass ModE, f.n. ?Raygrass **1** 29

ryge OE 'rye'. Rix(t)on, Ryall, Ryalls Cross, Ryall's Fm & Lane, Ryal's Plant., Ryeberry Hill, Rye Close, Rye Hill (× 2), Ryeland Coppice, Ryeland's Barn, ~ Coppice, Ryland Plant.; f.ns. *Ridon Hill*

2 53, Ryland **3** 253, *Rylandes* **4** 35, *Ryverlonges* **4** 411. In AS charters, *riȝ cumb* **4** 304

rygen OE 'growing with rye', f.n. ?*Reynakre* **3** 207

rymple lME 'a wrinkle', f.n. Rimple **3** 187

ryne OE 'stream, ditch'. *the myle ryme close* **4** 277; f.ns. Long Run **4** 299, *le Ryne* **1** 62, *Smalryn* **2** 216, *la Waterruyne* **1** 264

***rysce** OE 'rush-bed', f.n. Rush **2** 9

sack eModE 'large bag', f.n. ?Sack **4** 203

sacu OE 'lawsuit'. Sackmore Lane; f.n. *Sacmore* **1** 169

saddleback ModE 'saddle-shaped hill'. Saddle Back; f.n. Saddle Back **2** 9

sadol OE 'saddle-shaped ridge'. ?Saddle Street; f.ns. Saddle **4** 383, Saddle Cl **4** 330, 424, *Sadelburn* **3** 34

sǣ OE 'sea, lake'. Little Sea, Sea Barn, Seacombe Bottom, Sea Lane, Sea Mark Clump, Sea Hill, Sea Meadow, Sea Spray, Seatown, Sea Vale Fm, Sea View, Seaview Ctg, Seaway Lane, Seawell Knap. Also in f.ns., e.g. *Seelid* **1** 70. In AS charters, *æst Sæ* **1** 257, *sa, sce, Se* 1 32, *sæ* **4** 370, *sebergh* **3** 207, *west Sæ* **1** 272

sǣd OE 'sown grass', f.n. ?*Sedbourgh* **4** 476

sǣmann OE 'seaman', f.n. *Semenlond* **1** 238

sæppe OE 'fir-tree'. f.ns. ?*Sabeleye* **2** 271, ?*le Sappys* **1** 121

sǣt OE 'trap', f.ns. ?*Setecumforlang* **4** 183, ?Setty **3** 78

sǣte OE 'dwellers, settlers'. Dorset (county name)

safene, **safine** OE 'savine (a kind of juniper)', in AS charters, ?*safundune*, *saþendune* **1** 32

sāgol OE 'staff, club', in AS charters, *Saggelorð* **1** 272

sainfoin ModE 'the fodder crop sainfoin', f.ns. Cinque Foy **4** 342, Saintfoin Fd **4** 74, Singfoy **4** 383, 417

salt[1] OE 'salt'. Salt Box, Salt Hill Plant.

salt[2] OE 'salty, brackish'. Salterns Beacon, *Saltforde* **4** 395; f.ns. *Salteheygh* **3** 178, *Saltemede* **4** 11

salt-ærn OE 'building where salt was made or sold'. Saltings, Salterne, Salterns Beacon, ~ Copse. Also in f.ns, e.g. *Salterne* **1** 121, Salterns **2** 37

saltere OE 'salt-worker', f.ns. *Salteresthierne* **4** 283, Salters Plot **2** 32. In AS charters, *sealtera cumb* **4** 374

salthouse ME 'building where salt was made or sold'. *Salthouse* **1** 77

saltman ME 'maker or seller of salt', f.n. ?*Soutmannemede* **4** 218

salverei, **sauverei** OFr 'save the king'. *Sauveray* (a tower) 1280–5 (at Corfe C. **1** 12)

sanctuarie ME 'land endowed for the maintenance of a church sanctuary'. The Sanctuary; f.ns. *Le Sanctuarie* **4** 70, Gt & Lt Sanctuary **4** 370, *Sanctuary Close* **4** 70

sand OE 'sand'. Ashley Sand Gate, Sandbanks, Sandbury Batch & Moor, Sandford, ~ Ctg, Sandford Orcas, Sandhill Clump,Sandhills, Sandhills Coppice, Sand Holes, Sandley, Sandsfoot Castle, Sandway, Smallmouth Sand; f.ns. *Melcombe Sande* **1** 258, *Sandemore* **4** 11. In AS charters, *sanddoune* **3** 200, *Sandford* **3** 129, *sandhellesled* **3** 111, *sand mor* **3** 198

sandig OE 'sandy'. Sandy Barrow (× 2), Hr Sandy Coppice, Sandy Drove, Sandy Hill Fm, Sandy Hole Hanging, Sandy Lane (freq), Sandy Knap, Sandy View; f.n. *Sandy Lane* **2** 201

sandpit ModE 'sand quarry'. Old Sand Pit, Sandpits, Sand Pit(s) (freq), Sandpit (× 2), ~ Fm (× 2), ~ Lane

***sāpere** OE 'soap maker', f.n. Soapers Pool **3** 340

Sarsen 'Saracen', f.n. Sarsonage **4** 222

satan ModE. Satan's Square

sauge ME 'sage', f.n. Sage Grd **2** 234

sawmill ModE 'mill for sawing timber'. Friary Saw Mill, Saw Mill (× 2), Saw Mill Tavern

sawpit ModE 'pit for sawing timber'. Saw Pit (× 3), Sawpit Plant.; f.ns. *Sawe pyttes* **3** 238, *the Sawpit* **1** 354, Saw Pit Cl **2** 161, (Lt) Saw Pit Grd **2** 37, 200

scab ModE 'skin disease (in sheep)', f.n. Scab Acre **4** 391

scantling eModE 'small or scanty portion', f.n. Two Scondlings **4** 109

scar ModE dial. 'rocky cliff'. Scar Bank Ho

scarecrow ModE. f.n. Scarcrow Cl **2** 180

scēacere OE 'a robber'. ?Sugar Hill

sceacol OE 'shackle', f.n. Shacklewell **4** 249

sceadu OE 'shade, shelter', f.n. Shade Bush **4** 269

sceaft OE 'shaft, pole'. ?Shaftesbury

sc(e)aga OE 'small wood or copse'. Buckshaw Ho, Bushey, Cancer Drove, *Hartescath* **4** 494, Shave Cross & Fm, Shave Hill, Shave Lane; f.ns. *Middechave* **3** 33, *Shagh'ey* **3** 34, Shawford **3** 213, (the) Shave **3** 122

sceacga OE 'matted hair, etc', later 'tangled vegetation'. ?Shaggs, f.n. Shag's Mdw (*v.* Addenda **1** 128)

***sceald** OE 'shallow', in AS charters, *scealden forð* **3** 139

sc(e)amol, **scamela** OE 'market stall', also 'shelf of land, ledge'. *the Bochers Shambles* **3** 146, *schamellys* **2** 185, *Shambles* **1** 224 (a sandbank), *the Shambles* **1** 254, *le Shamell* **3** 366. In AS charters, *scamelen* **3** 136, *shamelen* **3** 112

scēap OE (***scī(e)p** WSax) 'sheep'. *Schipenedone* **2** 123, *Shaphegh* **4** 33, Shapwick, Shatcombe Fm, Sheepcroft Row, Sheep Down, Sheephouse, Sheeplands Copse, Sheep Lands Lane **3** 363, Sheep Linhay, Sheep Park Coppice, *Le Shepe Downe* **3** 125, Shipmoor Point, Shipstal Point, Shipton Gorge. Freq in f.ns., e.g. *Schephay* **4** 11, *Shapehegh* **3** 214, Sheep Croft **2** 222, *Shippelond* **2** 53. In AS charters, *Schepleg h'e* **3** 258

scēap-hyrde OE 'shepherd', f.n. *le Schyphurdes Gore* **2** 53

scēap-hūs OE 'shed for sheep'. Sheephouse, Sheephouse Bottom, ~ Plot, Shop House Plant.; f.ns. Shappis Md **2** 130, *Shepehouse Close* **4** 11, *la Shephouse* **4** 19, *Shephowse Meade* **3** 59

scēap-wæsce OE 'sheepwash'. Sheepwash (× 4), Sheepwash Copse, ~ Fm & Green, ~ Lane

sceard OE 'cleft, gap'. Bloody Shard Gate, Lawford Sheard Gate, Shade House Fm; f.ns. Broad Shard **2** 41, *Claye Sharde* **4** 26, *Gretesherde* **3** 32, Winshard **3** 64. In AS charters, *Scherdenberwe* **3** 258, *sherd* **2** 82

scearn OE 'dung, muck', f.ns. ?*Charncroft* **1** 183, Sharn Md **3** 253

scearp OE 'sharp, pointed'. ?Sharford Bridge; f.ns. Sharphays **3** 57, *Sherpecrofte* **3** 41. In AS charters, *scearpenhame* **3** 112

sc(e)aru OE 'boundary'. ?Sharford Bridge; f.ns. Shafford's Mdw **4** 19, *Sharforde* **4** 19, *Sharlond* **3** 34

scearu OE 'share', also 'shearing', f.n. *la Sherehey* **4** 11

scēat OE 'corner of land, projecting piece of land'. ?Bagshot, ?Bowshot Fm. ?Sheat Quarries, Shooter's Lane **3** 147 [*cf.* ***scīete**]

scelf OE (WSax **scielf**, **scylf**) 'shelf of land, ledge'. Shelves Plant., Shilvinghampton; f.ns. *Schulfacre* **4** 205, *la Schulue* **2** 216, Shelves **4** 424, Shilfe **4** 93, Shills **4** 34, (*lez*) *Shilues* **1** 238

scell OE (WSax **sciell**) 'shell'. Shell Bay; f.ns. ?Corn Scully **3** 29, Shill **1** 230, *le Shull(e)* **3** 29, *Shullourd* **2** 45. In AS charters, ?*tor scylget* **3** 102

sceoppa OE, **shoppe** ME 'shop, shed'. Flax Shop **4** 42, *the lytle Shopp* **4** 360, *le Shopp* **3** 180; f.ns. Shop Close **4** 40, Shop Fd **2** 66, *Shopp Mead* **3** 34

sc(e)ort OE 'short, small'. ?Chapman's Pool, *Shortallere* **3** 233, Short Coppice, *les Shortealres* **3** 268, Shortlake, Shortmoor, Short Moor, Short Wood, *Shortwood Common*; f.ns. *Schortclyve* **4** 11, *Scortstiche* **2** 72, *Shortelande* **4** 411, *Shorthey* **3** 34

***scēot**[3] OE 'steep slope'. ?Cowherd Shoot Fm, ?Gough's Shoot, ?Scite Coppice, Shatcombe Coppice, ~ Fm, ?Shot Lake Wood, ?White Sheet Hill (× 2), ?Whitesheet Hill; f.ns. *la Scheote* **4** 375, ?*Schutouere* **3** 34, ?Shute **3** 40 [*cf.* **scyte**, **shoot**, **shute**]

sc(e)ota OE 'trout'. ?Shotover Moor

school ModE, **scole** ME 'school'. Bryanston School, *the schole-house* **3** 366, *Schollane* **3** 149, Schoolhouse Fm, ~ Lane, School House (Lane), School House Quarter, School Lane (× 2), Sherborne School

schoting ME 'shooting, archery', f.n. *Showtinge Crofte* **3** 324, *Sotingestoke* **3** 34

scīd OE 'board, plank, beam'. ?Sedge Oak Coppice

sciell or **scill** OE 'resounding', in AS charters, ?*chellenberghe* **4** 173, ?*schilleburghe* **3** 258

sciering, **scering** OE 'shearing', f.n. ?*Sheryngdon* **1** 346

***scīete** OE 'corner or nook of land'. Evershot, ?Sheat Quarries; f.n. At Shoot **4** 9 [*cf.* **scēat**]

***scīewels** OE 'scarecrow'. f.n. *gamoldeschewole* **1** 95

scilling OE 'shilling' (?or an OE ***scil(l)ing** 'division, boundary'), f.ns. *Sciling* **3** 266, Shilling Plot **1** 182

scīr[1] OE 'shire'. Dorset (county name), Shermel Gate, Shire Lane, Shire Rack; f.n. ?*Schirewald* **4** 400

scīr[2] OE 'bright, clear'. Sherborne, Sherford, Sherford River (*v.* RNs); ?*Schirewald* **4** 400

scīr-(ge)rēfa 'sheriff'. Iwerne Courtney or Shroton; f.n. *Schireuesmoure* **3** 34

scite OE 'dung, shit'. Sedge Oak Coppice, Shadrach (Fm), Shadrach St. **4** 240, Shedrick (Bridge & Hill), Shit Yallery Hole; f.ns. *Shyteport* **1** 185

***scitere** OE 'stream used as a sewer, dirty stream'. *Sciterhuc* (*v.* RNs), ?Shitley, Shitterton; f.ns. *Schiterlynch* **1** 260, *Shiteroke* **4** 117, *Skitterlinch* **2** 181

scluse ME 'a sluice'. Sluice Gdns

scofl OE 'shovel'. ?Salway Ash; f.ns. Shoals **1** 228, *Le Shouelede lond* **4** 14 [*v.* **-ed**[2]], Shovels Copse **4** 166, ?Showell **4** 69

sconce eModE 'shelter', f.n. Sconce **1** 80

***scor(a)** 'shore, steep slope', f.ns. ?Sharkham **4** 453, *Shorland* **4** 35. In AS charters, *Soredich* **1** 238

***scoren**[1] OE 'a steep place'. ?Shorne Cliff, Shorn Hill

scoren[2] OE 'sharply cut off, precipitous'. ?Shorne Cliff

scoru OE 'boundary mark', f.n. *la Fryelandshor* **1** 326

scot OE 'tax, payment'. Scotland (× 2); f.ns. Scotland **2** 294, **4** 343, 499, Scotley Park **4** 209

scræf OE 'cave, hut, hollow', f.n. ?Shreave(s) **4** 116

scrag Do dial. 'crooked tree-branch', f.n. Scrags **3** 205

scratch ModE. Scratch Arse

scratchy ModE. Scratchy Bottom

scrēawa OE 'shrew-mouse', f.ns. ?Sherall **4** 69, Shrow Pits **2** 242

scrip eModE 'small bag', f.n. ?Long Scrips **4** 151

***scrubb** OE 'shrub'. Quarr Shrub; f.n. Shrubs **1** 168

***scrubbet** OE 'place characterized by brushwood'. Shrubbetts; f.n. Shrubbets **3** 253

scucca OE 'evil spirit, demon', f.n. ?*Sc(h)ochleham* **2** 283

***scydd** OE 'hovel, shed'. ?Shedbush Fm; f.n. Shud **3** 392

***scylen** OE 'stony', in AS charters, ?*scyleford*, *Scylenford* **1** 94

scypen OE 'cow-shed'. Shipney Lane; f.n. Godshepen **4** 234, *Shipenhegh* **4** 254, *Shippen Walle* **4** 35, Shippon Cl **2** 140

scȳr(e) OE 'hut, shelter', f.ns. *Lange-*, *Lytelschure* **3** 227

***scyrte** OE 'short piece of land', f.n. ?*Floteshorte* **3** 381

scyte OE '(steep) slope or hill'(also 'spring'?). Courthouse Shoot, ?Cowherd Shute Fm, ?Shot Lake Wood, ?White Sheet Hill (× 2), ?Whitesheet Hill; f.ns. *Schutouere* **3** 34, *Shute Barrs* **4** 26, Shutewell **4** 253, *Watersheete* **3** 382 [*cf.* **scēot**[3], **shoot**, **shute**]

s(e)alh 'a sallow, a willow'. Silton; f.ns. *Selbeddes* **4** 223, *Selwood* **3** 112, *Zeale Crofte* **2** 217; f.ns. Hosals **4** 217, Zales **4** 218. In OE charters, *ealde seale* **2** 82, *seales* **3** 138

seat ModE 'residence'. Eldon Seat, Minterne Seat Coppice

sēað OE 'pit'. ?*Horsyth*; f.n. *Horsithe* **3** 199. In AS charters, *chealc seðas* **3** 110, *soces seað* **3** 112

***seax** OE 'knife, sword', f.n. ?*Saxcemede* **1** 328

Seaxe OE 'Saxons'. Sixpenny Fm, Sixpenny Handley (Hundred). In AS charters, *lutlen seaxpennes suðeke* **3** 104

secg[1] OE 'sedge, rush'. Sedge Copse, Sedgecroft Common; f.ns. Sedgemoor **2** 172, *Seggebroke* **4** 271. In AS charters, *secg broc* **2** 158

secrestein ME 'a sexton', f.n. *Sextens Mede* **3** 98

secunda Lat 'second'. Netherbury **4** 230

sedgy ModE 'growing with sedge', f.ns. Sedgey Pce **1** 297, Sedgy Grd **1** 100

sege ME 'privy, excrement'. ?Sedge Oak Coppice

seld, *v.* **setl**

sele[1] OE 'a hall'. ?Seldown

***sele**[2], ***siele** OE 'willow copse'. ?Silly Hill; f.ns. *Selemer* **4** 308, Silly **3** 177, ?*Suyle* **3** 200

selvage ModE 'border, edge', f.n. ?*Sylfynche* **4** 223

***sengel** OE 'land cleared by burning, burnt place', f.n. ?Single Cmn **4** 373

***senget** OE 'land cleared by burning, burnt place'. ?Scent Close Plant.; f.n. *le sente* **1** 134

seofon OE 'seven'. Seaborough, Seven Acre Coppice, ~ ~ Plant., ~ ~ Withy Bed, Seven Ash Common & Fm, Seven Barrow Plant., Seven Barrows, ?Silverlake Fm; f.ns. *Seuethorn* **2** 276, 7 Acres **2** 51, *Sevenditches* **2** 271, *Seveneakre* **3** 207. In AS charters, *seofan æceras* **4** 374, *seuen diche* **3** 123, *seuen willes þry* **1** 32

seolfor, *v.* **silver**

***sēoluc** OE 'drain, gully'. Silklake Quarries; f.ns. Sel(ke)croft **3** 289, *Selkham* **4** 500, *Silke* **1** 379, Silk Md **3** 253, Sill Md **3** 166

serge ModE 'woollen fabric', f.n. ?Sarge **3** 162

serpentine ModE 'winding, sinuous'. Serpentine Plant.

session ModE. ?Sessions Gate

(ge)set OE 'fold, stable' or 'seat, dwelling'. *Chircheset* **3** 204, *Couschete* **4** 147, ?*Sete* (*v.* RNs); f.ns. *Algaressitte* **1** 260, ?*Setecumforlang* **4** 183, ?Setty **3** 78, *Wynneshamesete* **4** 271

seten OE 'plantation', f.n. ?*Sytenmede* **3** 264

setl OE 'seat'. Kingsettle Fm, *la Tolselde* **3** 150; f.n. *Selegyevessetle* **3** 34, ?*Setelyston* **3** 227. In AS charters, *biken settle* (in bounds of Sixpenny Handley **3** 113, *v.* Addenda)

seue ME 'sieve', ?*Seuemelne* **3** 73

seventeen ModE. Seventeen Acre Point

severell ME 'plot of privately owned land'. Plain Severals Plant.

sex OE 'six'. Six Acre Coppice (× 2), ~ ~ Copse (× 2), ~ ~ Plant., ~ ~ Wood, Six Cross Ways; f.ns. 6 Acres **2** 215, *Sixeakers* **2** 251. In AS charters, *sex þorn* **2** 58, *sixe acres* **3** 200

sextēne OE 'sixteen', f.n. *sexten'acres* **3** 342

sextry eModE 'sacristy', f.n. Sexty Md **3** 98

shady ModE. Shady Bushes

shag ModE 'shag'. ?Shag Looe

shale ModE. Manfield Shale Pit

shear ModE 'act of shearing'. Shearplace Hill; f.n. Sheep Shear **4** 414

shearing ModE, f.ns. Shearing Close **1** 249, ?*Sheryngdon* **1** 346

sheepwashing ModE. Sheepwashing Lane

sheer eModE 'steep', f.n. ?*le Sherepoyncte* **4** 174

***shingel**[2] ME 'shingle'. *The Shingles* **1** 224

ship ModE. E, N & S Ship Channel
shit-breech eModE, a derogatory term, f.n. ?*Shedbriche* **3** 34
shitten ME 'filthy', f.ns. Shitten Can **4** 40, Shitten Shard **3** 253, *Shitten Stile* **1** 185
shoemaker ModE, f.n. Shoemakers Close **4** 500
shood ModE dial. 'husk of oats', f.n. ?Shoud **4** 491
shoot, shute Do dial. 'steep slope or hill'. Shute Haye; f.ns Shoot Orchd **4** 424, Shute Md **4** 417 [*cf.* ***scēot**[3], **scyte]**
shot-hole ModE 'a small hole in a fortified wall through which to shoot'. Shothole
shovel ModE. Navvy Shovel
shrubbery ModE. Burn's Shrubbery, The Shrubbery (freq), Shrubbery Lane, Steepleton Shrubbery
shuttle eModE (**scyt(t)el(s)** OE) 'flood-gate, gate which bolts shut', f.n. *Shittles* **3** 376, *le Shuttle* **3** 298
sīc OE 'a water-course, a small stream'. Gussage Brook (*v.* RNs), Gussage All Saints/St Andrew/St Michael. In AS charters, *sic* **3** 258
sīd OE 'broad'. Fifehead Sydling, Sydling St Nicholas, Up Sydling; f.ns. *Sidberowe* **3** 298, Sidmans **3** 231, *Sidneham* **1** 346
sīde OE 'hillside'. Brookside (× 3), Downside, Elmside Coppice, Fernside, Forest Side, Hillside, Moorside, Riverside Works, Sunnyside Fm, Waterside, Waterside Fm, Weyside, Woodside (× 2), Woodside Ctg(s)
sideland ModE 'land alongside another feature', f.n. Sideland Cl **2** 161, **4** 399, Sidelands **4** 453
sidelong ModE 'sloping', f.ns. Sidelong Grd **4** 283, Sidelong Plot **1** 168
siding ModE 'short railway track for shunting'. Broom Ballast Siding, Carter's Siding
***sīdling** OE, ME 'land alongside a stream or other feature', f.n. Sid(e)ling **4** 391, 424, Sideling Bagwell **4** 470
signpost ModE, f.n. Signpost Close **2** 230
silk ModE. Westbury Silk Mills
silly-how ModE dial. 'a child's caul'. ?Silly How
silva Lat 'wood'. Mount Sylva
silver ModE (**seolfor** OE) 'silver'. ?Silverbridge Fm, *Silver Well* **4** 138; f.n. Silver Ho **4** 436, Silverlands **4** 410, Silver Md & Well **3** 67
sinder OE 'cinder'. Synderford
***singel** OE 'shingle', f.n. ?Single Cmn **4** 373
sister ModE. *The three Sisters* **4** 295
sixpenny ModE. Sixpenny Gate
skating ModE. Skating Pond

skilling Do dial. 'penthouse, shed'. ?Skilling Fm; f.n. Skilling **3** 67

skim Do dial. 'to mow', f.n. ?*Skimehayeslane* **3** 34, Skimming Close **4** 391

skinnere ME 'skinner'. Skinner St. **2** 43

skippet ModE (meaning uncertain) Hr Skippet Fm, Lr Skippet, Mount Skippet, Skippet Heath

slæd OE 'valley'. Slod Brook (*v*. RNs), Slodbrook Fm, *Trendle slade* **2** 156, Withy Slade Covert; f.ns. *Inslade* **2** 14, *Ocslad* **1** 260, *Odeslade* **3** 6, Slade Furlong **4** 34. In AS charters, *beteswirþe sled* **3** 104, *hecgan sled* **1** 31, *mere sled* **4** 173, *sandhellesled* **3** 111, *sledes* **3** 41, *sledes Northecge* **3** 123, *sledweie* **3** 112

***slæget**, ***sleget** OE 'sheep pasture'. ?Bonsley Common, *Hog Slight* **4** 179, Slait Barn (× 2), Sleight, Sleight Bldgs, South Sleight Coppice & Knap; f.ns. Harvest Slait **3** 27, *the North Sleight* **2** 14, The Sheep Sleight **4** 40, *La Slayitte* **3** 35. In AS charters, *slæget* **2** 162

***slæp** OE 'mud, mire, marsh'. ?Woolslope Fm; f.n. *le Louseslappe* **2** 289. In AS charters, ?*slaep* **3** 380

***slǣp** OE 'a slippery, muddy place'. Slape Fm & Ho, Sleep Bottom & Brook, Slepe, Slepe (Fm), The Slopes; f.ns. ?Cowslip **3** 55, *Sle*(*e*)*pe hedge* **2** 230

slāh OE 'sloe', f.ns. Sloe Plot **4** 436, Sloes **2** 63

slāh-þorn OE 'sloe thorn', f.n. Sloe Thorn **4** 436

***slāh-trēow** OE 'sloe tree', f.ns. Sloetree Cl **1** 151, *Slootry Close* **2** 53

sleck ModE dial. 'soft mud', ?Sleck **2** 105

sleech eModE (ME **sliche**) 'mud deposited in a river'. Sleech Wood

***slege** OE, ModE dial. **slay**, **sley** 'sheep pasture'. ?Bonsley Common, ?*Slay Strete* **4** 393; f.ns. Slaye Acre **3** 59, ?*La Sleyacre* **1** 42, *Le Sleye* **1** 366, *Sley Yates* **2** 262

***slicu** OE, **slyke** ME 'smooth', f.ns. Sleek Beard **2** 242, Slick beard **3** 64

slide ModE 'earthslip, slippery place'. ?Slide Acre Coppice; f.ns. ?East Slide Banks **1** 94, Slide Acre **4** 425, Slides **4** 399

***slīete** OE 'sleet, mud', f.n. Slitmoor **1** 143

sling ModE 'narrow curved piece of land', f.ns. Sling **2** 220, Sling Withey Bed **4** 110, Slyng **2** 149

slip(p)e ME 'narrow strip of land'. North Slip, South Slip; f.ns. Beach Slip **4** 39, Slip Close Md **4** 425

slippery ModE, f.n. Slippery Close **4** 249

slōh OE 'slough, mire, a miry place'. ?Mount Sloe, *Radeslow*, Sloes Hill, Slough Fm, Slough Lane; f.ns. *Horyslowe* **3** 334, Slough Croft

4 425, Slough Ditch **2** 230. In AS charters, *fulan slo* **4** 303, *horgan sloh* **3** 289

sloppe ME 'muddy place'. Slop Bog (Addenda **2** 228); f.n. Slop **2** 140

sloth(e) ME 'muddy place' (or ***slytte** OE 'muddy place'), *Sluth* **2** 162

slub, **slob** ModE 'thick mud', f.ns. Slobs **1** 228, *the Slubbe* **3** 227

slugger ModE 'sluggard' or **sluggardly** 'lazy', f.n. ?Sluggerly **4** 330

sluice ModE. Sluice Weir

slut(')-hole ModE 'place for rubbish', f.n. *Slutes hole* **1** 32

smæl OE 'narrow'. ?Smacam Down, ?Smacombe, Smallbridge Fm, Small Mouth, Smelland's Lane. Also in f.ns., e.g. *Smalelane* **3** 34, 149, *Smalemede* **2** 216, *Smalewey* **2** 14, *Smelecombesheved* **1** 42. In AS charters, *smalan hrucȝ* **4** 303, *Smalan pæþ* **4** 35, *smale broke* **3** 109, *Smalencumbe* **1** 32, *smal(e) þornes* **3** 41, 129

smēðe[1] OE 'smooth, level'. Smedmore Ho, Smithfield Coppice, ?Smitten Corner; f.ns. *Smethedone* **2** 53, 251, *la Smethemora* **2** 149, *Smethenhull* **4** 22, Smiths Md **3** 30, Smythe Hill Flg **1** 363

smicre OE 'fair, fine', in AS charters, *smiceran æc* **4** 454

smið OE 'smith', f.ns. Smithick Fd **3** 98, *Smythes lane* **3** 174

smiððe, **smeðe** 'smithy'. ?Smetherd Fm, Smithy (freq); f.ns. Smithie Cl **3** 133, ?*Bove smithelyngheye* **4** 181

smoke ME 'smoke money'. *Smokeham*, Smokeham Bottom; f.ns. Smock Grd **4** 411, *Smokaker* **3** 34, *Smokland* **2** 78

smoky ModE (reference uncertain). Smokey Hole; f.n. Smoky Hole **4** 425

smuggler ModE. Smugglers' Cave, ~ Lane

snaca OE 'snake', f.ns. Snake Land Flg **2** 148, *long-, shorte snake* **2** 175, Snakey Close **1** 286

snād OE 'a detached piece of land or woodland'. Snatford Bridge, ?Snodwell Fm

snægl OE 'snail'. Snail Creep, Snail's Bridge; f.n. ?Sneal Cl **4** 500

***snæp** OE 'boggy piece of land', f.ns. Snap **4** 203, Snape Cl **4** 500, Snapes **4** 331, *Snapis* **4** 100

snag ModE 'tree stump, sharp projection, obstacle, etc.' or **snag** Do dial. 'the small peabig variety or species of sloe'. Snag Fm, Snagharbour Wood, Snag Lane. Also in f.ns., e.g. Snag Coppice **1** 249, Snaghill **1** 134, Snag Md **1** 249

snāw OE 'snow'. ?Snow Hill; f.ns. Snaverland **1** 198, Snowdon **1** 378, Snow Hill **2** 259

snīte OE 'a snipe'. Snipe Gate, Snipe Moor [*cf.* **snype**]

snowdrop ModE. Snowdrop Cottage

snuff ModE 'thing of little value', f.n. Cole Hill Snuffhalf **2** 119

snug ModE '(that which is) comfortable'. ?Snug
snype ME 'a snipe', f.n. Snipe **4** 348 [*cf.* **snīte**]
soap-ashes ModE 'wood ashes used in soap-making', f.ns. Soap Ash **4** 411, Soap ashes **3** 101
soap-house, **soap-shop** eModE 'building for soap-making', f.ns. Soap House Md **4** 425, Soap Shop Orchd **3** 162 [*cf.* **sāpere**]
soc OE 'moisture, drainage', in AS charters, *soces seað* **3** 112
sockety ModE dial. '?boggy, marshy'. ?Socketty Hock, ?Sockety
sod eModE 'turf', f.ns. Sod Hay **3** 136, Sods **4** 453
sodom eMod 'a wicked place'. ?Sodern; f.n. Sodom **4** 92
soer OFr 'sister', f.n. *deux soers* **2** 161
sog, *v.* ***sugga**
soggy, *v.* ***suggig**
sogh ME, **sough** ModE dial. 'bog, marsh', also 'drain, trench', f.ns. ?Sowmead **3** 138, Surf **4** 331
soldier ModE, f.ns. Soldier **2** 26, The Soldiers Close, Soldiers Grd **4** 156
sop ModE 'wet ground', f.n. Sops **3** 101
sort ME 'lot, allotted share', f.n. *Sortelaunde* **4** 271
sōt OE 'soot', f.ns. *Southmestsotlond*, *Suteput* **1** 328, *Sutebarwe* **1** 346
sour dock ModE 'common sorrel', f.n. Sourdocks **1** 119
spa ModE. Nottington Spa, Spa Rd **1** 253, Spa Well **1** 206
spadu, **spada**, **-e** OE 'spade' ?*Spadeforde Bridge* **1** 18; f.n. *Spaddone* **2** 133
spanish liquorice ModE, f.n. Spanish Liquorice Grd **1** 281
spann[1] OE 'strip of land', f.ns. Span Wd **4** 269, Spanyard **3** 198
spear Do dial 'stem of a reed'. Spearpit Coppice; f.ns. Spear Acre Md **3** 30, Spear Bed **2** 180, Spear Grd **2** 37, Spere Bed **1** 51 [*cf.* **spīr**]
***spearca** OE 'brushwood', f.ns. *atte Sperke* **1** 216, *Sparke Meade* **3** 376
spear-grass ModE, f.n. Spear Grass Beds **1** 81
spearwa OE 'sparrow'. Sparrowbush Fm; f.n. Sparrowline **2** 66
spēc[1] OE 'brushwood'. ?Specket Ctg & Lane
speck eModE 'small plot', f.n. Green Specks **1** 93
spence ME 'a buttery or pantry'. Spence Fm
***speoht**, ***speht** OE 'the green woodpecker'. Spettisbury
spettle ME (OE **spǣtl**) 'saliva, spittle'. ?Spittle Pond, ?Spittles Lane
spice ME 'aromatic herb', f.n. Spice Croft **1** 271
spicket ME 'wooden peg or pin'. ?Specket Ctg & Lane
spinach ModE, f.n. Spinnage cl **1** 344
spine Do dial. 'turf, greensward', f.n. Spine Fd **4** 391

spinney ME 'copse, spinney'. Eweleaze Spinneys, Pit Bottom Spinney, Spinney (Pits) Coppice; f.n. Small Spiney **3** 44
spīr 'reed', f.n. *Spyreacre* **4** 19 [*cf.* **spear**]
spire ModE 'pointed structure'. ?Spire Hill & Fm
spitel ME 'hospital'. Spiller's Spittle **3** 150, ?Spittles Lane, *Sputel*; f.ns. *Spitelhaygh* **4** 3, *Spittill Corner* **1** 167, Spittle Fd **2** 32
spitting ModE, f.n. Spitting Box **4** 249
spitu OE 'spit or tongue of land'. ?Winspit
split ME 'a split, an opening, a gap'. Splitmead Coppice, Split Rock
splott OE 'small plot of land', f.ns. *Splatlackes* **2** 175, Splatts **3** 8, Splot **3** 258, *atte Splotte* **3** 329, Splotts **1** 94
spotte ME 'small plot of ground'. Pleasant Spot
spray ModE. Sea Spray
spraye ME 'brushwood'. Spray Coppice/Copse; f.ns. Long Sprays **4** 425, Spray **4** 209, Spray Md **3** 263, Sprays **4** 425, Spry Md **3** 300
sprēot OE 'pole, spear'. ?Spratley Wood
spring OE 'well-spring'. Bubble Spring, Castle Lake Spring, *Spring* (*v.* RNs), Spring Bottom (× 3), ~ Close Lake, ~ Ctg, ~ Coppice(s), ~ Copse (× 3), Springfield (freq), Spring Garden Barn, Spring Grove, Spring Head (× 2), Springhead, Spring Hill Ctgs, Spring Ho, Spring Plant. (× 2), Spring Pond, Stone Spring, Waddon Spring, West Spring Lake
spring ME 'young wood, copse'. ?Springs (a plantation)
spruce ModE. Spruce Road
***spura** OE 'spur of land', f.n. ?*Spurehey* **1** 199
spye ME 'one who keeps watch, lookout' ?Spyway (× 2), Spyway Barn
***spyrt** OE 'spirt (of water), spring', f.ns. Sperts' **3** 177, Spirtly Md **2** 105, Spurt Hill **4** 335
spyrte OE 'wicker-basket', f.n. *le Spurt* **1** 143
square ModE. Satan's Square, Square Clump (× 2), ~ Coppice (× 4), ~ Copse, ~ Covert, ~ Pit, (Part) Square Plant.; f.ns. the Square **4** 92, Carpenters Square **4** 280
squire eModE 'lord of the manor', f.ns. *Squire Meade* **4** 437, Squires Mead **4** 471
stab ModE dial. 'stump', f.n. Stabb Md **4** 411
stable ME 'stable for horses and cattle'. Staple Copse, ~ Croft; f.n. *Stabelborghe* **2** 53
staca OE 'stake, post'. *Bull Stake* **1** 168, Kingstag, Stake Fm, Stake Ford Cross, *Stakes*, Stakes Ford; f.ns. Stakeford **3** 293, *la Stakelane* **4** 357, *Stakkfurlong* **4** 22. In AS charters, *lippen stagen* **3** 102
stackyard ModE, f.n. Stack yd paddock **1** 331

stǣger[1] OE 'stair, steep ascent', f.ns. *Le Chapplestayers* **1** 252, *Haselstaire* **3** 58, *the Kyngges Stere* **3** 365, *le Racke stare* **3** 297
stǣger[2] OE 'steep'. ?Stair Hole; f.n. *Stayrok* **3** 34
***stǣne** OE 'stony place', f.ns. *Stene* **2** 216, *?Stenefeld* **2** 205
stǣnen OE 'stony', f.ns. *?Stenefeld* **2** 205, *Stenynhous* **1** 355
stǣner OE 'stony'. f.n. *Standerwoddes garden* **1** 169
stǣniht OE 'stony, rocky'. Stent Drove, Stintford Lane
stæpe OE 'stepping-stone'. Stepps Fm. Also in f.ns., e.g. Stepps **1** 185, ?Steps Pond Grd **1** 81
stær OE 'starling' (*cf.* Do dial. **stare**). ?Stair Hole, ?Starmoor Plant.
stæð OE 'river-bank', f.ns. *Stath'acre* **3** 200, *le Stathe* **3** 34, *les Stathes* **4** 11
stag ModE. Stag Gate
staircase ModE. Devil's Staircase
staked ME 'with a boundary marked by stakes', f.ns. *Stakedeham* **1** 328, Staked Gate Moor **4** 425
stall, **steall** OE 'stall for cattle'. Buckhill Stall, Channel Stall Coppice, Cow Leaze Stall, Grove Stall, Lower Field Stall, Pond Mead Stall, Redstall Copse, Shipstal Point, Stall Barn, Stall Coppice, Stony Down Stall; f.ns. Corshals **3** 262, Stall Close **4** 417, Stall Grd **3** 44. In AS charter, *?on æt steallesbeorh* 4 29 (*v.* Addenda)
stān OE 'stone'. Agglestone, ?Allweston, Ball Stone, Belstone Covert, *Blackstone* **3** 52, Breston Quarries, Broadstone, The Broad Stone, Cock-Crowing Stone, Dead Woman's Stone, Dimstone Hill, Durlston Bay & Head, Froxen Copse, Giant's Stones, Graston Ho, Greatstone & Littlestone Coppice, Hell Stone, *Hell-Stone* **4** 54, High Stoy, Hore Stones, ?Hurpston, Longstone Ope Quarries, *Lymestone* **3** 307, Magging Stone, †Nine Stones, Parkstone, Puckstone, ?†Rempstone, ?†Rempstone Gate, Stafford Park Copse/Fm, Stafford Row, West Stafford, Stallen, Stanbridge, Stanbridge Mill, Stanbrook, Stancombe Barn & Coppice, ?Stand Barrow, Standing Stone, Standpitts Lane, Standridge, *Stane Hundred*, Stanford Point, ?Stanley Court, Stanton St Gabriel, Stoborough, Stone, Hr & Lr Stonebarrow, Stonebarrow Hill, Stonecombe Fm, Stone Down Coppice, Stone Firs, Stonehill Down, Stone Hill Gate & Wood, Stone Hills Plant., Hr & Lr Stonehips, Stone Island, Stone Lane, Stone Pit(s), Stonerush Drove, ?Stone's Common, Stone Spring, Stone-way Coppice, Storridge Fm, ~ Hill & Lane, Valley of Stones, Wardstone Barrow, ?Wytherstone Fm. Freq in f.ns., e.g. *Hurn'ston* **3** 226, *Lang(ge)ston* **3** 226, Standledge **4** 43, Standly Flg **3** 205, Standpit **4** 269, *Stanfurlang* **1** 260, †Staunchester **4** 203. In AS

charters, *blake ston* **1** 271, *haran stanas* **2** 161, *staford* **4** 304, *stan* **3** 289, *stan*, *stan tor* **1** 32, *stanbroc* **3** 258, 289, *stan castel*, *stancysten* **3** 208, *standene* **2** 58, *stanhecheres ande* **2** 58, *stan istel* **3** 208, *stan wal*, *stanweal* **1** 32, *stan wæg* (*v*. Addenda **3** 267), *stan wege*, *stanwei* **1** 32, *stanwey* **3** 200, 258

stande OE 'pond', f.n. ?*Scondmulle* **2** 216

standing ME 'a hunter's stand'. *Chepmans Standinges* **3** 146, King's Standing; f.n. Kings Standing **1** 301

***stānen** OE 'made of stone'. *stoninge bridge* **3** 80, ?*Vyes stonen still* **1** 171; f.n. *Stoning Close* **1** 271. In AS charters, *stanene bregge* **1** 32, *stanen wal* **1** 32

stānig OE 'stony, made of stone'. Stone Heaps Plant., Stonerush Drove, Stony Bottom, Stony Down (Lane & Stall), Stony Head, Stony Lane (× 2), ?Stony Lawn, Stony Post Ctgs, Stony Weir; f.ns. *la Stoniestrete* **3** 35, *Stonybrigg* **3** 227, *le Stonye Lake* **2** 28, *Stonyland* **4** 19. In AS charters, *stanegan crundel* **2** 122, *staniȝan weȝ* **4** 304

stān-scilig OE 'stony (ground)', f.ns. *Hydiestansilie* **1** 113, *Overstonschulye* **1** 328, ?Stone Shell **3** 45, *paruam Stonschyly* **3** 227, *Stonschulye* **1** 230

stapol OE 'post, pile, pole, boundary post'. *Puttes staple* **4** 94, Stable Barrow, *Stapleford*, Stalbridge; f.ns. Staple Flg **3** 177, *Staples* **2** 53. In AS charters, *stapel* **3** 238, *þigerðes stapel*, *Wigheardes stapele* **3** 112, 178

starve eModE. Starve Acre Clump, Starveacre Lane; f.ns. Starve Acre **3** 216, 310, **4** 249, 331, Starveall **3** 213, Starve Farthing **4** 436, Starve Land **3** 88

stay ModE 'prop, support', f.ns. Barn Stays, Plovers Stays **1** 206

stēap OE 'steep'. Stapehill

stede OE 'place, site', f.ns. Brinstead **2** 220, Hookstead(s) **3** 44, *Laxted forlong* **2** 53, *Nellstede* **3** 186, *Podingstede* **4** 139. In OE charters, *mylen stede* **3** 117, *olde wested'* **3** 201

stell OE (**stiell** WSax) 'enclosure, fish weir'. *Stones Still* **4** 115, ?Vyes stonen still **1** 171; f.ns. Short Stills **4** 116, ?*Stillehey* **3** 158. In AS charters, *ston istel* **3** 208

stēor OE 'steer', f.n. Starley **2** 172

steort OE 'a tail or tongue of land, a projecting piece of land'. Steart Lane, ?Sterte, Sturt Coppice & Fm, Sturt Lane, ?Sturt's Fm, Stut Lane, Woodstreet Fm; f.ns. Ste(a)rt **2** 119, *la/le Sterte* **4** 228, *Sterthull* **3** 227, Stiert **3** 236, *le Stort* **2** 14, Gt Sturts **3** 47

***steortel** OE 'projecting or pointed piece of land'. Sturthill

step ModE (OE **stæpe**) 'step, stepping stone'. Freshwater Steps, Steps; f.n. Stepps **1** 185

stēpel OE (**stīepel**, **stȳpel** WSax) 'a steep place, a steeple'. Steeple, Steepleton Iwerne, Winterborne Steepleton

stepping stone ModE. Stepping Stone Bridge, Stepping Stone(s) (× 6)

sticca OE 'stick, boundary post'. ?Sticklehill Barn; f.n. *Merestike* **1** 199

stiche ME 'piece of ploughing land', f.ns. *beane stiches* **1** 293, *Cattestich* **4** 270, Friars Stitch **2** 119, Gapstitch **3** 101, *Scortstiche* **2** 72, *Stiche* **1** 346

stichen ME 'piece of ploughing land', f.ns. Miss Stichings **2** 241, Stiching(s) **2** 37, **3** 288, Stitching **3** 64

sticol OE 'steep'. ?Stickland Coppice, Stickley Coppice, Winterborne Stickland; f.ns. Stickelewell Md **3** 57, Sticklands Hill **3** 30, Stickway **2** 249, *Stikelclif* **4** 142. In AS charters, *sticelan ford* **4** 454, *sticelen lane* **3** 208, *Sticholnelinche* **3** 246

***sti(e)lling** OE (WSax) 'cattle shelter', f.n. Stillings **2** 98

stīg OE 'path'. ?Stivvicks Bridge [but now *v.* Addenda **2** 103]

stigel OE 'a stile, a steep ascent'. Birch Hill Stile, Black Man's Stile, *Brodemeyde Style* **4** 376, Buck's Stile, *Churchstile* **4** 342, *Church'style* **4** 326, Churchstyle Fm, Copse Stile, Cripplestyle, Deadman's Stile, Milborne Stileham, *Mylstile* **4** 379, *Stigheleslane* **3** 149, Stile Way, Stockstyle; f.ns. Amen Style **2** 25, Clay Stile **3** 8, *Jorden Style* **2** 180, Stile Grd **2** 200, Stile Stones **3** 30. In AS charters, *stigele* **3** 112, *stiȝele* (*v.* Addenda **4** 193), *stigel hege* **3** 139, *stigla pæþ* **2** 136

stink ME 'offensive smell', f.n. *Stynkbargh'wysway* **3** 227

stinking ModE, f.n. Stinking Hill **4** 436

***stint** OE 'the sandpiper or dunlin'. Stinsford

stirc OE young bullock', f.n. *Sterkforlang* **3** 200

stīð OE 'hard', in AS charters, *stiþeleighe* **3** 258

stiward ME 'steward (of a manor)'. ?Steward's Lane; f.n. Stewards Craft **3** 156

stoc OE 'outlying farm buildings, a secondary settlement'. Burstock, Cattistock, Chardstock, *Crockern Stoke*, East Stoke, Halstock, Heathstock, Hogstock, Mandeville Stoke Fm, Netherstoke, Laverstock Fm, *Osanstoke* **4** 338, Powerstock, ?Prystock Fm, ?Stock Fm & Hill, Stockford Ctg, Stock Gaylard Ho, Stockland, *Stockland*, Stockwood, Stoke Abbott, *Stoke Atram* **4** 426, Stokeford, Stoke Wake, *Turberville Stoke*, Wild Stoke or Buckett's Fm, *Wulfheardig stoke* **4** 338. In AS charters, ?*holne stoke* **1** 31, *niþerstoce* (*v.* Addenda **4** 193)

stocc OE 'tree-trunk, stump'. Hr & Lr Stockbridge Fm, Stock Coppice, ?Stock Fm & Hill, ?Stockfield Fm, Stockham Dairy & Lane, Stock Hill, Hr & Lr Stockley Fm, ?Stocks Fm, Stockstyle; f.ns. *la Schetyngstokk* **3** 34, Stock Md **2** 230, *Stockmead* **2** 216, *Stockume* **2** 95. In AS charters, ?*holne stoke* **1** 31, *stocchylle* **4** 301

stoccen OE 'made of logs'. Hr & Lr Stockbridge Fm, Stocken Bridge **3** 67, Stockham Bridge; f.n. *Stokkenebriggelake* **4** 386

***stoccet** 'place with tree stumps', *Stockett Mylne* **4** 437

stock-house eModE 'prison where offenders were set in stocks', *le Stockehous* **4** 3, *the Stockhouse* **4** 255

stōd OE 'herd of horses'. Compton Stud, Stud Fm, Stud Ho, Studland, *Studley*; f.ns. *Stodelane* **3** 34, ?*Stodenhull* **2** 271, *Stoudlond* **2** 216, *Studburgh* **4** 387. In AS charters, *stod dic* **1** 30

stōd-fald OE 'stud-fold', f.ns. *Stodefold* **3** 227, ?Stoutfold **4** 491. In AS charters, *alde* ~, *ealden stodfald* **1** 30

stōl OE 'stool, seat, tree-stump'. Bussey Stool Fm

stoop, **stoup** eModE 'post used as boundary mark', f.ns. Stoop Hill **4** 331, *Stope* **4** 386, ?*La Stoupe* **4** 133

stōpel OE 'stepping-stone', f.n. Stoples **3** 258

storehouse ME 'storehouse', *le Storehous* **4** 3

stote ME 'a stoat'. ?Stoat's Hill

stott OE 'an inferior kind of horse, an ox'. ?Stottingway; f.n. Stot Fd **3** 353, Stotfield **2** 19

stōw OE *Edwardistowe* **3** 141, Instow. In AS charters, *halgan stowe* **3** 359

straddle ModE 'act of striding over', f.n. *Strad*(*d*)*lehole* **1** 230

strǣt OE 'a Roman road, a paved road', later (as ME **strete**) 'a street of houses, a hamlet, a straggling village'. Abbott Street, *Baggeridge Street* **2** 285, Belchalwell Street, †Brog Street, Brown's Street, Cole Street Fm & Lane, Hammond Street Fm, Higher Street Lane, Hill Street, *Knight Street*, Lower Street, *Middlestreet*, Mutton Street, New Street (Ctgs &) Lane (× 2), *Penstret* **3** 195, Saddle Street, ?Sterte, †*Stratford*, †*Stratforde* **4** 275, †Stratton, The Street **4** 372, Street Fm, Street Lane, Streetway Lane, †*Stretford* **1** 275, †Week Street Down. Freq in modern urban street names as ~ Street. Also in f.ns., e.g. *Cleystrete* **1** 183, †*Old Strete* **2** 53, *Pleistret* **3** 34, Press Street **3** 177, *la Stoniestrete* **3** 35, Stratfield **4** 217, Street Md **3** 231, *Stretacre* **2** 216, †*Stretforlange* **2** 72. In AS charters, *elþen stret* **2** 57, *strœt* **4** 35, *strœt forda* **4** 454

strait eModE 'a narrow lane'. Oakley Straight, ?Straight Lane, Straits

strand OE 'shore'. Strand St. **2** 43, *the Strend* **1** 259; f.n. *Strond Close* **1** 169

strang OE 'strong'. Strongate Fm

strap OE (later also **strop** ME) 'long, narrow piece of land'. Strap Bottom, Strap's Coppice; f.ns. Beckfords Strap **3** 388, Long Strap **4** 40, Strap Orchd **4** 253, Strop **3** 5, *le Sttrop* **4** 14

strawberry ModE. Strawberry Coppice, ~ Wood

strēam OE 'stream'. Bere Stream (*v.* RNs), Combe Stream (ib), Ebblake Stream (ib), Engine Stream (ib), Holy Stream (ib), *Knollton Streame* (ib), *Lake Mill Stream* (ib), Mill Stream (× 3, *v.* RNs), *le Streme* **2** 44. In AS charters, *pidelen streame* (*v.* RNs), *stream* **3** 163, **4** 305, *streame*, *stremes* **1** 32, *streme* **3** 253, 258, *wicum streames* (*v.* RNs)

strēaw OE 'straw'. ?Straw Barrow, *Strawbarrow*; f.ns. Strawberry **1** 183, *Strowbrege* **4** 35, *le Strowhouse* **3** 380

***strecche** ME 'stretch of land', f.ns. High Stretch **4** 14, Strech Cl **4** 269

string(s) ModE 'long, narrow piece(s) of land', f.ns. Strings **3** 30, 57, **4** 425

strip eModE 'strip of ground'. Big Oak Strip, The Strip, Tunnel Strip Plant. Also in f.ns. e.g. *Gatstrippe* **1** 128, Strip Md **4** 331, ?*la Strippam* **4** 375

strōd OE 'marshy land overgrown with brushwood'. Strode Manor Fm & Place, Stroud Bridge, ~ Common, ~ Fm (× 2); f.ns. *Strodebriggeford* **2** 11, *Strode Lane* **3** 149, *Stroode* **2** 20, *Stro(w)de* **3** 382

***strūt** OE 'dispute', f.n. *Strouthirste* **4** 117

stubb OE 'tree-stump'. Cutty Stubbs, Pimperne Stubs, Stivvicks Bridge (*v.* Addenda **2** 103), The Stubbs, Stubb's Coppice ?Stubb's Fm; f.ns. Elder Stubs **2** 62, *Stubbelese* **2** 181, *le Stubbes* **2** 70, Stubland **4** 425. In AS charters, *ellenstubbe* **1** 237, *þornstub* **3** 208

stubbed ModE 'with tree-stumps removed', f.n. *Stubbed Close* **4** 223

***stubbig OE** 'with tree-stumps', f.ns. Stubbey Down **2** 81, Stubby Beach **4** 340, *le Stubbymerssh* **1** 287

stubbil ME 'stubble', f.ns. *le Stubble field* **1** 134, *le Stubble Fildes* **4** 26, Stubble Leaze **3** 133

***stubbing** OE 'place with tree-stumps removed', f.ns. Stubbing **4** 425, Stubbin Mdw **4** 152

stump ME 'tree-stump'. Thorn Stump

stunt OE 'steep', f.n. ?*Stuntecnoll* **1** 185

***Stūre** r.n. of uncertain origin. R. Stour (*v.* RNs), giving name to E & W Stour, Stourpaine, Stour Provost, Sturminster Marshall & Newton

stut[1] OE 'gnat'. Stutcombe Bottom; f.n. *Stotecomb croft* **4** 365, ?*Stowtlondes* **3** 342

stūt[2] OE 'stumpy hillock'. ?Hounstout Cliff; f.ns. Stout **4** 69, ?*Stowtlondes* **3** 342. In AS charters, *stut* **3** 200

stybb OE 'tree-stump'. Stubhampton. In AS charters, *hnottan stybb* **4** 454, *styb* **3** 129

***Stȳrcel** r.n. of uncertain origin. R. Stirchel (*v.* RNs)

sugar ModE, f.n. Sugar Md **4** 217

***sug(g)a** OE 'swamp, bog' (**sog** ModE dial.), f.ns. Sog Md **4** 425, Sogmire **4** 391, Sog Plot **4** 283, Sogs **4** 453

***sug(g)ig** OE 'swampy, boggy' (**soggy** ModE dial.). Soggy Covert; f.ns. Soggy Bottom **4** 391, Soggy Breach **4** 425. In AS charters, *sucgimade hauede*, *sugging made* **3** 136, 138

sulh OE 'plough', f.ns. ?*Soulescumbe* **3** 59, *Sulaker* **3** 200, Sull Acre Way **1** 363

sumor OE 'summer'. Somerleigh Court, *Summer Beer* **3** 100, Summer Ctgs, ~ Hill Ho, ~ Ho (× 2), The Summer Ho, Summerlug Hill; f.ns. *Somerfeld* 4 43, *Somerlynch* 4 19, Summer Grd **3** 30, Summer Lane Cl **2** 230, Summerlear(s) **4** 425

***sumpt** OE 'marsh, swamp', f.n. *Sunptun' Alres* **3** 258

sundor OE 'apart', f.n. *Sunderton* **2** 14

sunny ModE. Sunnyside Fm

superior Lat 'higher'. Upbury Fm **4** 230

sūr OE 'sour', f.ns. ?(*le*) *Soure* **1** 238, 199, *Sourehams* **3** 376, (*la*) *Sourlond* **2** 242, **3** 208, **4** 11. In AS charters, *suran apældran*, ~ *apeldran* **4** 375

sūre OE 'sorrel', f.ns. ?(*le*) *Soure* **1** 238, 199

sūtere OE 'shoe-maker', f.n. ?*la Soutereshurne* **3** 200

sūð OE 'south'. *Soddon*, ?Sodern, Sodom Lane, South Admiston, *Southam* **2** 20, South Bowood, South Bridge (× 2), South Buckland, *Southcombe*, Southcombe Fm, ~ Plant., South Common (× 2), Southcot, South Court, South Deep, South Dibberford, *South Down*, South Down, South Down Ctg, South Down Fm, South Drove, Southeast Ctgs, South Eggardon Fm, *Southeparke* **4** 225, South Ewe Leaze, South Fm, South Field Down & Hill, Southfield Ho, ~ Lane, South Ground Plant., South Hanging, South Haven Point, South Heath (× 2), South Hill Estate, South Holworth Barn, *South Hungerhill*, South Lake, Southley Fm, South Lodge (freq), *South*

Lym **4** 392, South Mill, Southmoor, Southover (× 2), Southover Fm & Ho, South Perrott, South Pier, South Plant., South Poorton, South Sleight Coppice & Knap, South Slip, *Southway*, Southwell, South Winterborne (*v.* RNs), *South Wooth* **4** 322, ?Sovell Down, Suckthumb Quarry, *Sugestone*, Sutcombe Wood, *Suthbury* **4** 123, Sutton, Sutton Fm & Hill, Sutton Poyntz, Sutton Waldron; f.ns. *Southebroke* **4** 22, *Southecombe* **1** 264, South Fd **2** 51, *Suthovre* **3** 35. In AS charters, *seuen diche suðende* **3** 123, *sudwde* **3** 163, *suðbeara* **2** 162, *suðeke* **3** 104, *suðheal* **2** 162

sūðan OE 'south, south of'. Southbrook

sūðerne OE 'southern'. Southern Hill

sūðerra OE 'more southerly', f.n. Sowtherhaye **1** 62

sūðmest OE 'southmost', f.n. *Southemostefurlang* **3** 208

swalwe[1] OE 'swallow', f.ns. Swallow Close **3** 109, Swallow Coombe **4** 344

***swalwe**[2] OE 'whirlpool', ***swalg** OE 'pit, pool', *Swalwe* 1280 (prisons at Corfe C. **1** 11); f.ns. *Blakeswelew* **4** 132, Water Swallow (Fd) **2** 57, 66

swan[1] OE 'swan'. Sandhills, ?Swanage, Swan's Pond, ?Swan Way; f.ns. (*le*) *Swan*(*n*)*esneste* **1** 329, Swans nest **2** 9

swān[2] OE 'herdsman, peasant'. Swalland Fm, ?Swanage, ?Swanderbury

swannery ModE. (Abbotsbury) Swannery

swash ModE 'flooded land', *the Swatchway* **1** 76

sweet-briar ModE. Sweetbrier Drove

swampy ModE, f.n. Swampey Plot **4** 453

(ge)swelg OE 'abyss, whirlpool, pit', f.ns. *La Suelghesheude* **2** 271, *la Swelelake* **4** 157, *la Swelie* **3** 315, Swelly Cl **3** 109

***swelle**, **(ge)swell** OE 'swelling', 'steeply rising ground'. ?Poxwell; f.n. *le Swell* **1** 185

swēora OE (late WSax **swīra**, **swȳra**, **swūra**) 'neck of land, col'. *Swerebrok* **3** 188, *Swire Hill*, Swyre, Swyre Bottom & Hill, Swyre Head (Chaldon Herring, Corfe Castle); f.ns. *Swurefurlang* **4** 133, Swyer **3** 8, 98. In AS charters, *swuran* **1** 19

swēte OE 'sweet, pleasant'. Sweet Apple Copse, Sweet Hill, *Sweet Hill*; f.ns. Sweet Moor **2** 19, ?*Swetenham* **2** 175, *la Sweteplek* **1** 329, *Sweytacre* **3** 258, *Swoteham* **4** 133, *Swotemour* **2** 289

swice OE 'trap', f.ns. Sweek **1** 228, *Swykaker* **1** 216, Swyke **1** 258

swimming ModE, f.n. Swimming Mdw **2** 9

swīn[1] OE 'swine, pig'. Toller Porcorum, ?Swindale, ?Swineham; f.ns. *Swynehams* **1** 281, *Swyneslane* **3** 149. In AS charters, *swindone*, *-dune* **3** 208

***swīn**[2] OE 'creek, channel'. ?Swineham; f.ns. *la Swyne* **1** 329, ?*Swyneswell* **1** 32

swylle, ***swille** OE 'sloppy mess, place where water flowed freely', f.n. *Swilyate* **1** 185. In AS charters, *swylles* **3** 129

sycamore ModE. Sycamore, *Sycamore...Mead* **1** 248, Sycamore Pond; f.ns. Sycamore **3** 198, **4** 368

syle[1], **sylu** OE 'bog, miry place', f.n. ?*Suyle* **3** 200. In AS charters, ?*sulangraf* **3** 112

table ModE. ?Table Plant.

***tacca**, ***tacce** OE 'young sheep'. ?Acton

tadde, **tāde** OE 'toad'. Taddle Fm, ?Tadnoll Dairy & Mill, Tatten, Tibble's Copse & Lane; f.n. Tadbrook **3** 288, *Tadmeade* **1** 63, Tadmoor **2** 9, ?Tatcombe **1** 305

tægl OE, **taile** ME 'tail of land', f.ns. Beer Hill Tail **4** 233, Clappentail Lane **4** 393, Mallardes Tail **2** 72, *Tailecopice close* **2** 216, *Taylepytte* **1** 122, *le Westmylle Tayle* **4** 362

tǣse OE 'useful', f.n. ?*Teswell* **3** 59

tǣsel OE 'teasel', f.n. Teazle Plot **4** 234

tag ModE 'tail-piece', f.n. Long Tags **3** 5

***tagga** OE 'young sheep', f.n. Tag Hill **2** 57

tamarisk ModE. Tamarisk Fm (*v.* Addenda **4** 24)

tān OE 'twig, shoot'. ?Tan-hill Copse

tan ModE 'material used for tanning'. Tan Acre Plant., ?Tan-hill Copse

tang OE 'tong, forceps', f.n. *Tanglond* **1** 138

tanhous ME 'a building in which tanning is carried on'. *Tanhous* **1** 355, Tan Lane

tannere ME 'tanner'. Tanners Lane **1** 156

tannery ModE. Tannery

tanpit ModE 'pit where tanning is carried on'. Tanpits Bridge & Coppice

tansy ModE 'tansy'. Tansey Close Coppice; f.n. Tansey Cl **4** 500

tanyard ModE 'yard where tanning is carried on'. Tanyard Fm, Old Tanyard; f.n. Tanyard Lane **3** 147

tap-house ModE 'an alehouse'. Taphouse Fm

tar ModE. ?Tar Rocks

tavern ModE. Saw Mill Tavern, Tavern Fm

telegraph ModE. Telegraph (× 3), Telegraph Clump, ~ Ctg (× 3), ~ Hill, ~ Plant.

temple ME 'land of the Knights Templar'. Temple Brook (*v*. RNs), Temple (Brook) Fm; f.n. (*le*) *Templelond* **4** 500

tēn OE 'ten'. Ten Acre Coppice/Copse (× 3), ~ ~ Gate, ~ ~ Plant., ~ ~ Row; f.n. *ten' acres* **3** 208

tenant ModE. Tenant's Bottom & Hill, Tenants Hill, *Tennants Downe* **2** 202; f.n. Tenanten Fd **1** 87

tenantry ModE 'land let out to tenants'. Tenantrees, Tenantry Down, ~ Wood

tender ModE 'soft ground', f.n. ?Tenders **4** 409

tenement ME 'tenement, dwelling'. Jarvis's Tenement; f.ns. *Blymundestenement* **2** 136, *Skutts tenement* **2** 45, *Thomelynstenement* **2** 251, *Weryngesten't* **3** 35

tennis court eModE. *tenys cowrt* (1551) **4** 360

***tēo** OE 'boundary', f.n. *Overtebourneslake* **3** 274

tēoða OE 'tenth'. Tithe Barn; f.n. Tithewell **4** 234

tēoðung OE 'tenth, tithing'. *Free Tithing* **3** 10; f.ns. Titheing plot **4** 209, Tithing Md **4** 476, Tithing Plat **3** 5, Tything Plot **4** 344

tēoðung-mann OE 'the chief man of a tithing', f.ns. *the tythinge mans acre* **2** 95, *Tythingman's Ld* **2** 114

terig Welsh 'ardent, rough'. ?*Ter*(*r*)*ig* (*v*. RNs)

terra Lat 'land, tenement'. *Netherbire in Terra* **4** 311

terrace ModE. Marquis Terrace, Peacemarsh Terrace, The Terrace

tetchy ModE 'irritable', f.n. ?Titchy Md **4** 425

þæc, *þæcce OE 'thatch', f.n. *Thachehowe* **2** 230

thatched ModE. Thatched Ho

þēof OE 'thief', f.ns. ?*Theven'howe* **2** 230, *Thevenwell* **4** 250

þersc OE 'threshing', f.ns. *Thersshefoldes* **2** 78, *le Thresshfolde* **3** 158

þicce[2] OE 'thick, dense'. Heckford Rd, Thickthorn Down/Fm/Wood, Thickthorn Lane; Thick Thornes **3** 387

thicket ModE 'thicket, dense bushes'. Briar Thicket; f.ns. *Thicket* **4** 454, Thickets **2** 256, Thickett Md **3** 57

thieving eModE 'infested by thieves'. *Thieveing Way* **3** 307

þing OE 'property', f.ns. *Cosynesthyng* **2** 250, *Haukynsthyng* **1** 128, *Pudelthynk* **1** 113, *Sewardesthyng* **1** 287

þistel OE 'thistle'. Thistlegate Ho, Thistle Hill; f.ns. Thistle Close **2** 220, Thistley **4** 414

thistl(e)y ModE, f.ns. Thistley Fd **3** 225, 321, Thistly Cl **4** 88

þorn OE 'thorn-tree'. Colston Fm, Crawthorne Fm, Crowthorne Lane, Godderthorn Hd, Long Thorns, Longthorns, Mosterton, *Mulethorne* **4** 179, Nyland Thorn Covert, Thickthorn Down/Fm/Wood, Thickthorn Lane, Thorn Barrow, Thorncombe (× 4), *Thorncombe* **4**

378, Thorncombe Beacon, ~ Thorn, ~ Wood, Thorn Coppice, ~ Ctg, Thorncroft, (*La*) *Thorne* **1** 42, **4** 388, Thorney Down Fm, Thornford, Thorngrove, Thornham, Thorn Hill Fm, Thornhill (Fm), Thornhill's Coppice, Thorn Ho, Thorn Stump, ?*Thorrington* **2** 102, Thorton Fm, *Torne* (1086 DB) **1** 5, 35, 42, Turnworth; freq in f.ns. e.g. *Bareleggethorn* **2** 13, *Lenkethorne* **2** 70, *Thorendon* **2** 125, *Thornehill* **3** 35. In AS charters, *an lipien þorn* **3** 122, *cockes þorne* **2** 122, *ham woldes þorne* **2** 158, *Ruþanþorn* **2** 58, *smal*(*e*) *þornes* **3** 41, 129, *sex þorn* **2** 58, *þorn*(*e*) **1** 32, **3** 238, *thorncombe* **3** 246, *þorndunes cnep* **3** 208, *þorn rewe* **3** 200, *þorn þiuel* **2** 58, *þorþivel* **3** 208, *þornstub* **3** 208, ?*þorthn* **3** 258, *twam þornen* **3** 139

þornig OE 'growing with thorn-trees'. Thorney Down Fm, Thorny Barrow, ~ Cleeve, ~ (Close) Copse, ~ Ham; f.ns. Thorney Down **2** 270, *Thornyham* **2** 216

thousand ModE, f.n. Thousand Acres **2** 19

three legged ModE 'having three legs'. Three Legged Cross; f.n. the Three Legged Cross **4** 400

þrēo OE 'three'. Three Ash Cross, †Three Barrow Clump, Three Barrows, The Three Clumps, Three Cornered Clump, ~ ~ Coppice, ~ ~ Plant., Three Elms Inn, Three Gates, Three Guinea Bridge, Three Hedges, Three Lanes End, Three Lords' Barrow, Three Oaks, *The Three Sisters* **4** 295, Three Ways; f.ns. *three crosse* **3** 112, *threlansshorys* **3** 227. In AS charters, *thre landschere* **3** 258, *þrem land sharen* **3** 139, *þreora gyrda* **3** 205 (now **1** 316, *v*. Addenda), *þry akeres* **3** 139

threshing eModE, f.n. *le Thressyngplace* **3** 342 [*cf.* **þersc**]

thritty ME 'thirty'. Piddletrenthide

throck, drock ModE dial. 'a covered drain, a flat stone across a ditch'. Drow Coppice

þrop OE 'outlying farm, secondary settlement'. Droop, Throop, *Throop*, Throop Dairy Ho, Whinthorpe; f.ns. Droop Fd **1** 226, *Edmondesthorp* **3** 353, *Ravenestorp* **3** 280, *atte Thrope* **4** 45, *Thropp Close* **4** 305

þrote OE 'throat, passage'. *Wydecombesthrote* **3** 246, *Yolowhamthrote* **1** 320; f.ns. Throat **3** 67, *la/le Throte* **1** 138, 346

þrūh OE 'trough', in AS charters, *seuen willes þry* **1** 32

þrysce OE 'thrush', f.n. Truss Fd **3** 178

thurgh ME 'through', f.n. *Thorowaye* **3** 6

thurghfare ME 'a through road', *the Thorough Fair* **1** 354, Thoroughfare **1** 151

thwart ME 'athwart, lying across', f.ns. Thirt the Path **2** 119, Thwart Ld **1** 313
thwart-over ME 'lying across', f.ns. Thirt Over Comb **2** 119, Thurtover Burclinche **1** 313
þwēores OE 'across', in AS charters, *þwers dich* **1** 32
þȳfel OE 'a bush, a thicket'. Holm & Ivy Fm; f.n. *Whytethifel* **4** 271. In AS charters, *hordþiuel* **3** 207, *þorn þiuel* **2** 58, *þorþivel* **3** 208, *wiþig þefele* **1** 33, *wiþig þeuel* **2** 69, *Wiþig þeuel* **2** 69
þyrel OE 'hole'. Durlston Bay & Head; f.n. *la Thurleston* **3** 228
(ge)þyr(e)lod OE 'pierced', ***þyr(e)lede** OE 'having a hole'. Durdle Door, ?Durdle Pier
þyrelung OE 'piercing'. ?Turlin Fm
þyrne OE 'thorn bush'. Thornford, Turnworth, *Tyrryestherne* **3** 74; f.ns. *la Blakethurne* **2** 52, *Salteresthierne* **4** 283, (*le*) *Thurne* **3** 214, (*le*) *Thurnecott* **4** 228. In AS charters, *þyrnen* **3** 208, ?*Wirnen* **3** 258
***þyrnet** OE 'a thorn copse'. Thurnwood Dairy
þyrs OE 'a giant, a demon', f.n. *Thursdyche* **1** 238
ticce(n) OE 'kid, young goat', f.n. Tichford **2** 110
tide ModE 'flood-tide'. Tidmoor Point; f.ns. Tidemoor **1** 81, Tide Moors **1** 84
tīedre OE 'thin, tender'. Tytherleigh
***tige** OE 'goat'. Tyneham
tigel, **tile** OE 'tile'. Tile Barn, Tiley
tiled ModE 'roofed with tiles'. Tile House Fm
tilekiln OE 'kiln for tile-making'. Tilekiln Copse
tilage ME 'land enclosed for arable use'. *Clan down Tillidge* **4** 262; f.ns. Garden Tillage **2** 199, Tillage **2** 200
timber OE 'timber'. Timber Hill; f.ns. Timber Hills **4** 311, *Timberhurste* **3** 274. In AS charters, *tymbercumb* (*v.* Addenda **4** 193)
tiny ModE. Tiny Coppice
tippet eModE 'long scarf or band of cloth', f.n. Tippetts **3** 340 (*v.* NDEFN 428)
tithe, *v.* **tēoða**
tithing, *v.* **tēoðung**
***titta** OE 'a tit', f.ns. *Titcombe* **4** 25, 43, Titfurlongs **4** 337, ?Tudcombe **4** 26
toffee ModE, f.n. ?Teoffee Mdw **3** 5
toking, *v.* **tucking**
tol ME 'tax, toll', f.ns. *Tolehammede* **4** 12, Toll Goose **4** 436, *la Tolselde* **3** 150

toll-gate eModE 'gate where tolls are collected'. *Toll Gate*; f.n. Toll Gate Pdk **3** 57

toll-hall eModE 'hall where tolls were collected'. *La Tolhalle* **4** 360

tolhous ME 'house where tolls were collected'. *Drimpton Tollhouse* **4** 263, Tollhouse Ctg

topmost ModE. Topmast Quarry

topp OE 'top'. Fore Top, Hilltop Bungalows, Tophill, Top o' Town, Whitey Top; f.n. Tophill Plot **1** 81, *le Toppe* **1** 184

torr OE 'a rock, rocky outcrop'. Baiter, High Tor, ?Tar Rocks, ?Tor, ?Turford. In OE charters, *haren torre* **3** 206, *stan tor* **1** 32, ?*þoure* **3** 238, *tor scylget* **3** 102

***tōt(e)** OE 'a look-out, a look-out hill'. Hambury Ho & Tout, Honeycomb Tout, ?Hounstout Cliff, Marm Tout, Nettlecombe Tout, ?Todber, ?Totnell, Tout, Tout Hill, Tout Quarries, The Toyte, Worbarrow Tout; f.ns. *Totemede* **3** 274, Totter Hills **4** 487. In AS charters, ?*toten berg'* **3** 123

***tōt-hyll** OE 'look-out hill'. Tout Hill (× 2), Tut Hill; f.ns. Toothill **3** 109, Tout Hill **4** 253

tower eModE. Charborough Tower, Clavel Tower, Horton Tower, Roman Tower, *le Towar* **2** 193, Tower Hill, The Towers

trade ME 'path, way'. Trade Quarries

tramway ModE. Portland Tramway

trapdoor ModE. Trapdoor Fm

tray ME 'shallow open vessel', ?*le Tray* **3** 9

trenche OFr, ME 'ditch, trench', f.ns. *Trenches* **3** 376, *La Trencheye* **1** 42, Lt Trench **3** 88

trenched eModE 'with a ditch or trench', f.n. Trenched Md **4** 287

trencher eModE 'a plate'. Giant's Trencher

trendel, **tryndel** OE 'a ring, something circular'. Trendall's Copse, *Trendlecombe*, *Trendle Hill* **4** 137, Trendle St. **3** 363; f.ns. *Trendele* **2** 133, *Trendel* **1** 298, The Trendle **4** 43, Trendles **4** 92.

trente OFr 'thirty'. Piddletrenthide

trēow OE 'tree' also 'beam, post'. Boundary Tree, Cherry Tree Copse, Cross Tree Fm, Culliford Tree (Hundred), Elm Tree Fm, Fifteen Trees, Fir Tree Ctg, Firtree Ctg, Fir Tree Inn, Larmer Tree, Little Dunster Fm, Loxtree Fm, *Millhays Treen* **4** 421, Pear Tree Fm, Pepperclose Trees, *Trehurne* **4** 154, Twelvetrees Place, Willow Tree Fm, Yew Tree Ctg, Yew Tree Fm (× 2), Yewtree Ctg, ~ Fm, ~ Pit; f.ns. *Cutebolstre* **1** 345, Tree Grd **3** 45, Treenwoods **4** 453. In AS charters, *ecgulfes treow* **3** 310

triangle ModE. Gorse Triangle, The Triangle (× 2), Triangle Plant.; f.n. Triangle **4** 124

***trifel** OE 'grinding' or **trifle**, **trufle** ME 'trifle, thing of little value', f.n. *Trifulmede* **3** 274

trigon ModE 'triangle'. ?Trigon Fm/Ho/Hill

***Trihanton** PrWelsh 'river liable to floods'. The Tarrant (*v.* RNs, giving name to Tarrant Crawford, ~ Gunville, ~ Hinton, ~ Keynston, ~ Launceston, ~ Monkton, ~ Rawston, and ~ Rushton), Trent Brook (*v.* RNs, giving name to Trent)

trim-tram eModE 'ornament of little value', also 'kissing gate', f.n. Trim Trams **3** 30

***trind**, ***trynd** OE 'circular', f.n. Trinleys **4** 287. In AS charters, *tryndlea weȝe* **4** 305

trip Do dial. 'culvert over ditch', f.ns. Trip **2** 172, Trip Cl **1** 143

trocle, **trokel** lME 'a truckle, a pulley, a hoist for moving loads'. Truckle Field Lane

trog OE 'valley, trough', f.ns. ?*Kenmynges trowe* **2** 174, *Subtrehays* **3** 293, *le Trowe* **3** 238. In AS charters, *hunda troh* (*v.* Addenda **4** 193)

troy town eModE 'a maze'. Tray Town, Troy Town Copse; f.n. Troy Town **3** 373

truht OE 'trout', f.n. ?Trout Mdw **1** 100

***trun** OE 'round', f.ns. Turn Bridge **3** 128, Turvill Ld **3** 236

trunk ModE 'a perforated floating box in which live fish are kept'. ?Trunk Hole; f.n. Trunk Md **3** 216

***trynd** OE 'circular', in AS charter, *tryndlea wege* **4** 305

tryndel, *v.* **trendel**

tub ModE 'wooden vessel', f.n. Tubb **4** 58

tucker eModE 'fuller of cloth', f.ns. Tuckers Grd **3** 79, Tuckers Hams **3** 64

tucking eModE 'the fulling and stretching of cloth'. Tokenhills Coppice

tucking-mill eModE 'mill for fulling and stretching cloth'. Bindon Mill, *Le Tokyngmyll* **4** 476, *the tooken mill* **4** 58, *la Toukyng Mull* **1** 297, Tucking Mill, *Tucking Mill*

tuffe, **tufte** ME 'tuft, cluster of trees or bushes', f.ns. *Hownystuff* **3** 207, Tuft Plot **4** 269

***tull** PrWelsh 'a hole'. ?*Toller* (*v.* RNs), giving name to Tollerford, Toller Fratrum and Porcorum, Toller Whelme

tumbledown ModE 'steep or rough ground', f.ns. Tumble Down **1** 207, 152, 249

tūn OE, **toun** ME 'farmstead, estate, village'. *Abbatston* **1** 307, Abbott's Wootton Fm, Afflington Fm, ?Allington, ?Allweston, Als(t)on, Alton Mead Lane, Alton Pancras, Anderson, Ashington, Ashton Ctgs & Fm, Ashton Fm & Wood, Ashy Coppice, Athelhampton, Baltington, Bardolfeston, Barnston Fm, Bexington Fm, Bhompston Ctgs & Fm, Blackmanston Fm, Hr & Lr Bockhampton, Bovington, Braytown, Brockington Fm, Bryanston, *Bubbeton* **4** 210, *Buchœmatune* **4** 242, Buckhorn Weston, Burleston, Burton Bradstock, Caplestone Ctg, Castleton, Castletown, Catherston Leweston, (Hr) Chalmington, Charlton Dairy Fm, Charlton Hr Down, Charlton Marshall, Chaston Fm, ?Chedington, *Church(s)ton* **1** 306, Clapton Bridge, Clifton Maybank, Colliton Ho, Compton Abbas, Compton Acre(s), West Compton, Compton Valence, Corton Fm, Cripton Barn, Cruxton, Dodding's Fm, Drimpton, Durweston, Eastington Fm, Easton, Egliston, Ellston Hill, *Eston*, *Estretuna* **3** 131, Filcherton, Fordington, Forston, Frampton, Galton, Gatemerston, Godlingston Fm & Manor, Godmanstone, Grimstone, Hampreston, ?Hatherington, Herrison, Hethfelton, Hilton, Hinton Martell, Hinton St Mary, E & W Holton, Horton, ?Hurpston, Ibberton, Ilsington, Kingston (× 2), Kingston Lacy, Kingston (Maurward), Kingston Russell, Kington Magna, Kinson, Knighton (× 2), East Knighton, Knighton Ho, West Knighton, Knitson, Knowlton, Langdon, Langton Herring, Langton Long Blandford, Langton Matravers, Lazerton Fm, Leeson Ho, Leweston, Lillington, ?Linnington, Littleton, Litton Cheney, Lorton, Luton Fm, Lutton, Madjeston, Magiston Fm & Hill, Maiden Newton, Mangerton, Mannington, Manston, Mapperton (× 2), Martinstown, *Milborne Michelston*, *Milborne Symondeston*, Milton, Milton Abbas, Milton on Stour, West Milton, Minchington, Moncton Bushes, Monkton Up Wimborne, Moortown Fm, Moreton, Muston Fm, *Myleburn Munketon* **1** 282, Netherton (Fm), Newton, Newton Fm (× 3), Newton Manor, *Newton Montacute*, Newton Peveril, New Town (freq), Newtown Fm, Newtown (Hill), Nottington, Notton, Osmington, Pallington, *Pentringtone* **2** 235, *Philipston*, Piddlehinton, Pierston Fm, ?Pixon Barn, N Poorton, Povington, Poyntington, Preston, Preston Fm, Preston Hill, Prytown Fm, Puddletown, Pulston Barn, Putton, Quarleston Fm, Ranston, ?Rempstone, Rollington Fm, Rushton, *Rushton* **2** 232, Seatown, *Selavestune*, Shillingstone, Shilvinghampton, Shipton Gorge, Shitterton, Shroton, *Sidelyntone* **4** 175, Silton, Stalbridge Weston, Stanton St Gabriel, Steepleton Iwerne, Stratton, Stubhampton,

Sturminster Newton, Sutton Poyntz, Sutton Waldron, Tarrant Hinton, Tarrant Keynston, Tarrant Launceston, Tarrant Monkton, Tarrant Rawston, Tarrant Rushton, Tatton Fm & Ho, *Thorrington* **2** 102, Thorton Fm, Ton Bridge, Top o' Town, Town Bridge (× 2), *le Townehowse* **3** 180, Town Fm (× 2), Town Hill, Town Lane, Town Mills, Town's End (Fm) (× 5), Townsend, Uppingtown, Upton (× 2), Upton Manor Fm, Walton Elm, Waterston, Watton, ?Watton Hill, *Wayhoughton*, Waytown, *Westmyngton*, Weston (× 2), *Weston*, Weston Fm, Westown Estate, *Westretun* **3** 131, White Lackington, White Lovington, Winterborne Clenston, ~ Herringston, ~ Houghton, ~ Kingston, ~ Monkton, ~ Muston, ~ *Nicholston*, ~ *Phelpston*, ~ Steepleton, ~ Tomson, ~ Zelstone, Witchampton, Wolfeton Ho & Manor, Wonston, Wood Town Fm, Woolgarston, Woonton Fm, Wootton Fitzpaine, ~ Glanville, North Wootton, ?Wytherstone Fm; freq in f.ns. e.g. *fordetune* **4** 17, Horsingtons **3** 27, *Monketon* **3** 227, *Touneshende* **2** 63. In AS charters, *tunes hirne* **2** 82, *Winteintune* **3** 113

tunge OE 'tongue of land'. Tongue Copse; f.n. Tongue **1** 168

tunman ME 'villager', f.n. *le Tounemenfeld* **1** 122, *Le Townemensfeild* **4** 305

tunne OE 'cask, tun'. Tun Coppice; f.n. *Tunne* **1** 259

tunnel ModE. Tunnel Head, Tunnel Ho, Tunnel (Strip) Plant.; f.n. Tunnel Plot **4** 249

turbarye ME 'turf land, peat bed cut for fuel'. *Featherlake Turbary* **4** 420, *Quantocks Turbary* **4** 421, *Shore Turbary* **4** 422, Turbary Allotment; f.ns. Turbary Plot **4** 425

turf OE 'turf, peat', f.ns. *the Turffe Howse* **2** 78, *la Turfput* **4** 269, *Turuesethe* **3** 35, *Turveclose* **4** 438, *Turveshethe* **3** 35

***turk** (gen.sg. ***tïrk**) PrWelsh '(wild) boar'. Pentridge

turning ModE 'where a plough can be turned', f.n. Turning Fd **4** 414

turnip ModE, f.ns. Turnip Fd **2** 158, **3** 5, Turnip Grd **1** 100

turnpike ModE '(road with) a barrier or tollbar'. Old Turnpike, Turnpike Ctg (× 2), ~ Wood; f.ns. Turnpike Grd **4** 301, Turnpike Md **1** 100

twēgen OE 'two'. The Two Barrows, Two Droves, Two Gates (× 4), Two Mile Coppice; f.ns. *Bytwene Tweylynches* **2** 133, ?*Tweygryn* **1** 366, *Twomslade* **4** 183. In AS charters, *twam þornen* **3** 139, *twam wicon* **4** 454, *twegen æcras* **2** 136, *tweie pettes* **3** 208

twelf OE 'twelve'. Twelve Acre Coppice, The Twelve Apostles, *the Twelve Men Way* **1** 354, Twelvetrees Place; f.ns. *Twelfeacres* **3** 342, 12 Acres **3** 30

twēntig OE 'twenty'. Twenty Acre Coppice (× 2), Twenty Acres Altmt Gdns; f.n. 20 Acres **2** 72

twī- OE 'double'. Twofords Bridge, Twyford; f.n. *Twycroft* **4** 500

twicen(e) OE 'fork of a road, cross-roads, narrow lane'. Twitchings Copse & Ctg; f.ns. Twiching Cl **3** 279, Twitchams **2** 10, *Twychene* **2** 217, *Twytchynge* **2** 95

twīfeald OE 'double', in AS charters, *twifelde dich* **3** 208

twinn OE 'double', f.n. *Twynelde* **1** 70

twinnyng ME 'separation, division', f.n. *Twynnynglane* **1** 122

(ge)twis OE 'forked, bifurcated'. Twisting Alders Coppice

twisla OE 'fork', f.ns. Tweazels **4** 383, Twissell Barrows **2** 127. In AS charters, *weie itwislen* **3** 129

***twist** OE 'something twisted'. Twist; f.ns. (Broad & Long) Twist **3** 254, **4** 464, *Twiste* **4** 183, *La Twyste* **3** 238

twite eModE 'species of linnet', f.n. ?Twite Hill **4** 425

two, *v.* **twegen**

***tydd** OE 'brushwood', f.ns. ?*Tiddon* **1** 364, *Tudpytt* **4** 387

***tȳnincel** OE 'small farm or estate'. Tincleton

***tȳning** OE 'fencing', f.n. Tyn(n)ings **3** 254

tynkere ME 'tinker'. ?Tinker's Barrow, ~ Copse, ~ Lane **1** 156; f.ns. Tinkers Close **2** 249, Tinkers' Drove **2** 283

***tyrl** OE 'that which turns or rolls along'. *Tril* (× 3, *v.* RNs), Trill Bridge, Trill Fm & Ho; f.n. Trilcombe **4** 181

***ariā** Brit (meaning obscure) Dorchester

uferra OE 'higher, upper'. Higher Kingcombe, Higher Sandford, Melcombe Horsey, ?Overcombe Beach, Over Compton, Overcoombe, *Overdownes* **4** 220, Overleaze Coppice, Over Newland Fm, Upbury Fm, *Vuerȝep* **4** 427; f.ns. *Ouerclose* **3** 33, Overwood **2** 9

***uindo-** Brit 'white, bright'. *Vindocladia* **2** 177

ūle OE 'owl'. Ulwell, ?†Woolsbarrow

under OE 'under, below'. *Hunderhendone* **4** 178, Underbarn Walk, Under Cross Plant., *Under Curredun* **4** 175, *Underdown*, Underdown Hollow, Underdown Lane, Underhill (Fm), *Undermildone* **4** 130, *Underukepen* **4** 158, Underwood Fm; f.ns. *Under Cliffe* **4** 43, Underhills **3** 5

upp(e), **uppan** OE 'up, upper, higher up, upon'. *Lopegarston* **3** 174, Melcombe Horsey, Piddletrenthide, Toller Porcorum, Upbury Fm, Up Cerne, Upcot, *Upehendone* **4** 178, Uphall, *Uphay*, Uphill, *Uphill*, Uploders, *up(pe) Grene* **3** 360, *Uppeheye* **4** 107, Uppingtown, Up Sydling, Upton (× 3), Upwey, Upwood, Wimborne

St Giles; f.ns. *Hupehulle* **3** 178, *Huppe Haukeshulle* **1** 137, Up Marsh **2** 13, *Uppehulle* **2** 38, Upp Hurst **1** 137

upper ModE. Upper Barn, ~ Fm (× 2)

ūt, ūte OE 'outside, outer, more distant'. *le owte Downe* **4** 294, *le Out Hundred* **1** 6, *le Outewod* **2** 193, ?*Outrack Pryme* **4** 197, *Out Ryme* **4** 53; f.ns. *the Outhaies* **2** 90, Out Md **2** 66, *Outorchard* **2** 157, *Owtfild* **2** 216

ūterra OE 'outer'. Outer Bracket's Plant., Outer Heath, Outward Lardenhall; *lez utter Batelease* **3** 175, *le utter closse* **4** 167

utlete ME 'outlet, channel', f.ns. Outlet **2** 200, Outlets **2** 220, *le Owte lett* **1** 137

vacherie OFr, ME 'dairy farm', *vacherie* **3** 88

vale ME 'wide valley'. Blackmoor Vale, Halcombe Vale, Marshwood Vale, Sea Vale Fm, Vale Fm (× 2), Valehouse Fm, Whitcombe Vale

valley ModE. Baker's Well Valley, Bourne Valley, Brookman's Valley, Coombe Valley, Happy Valley, Luscombe Valley, Valley of Stones, Valley Rd; f.n. Valley Mdw **3** 178

veeir, **v(i)eue** OFr, ME 'view' (eModE **view**). Belle Vue (× 2), Belle Vue Barrow, Bellevue Plant., Hill View (× 2), Park View (Point), Sandy View, South View Nursery, Stour View Ho; f.n. Beauvoir **4** 138

vermin ModE 'animals of a noxious or offensive kind'. ?Vermin Lane

vetch ModE 'vetch', f.ns. Vatch Hill **4** 331, Vetch Cl **3** 45, **4** 217, Vetch Hams **3** 387, Vetchy Cl **4** 436

vicarage ModE. Vicarage (freq), Vicarage Fm (× 2), *the Viccaridge house* **4** 367

vicarie, **vicare** ME 'vicar'. Vicarage Coppice; f.ns. *Vicar(ie)s Breade/bredde* **3** 299, *Ye Vycary Leez* **2** 54

villa ModE. Barton Hill Villa, Bindon Villa, Boyne Villa, Cerne Villa, Coombe Villa, Laburnum Villa, Orchard Villa, Peacemarsh Villa, Providence Villa, The Villa

village ModE. Talbot Village

vine ME 'vine', f.n. Vine Md **4** 139

vinery ModE 'vineyard'. Leigh Vinery, Vinery (Hill)

vineyard ModE. The Vineyard (× 2), Vineyard Coppice, ~ Copse; f.ns. Furzy Vineyard **1** 79, *the Vyneyarde* **3** 311

vinny Do dial. 'mouldy, mildewed'. Veiny Cheese Pond, Vinney Cross; f.ns. Vinney Md **4** 453, ?Winywick **3** 315

vinye ME 'a vine', f.n. *Vyni* **2** 251

violet ModE. Violet Fm

voyde ME 'vacant, useless', f.ns. *Voydehowen* **2** 26, *le voyde howze close* **1** 346

***wacu** OE 'a watch, a wake'. Wakeham

wād OE 'woad'. (Little) Waddon, Waddon Fm, ~ Hill (× 2), Wadhill Coppice, ?Watton Hill; f.ns. *Wadbarough* **3** 228, Wadden, Waddon **2** 98, *Waddon* **2** 14, Wade Lane Cl **1** 230, Wadmore Coppice **4** 425. In AS charters, *waddene* **3** 129, *wadeleighe* **3** 200

(ge)wæd OE 'ford'. Shotover Moor, ?Wadbrook, Waddock Copse; f.ns. *la Wade* **2** 181, *Wade*, *Wath* **1** 230, Wade Md **1** 337, *Wathe* **4** 400 [*cf.* **wathe]**

wæfre OE 'unstable, wandering', ***wæfer** OE 'that which wanders', f.n. *Waverland*(*e*) **4** 43, 236

wægn OE 'cart, wagon', f.ns. One Sheard **2** 283, *le Wayneway* **4** 285, *Weaneway* **4** 93, *Wenesherd* **2** 78, Winchard Cl **3** 110, Winshard **3** 64

wǣl OE 'deep pool, deep place in a river'. ?*Walbridge* **4** 57

wælcyrige OE 'witch, sorceress', f.n. *Valgirimede* **4** 12

wærna[1] OE 'wren', f.n. ?*Wernehull* **2** 14

***wǣrna**[2] OE 'stallion', f.n. ?*Wernehull* **2** 14

(ge)wæsc OE 'flood, shore flooded at high tide', f.n. *In the Wassch* **1** 216, Wash **1** 51, Washford **1** 258

wæsce OE 'sheepwash, place for washing'. Cotley Wash, Wash Coppice, Wash Hill Coppice; f.ns. Morlis Wash **4** 152, Wash (Md) **2** 19, **4** 425, *Wayshemeade* **4** 223, *Wesh* **4** 287, *Wesshewer* **1** 346

wæsp OE 'wasp', f.n. Wopses Castle **1** 325

wǣt, wēt OE 'wet'. ?Wadbrook; f.ns. *Wetfurlang* **4** 472, *Wet'mede* **3** 178

wæter OE, **water** ME 'stream'. Backwater, Back Water Channel, Blackwater, *Blackwater*, Blackwater Bridge, ~ Fm, ~ Ford, ~ Lane, Blackwater River (*v.* RNs), Bowering's Water, Burton Freshwater, Burt's Water Ctgs, *Chelwell Water* (*v.* RNs), Common Water, Coombe Water Lane, Cox's Water, Dead Well Water (*v.* RNs), Empool Water (ib), Fair Water (*v.* RNs), Fairwater Head, Fordwater, Freshwater Bay, Freshwater Steps, Holebrook Water (*v.* RNs), *Hollish*, *Hollwysshe Water* (*v.* RNs), Lecher Water (ib), Mash Water (ib), *Maynewater* (ib), *Meerehay Water* **4** 243, *Milbyri Water* (*v.* RNs), Millwater Fm, *Newell Water* (*v.* RNs), Orchard Water Bridge, Park Water Ctgs, Rye Water Fm, Shreen Water (*v.* RNs), Stock Water (ib), Stoke Water (ib), Sydling Water (ib), Uddens Water (ib), Water Barn (× 2), Water Barrows, Watercombe, Watergates Ctg & Lane, Water House Fm **4** 351, Water Lake, Water Lane (freq), Water

Lane Fm, Waterley Wood, Water Point, Waterside, Waterside Fm, Water Wheel (× 2), Watery Lane **3** 308, Westhay Water, Winscombe Water Bridge, Woodswater Ctg, *Woodwater* (*v*. RNs), Woodwater Fm, Woolford's Water. Freq in f.ns, e.g. Dun Water **3** 55, *Freshwater lake* **2** 38, *Heldwater* **4** 40, Over the Water Md **2** 215, Water Md **2** 214, *la Waterruyne* **1** 264. In AS charters, *wæter pytt* **4** 305, *water pet* **2** 69

wæter-scipe OE 'conduit', f.n. *Watershippys* **3** 80

***wagge** OE 'quaking bog, marsh', f.n. *Waggefryth Copse* **2** 289

wain-house ModE 'wagon house, cart shed'. *Wainhouse* **4** 465, Wanehouse Fm

waite ME 'a watch, a lookout place', f.ns. Wa(y)thay **2** 99, *Waytehulle* **3** 158, *la Weyte* **2** 271

walh, **wealh** OE 'Welshman, serf'. ?*Walgrove*; f.ns. *Waleland* **1** 199, *Waleweye* **4** 271, *Wallcote Lake* **1** 52, ?Wall Moor **1** 137

walke ME 'division of a forest', also 'sheep-walk, path for walking'. Blind Walk Plant., *Busseystool walk* **2** 244, Chetterwood, *Clear-walk* **3** 52, *Cobley Walke* **2** 236, Cock Walk, Cottage Walk, Dairy Walk, The Dark Walk, Forest Fm & Side **3** 13, Jubilee Walk, *Laund-walk* **3** 21, Long Walk (Plant.), Underbarn Walk, Walkway, West Walk, Woodsend Fm; f.ns. *the Baleys Walke* **3** 31, The Walk **2** 161, *the Weste Walke* **1** 84

walking ModE. Ladies' Walking Field

walnut ModE, f.ns. Walnut Acre **2** 119, Walnut Grove **1** 313

walu OE 'ridge, bank'. Holwell, ?Walditch, ?*Walgrove* , ?†Wall's Plant.; f.ns. *Damfretyswale* **3** 226, *Holewal* **3** 58 Long Whale **3** 379, *Oldewale* **4** 142, *Wale* **3** 226. In AS charters, *ealdan wale* **4** 34

wamb OE 'womb, belly'. Wambrook

***wanoting** OE 'intermittent stream'. *Wanetinge* (*v*. RNs)

wante ME 'mole'. ?Wantsley; f.n. Want Hill Close **2** 19

-ware OE 'dwellers, inhabitants'. Dorchester

ware Do dial. 'sheep walk'. Dancing Ledge, E & W Weare, Lower Ware, Penns Weare, Scratch Arse Ware, Tilly Whim Caves, Wear Cliffs. Also in f.ns., e.g. (the Common) Ware **1** 61, Ware, Gt Wear Grd **4** 383, W(e)ar **4** 425, Wytch Wares **1** 30

wareine ME 'game preserve, warren'. Harry Warren Ho, Higher Warren, Newlands Warren, Old Warren (Ctgs, ~ Hill), Pimperne Fox Warren, Rabbit Warren, *Warren*, (The) Warren (freq), Warren Barn, ~ Close Plant., ~ Fm, ~ Heath, ~ Hill (× 3), ~ Ho (× 2), ~ Plant., ~ Road, ~ Withy Bed, The Warren (Wood), Wyld Warren; f.n. Home Warren **3** 236

warod OE 'shore meadow', f.ns. ?*Warddeclose* **4** 387, Ward Mdw **1** 344, *East Ward Meddow* **4** 392, Wared **4** 70. In AS charters, *warrode* **3** 258

wāse OE 'mud', f.n. Ouse Holes **4** 268

washe, *v.* **(ge)wæsc**, **wæsce**

washel ME 'washing place', f.n. Washalls **1** 61

washing ModE. Clay Washing Pits, Washing Ledge, Washingpool Fm; f.ns. Washingpool **4** 26, *Wasshyng Place* **3** 311

waste ME 'wasteland'. Ham Waste; f.n. *the wast grownd* **4** 387

watch ModE, **wache** ME 'watch, lookout'. *Wacheknappe* **4** 397, Watch Rock; f.n. Watch Oak Coppice **2** 238

watch-house ModE 'house in which a guard is placed' *the watchehouse* **1** 254; f.n. Watchhouse Fd **2** 172

waterfalle ME 'waterfall'. Waterfall (Ctg); f.n. *Waterval* **4** 181

water-pot ModE 'vessel for holding water'. Waterpot Hill

watery ModE. Watery Lane (× 4), Watery Way Coppice

wateryng ME 'watering place for horses and cattle'. St Edward's Bridge; f.ns. Watering Place & Pond **2** 32, *Watringe place* **1** 366, *the Watteringe place* **4** 236

wathe ME '?sheep walk', f.ns. *Bitakewathe*, *Wath* **1** 42 [*cf.* **(ge)wæd**]

watreles ME 'waterless'. Winterborne Abbas

w(e)ald OE 'woodland'. ?Walton Elm; f.n. *Schirewald* **4** 400

w(e)alh OE 'Briton, Welshman'. ?Walton Elm

w(e)all OE 'wall, rampart'. †Bestwall Fm & Ho, Buddlewall, Green Wall Plant., North Wall, †The Town Walls **1** 353, Walditch, ?*Walgrove*, ?†Walls Coppice, †Wall's Ctgs, Wallsend Cove, †Wall's Plant., ?Walton Elm, Works Wall; f.ns. *Shippen Walle* **4** 35, †Walls Fd **1** 342. In AS charters, *Elde Waldich* **3** 256, *stanen wal*, *stan wal*, *stanweal*, *walle* **1** 32, *Walle dich* **3** 238, *weal dic* (*v.* Addenda **3** 267)

***w(e)alt** OE 'shaky, unsteady'. ?*Walbridge* **4** 57, Walford Bridge & Fm

weam eModE dial. 'belly'. Weam Common Hill

weard OE, ward ME 'ward, watch, division'. *Firste Warde* **1** 12, ?Wardon Hill, Wardstone Barrow, Wor Barrow, ?Worbarrow; f.n. ?*Wyreberge* **2** 11

weard-setl OE 'watch house', f.ns. ?Kite Whistle **2** 19, ?*Wastle meade* **2** 19

wearg OE 'felon, outlaw', f.ns. ?Waryhall **1** 145, ?*la Werawsdich* **1** 329

wearg-rōd OE 'the gallows'. Worgret

wearm OE 'warm'. Warmwell, Warmwell Fm
weathercock ModE. Weathercock Coppice
weax OE 'wax', f.ns. Wax Md **4** 301, *Wexhayes* **3** 342, *Wexihamme* **2** 284
webbe ME 'weaver', f.n. *Webcliffe* **4** 465
wedge ModE. ?Wedge Hill; f.n. Wedgeditch **4** 391
weg OE 'way'. *Alreweye* **2** 239, *bithechurcchweye* **4** 216, Blood-way Coppice, Bowleaze Coveway, ?Challow Fm & Hill, *Churchewey* **3** 221, **4** 21, *Churichwey* **4** 32, Colway Gate & Lane, Coombe Bottom **1** 10, Cornerways, Crockway Fm, Cross Ways, Dolway Ctgs, *Douneweye* **4** 232, Hr & Lr Farway Fm, Ferney Way, Fishway Lane, Five Ways, Fuller's Way, *Greenway*, Greenway (Barn, ~ Head, ~ Lane), High Stoy, Hillway Copse, Hodway Lane, Hollis Hill, Holloway Lane, Holloway's Wood, Hollow Way, Holmes & Puddleway Coppice, Holway (Fm, ~ Hill), *Kathilleweia* **4** 289, Millway, Millway's Lane, Moorway Lane, *Mudwaye* **2** 203, Norway Lane, Nutley Way, Priest's Way, *Radway*, Ridgeway Hill, Ridingway, Rodway Fm, Salway Ash, Sandway, Seaway Lane, ?Silly Hill, Six Cross Ways, Spyway (× 2), *Stentewaie* **4** 247, Stile Way, Stone-way Coppice, Streetway Lane, Swan Way, Three Ways, Tweenways, Waddon Way, Watery Way Coppice, Waverlawns, Way Close Plant., ?Waycroft Quarries, Way Fm, Wayland, ?Waytown, White Way, Whiteway, Whiteway Fm, White Way Hill, Whiteway Hill, Whiteway Hundred, Whitey Top, Willow Way, Woodward's Way; f.ns. Kingway **2** 57, Rogeryswey **3** 227, *Sutteweye* **1** 199. In AS charters, *boddingc weg* **2** 136, *ealdan weg*(*e*) **2** 136, **3** 289, **4** 35, *ealde weg* **1** 31, *earna lea weȝe* **4** 290, *eastemestan holan weg* **3** 380, *ȝrenan weȝ* **4** 303, *grenan weges* **4** 454, *grenwai* **3** 128, *gyrd weg* **4** 374, *higeweges ande* **3** 135, 137, *holan wege* **4** 454, *holewei* **3** 111, *hriycsþeg* **3** 111, *imeren hole weg* **3** 102, (*i*)*richt wege, -wei* **1** 31, *meairweie* **2** 82, *merewege* **3** 129, *radenwey* **3** 257, *readan weg* **2** 135, *readan weȝ* (*v*. Addenda **4** 193), *riȝtwey* **3** 257, *sledweie* **3** 112, *snelles hamme weghe* **3** 112, *staniȝan weȝ* **4** 304, *stan wæg* (*v*. Addenda **3** 267), *stan wege*, *stanwei* **1** 32, *stanwey* **3** 200, 258, *struthherdes wege* **1** 110, *tryndlea weȝe* **4** 305, *weg, wei, weilaite* **1** 32, *weilete* **3** 258, *wege* **3** 123, *weges* **2** 136, **2** 158, *weie* **2** 58, **3** 41, *weie itwislen* **3** 129, *weies* **2** 58, **3** 258, *White Wey* **3** 259, *wicweie* **2** 82, *wines weie* **3** 112, *Wistonwey* **3** 259
weg-gelǣt(**e**) OE 'cross-roads', f.ns. *la Wendlete* **1** 366, *Wenelete* **4** 487. In AS charters, *weilaite* **1** 32, *weilete* **3** 258
weigh ME 'a balance, a pair of scales', *la Weygh* **4** 153

well(a) OE (**will(a)**, **wyll(a)** WSax) 'spring, stream'. Armswell Fm, Askerswell, *Attewell* **2** 21, Baker's Well, Belchalwell, Berie Well, Blashenwell Fm, *Bowell Lane* **3** 148, Bridewell (Plantations), Brittwell Ho, Castlewell, Caswell Fm & Wood, ?Challow Fm & Hill, Chelwell Water (*v*. RNs), Claywell, *Clingerwell* **3** 242, Cockwell Coppice, ~ Fm, Cookwell Brook (*v*. RNs), Coverwell Coppice, Crabb's Well, Culver Well, *Culverwell*, Dead Well Water (*v*. RNs), ?Dogwell Fm, Eastwell Lane, Elwell (Fm), Fortuneswell, Furzy Woolly Coppice, Gorwell Fm, Holwell, Holwell Fm, Holy Well, Holywell (× 3), *Houndeswellestrate* **3** 148, ?Kitwells Cliff, Knaveswell Copse & Fm, Lady Abingdon's Well, Lady's Well (× 2), Leper's Well, Loscombe's Well, Mandeville, Mary's Well, Morning Well, Newell **3** 362, Old Smuggler's Well, *Parkwellelake* **4** 165, Pellwell Fm, Pen Well Hill, Pill Well, Pillwell, ?Pink Knoll Hollow, Potwell Covert & Fm, Poxwell, *Presteswlle* (*v*. RNs), Pyle Well, Rodwell (× 2), St Andrew's Well, St Austin's Well, St Eustachius's Well **3** 215, St John's Well, The Saint's Well, Seawell Knap, ?Silverlake Fm, *Silver Well* **4** 138, Snodwell Fm, Southwell, Stallen, *Stentwill* **4** 247, Ulwell, Wall Fm, Wanderwell, Warmwell, Warmwell Fm, Welcome Hill, The Well, Well Bottom (× 4), Wellclose Plant., Well Coppice/Copse, Wellen Ctg, Wellfield Fm, *Wellhead* **3** 107, Well Ho (freq), ~ Plant., ~ Wood, Wells Fm, Wellwood, Whitefield, Whitewell Ctg, Whitfield Barn & Fm, Wills Wood, ?Willwood Ctgs, Wishing Well, Wolfridge Fm, Woodwalls, Wool, *Wooland*, Wool Bridge/Woolbridge, Woolcombe (Fm), ?Woolhouse Fm, Woolland Grove, Wools Bridge, Woolslope Fm. Freq in f.ns., e.g. Bannell Mdw **2** 66, Blindwell **4** 433, *Boreswill* **3** 31, *Melkwell* **4** 19, Well Grd **3** 31, Well Hole **2** 10, *Willeheye* **2** 148. In AS charters, *aettan wylle*, *aettan wylles heafde* **3** 310, *Alfricheswelle* **1** 30, *Bertes welle* **3** 205, *blehildewelle* **3** 255, *bogen wylle* **3** 148, *fyldena wylle* **3** 289, *halgan wyl* **4** 35, *he(f)deswelle* **3** 162, *holenwelle* **3** 102, 111, *holenwylle* **3** 102, *hwitan wyllon* **3** 207 (now **1** 316, *v*. Addenda), *lillanwelle* **3** 257, *murter* [?*murtes*] *wyll* **4** 304, *pimpernwelle* **2** 110, *ran wylle* **3** 380, *seuen willes þry* **1** 32, *wal*, *welles* **1** 32, *welle* **3** 259, *welle siluen* **3** 110, *wung wylle* **3** 40, 129

well-head ModE 'source of a spring'. Newell **3** 362; f.n. Well Head Cl **1** 193

welm OE 'spring', f.n. *la Nywelme* **3** 362

welsh-nut ModE 'walnut tree', f.n. Welch Nut Plot **2** 37

***wemm** OE 'dirty or muddy place', f.n. *Wemb'furlang* **4** 133

***wende** OE 'bend', f.ns. *North-*, *Southwendlond* **3** 208

wenn, **wænn** OE 'a wen, a tumour'. Wancombe Bottom & Hill; f.ns. *Pilewen* **2** 90, *Wen Close* **3** 228, *Wencome* **1** 230

wēnung OE 'expectation, hope', f.n. *Wenyng* **1** 238

wēod OE 'weed', **weedy** ModE, f.ns. *Wedylond* **1** 216, *Weede crofte* **2** 217, Weedy Plot **2** 26

(ge)weorc OE 'building, earthwork'. ?Wardon Hill, *Wichwerkes*, ?*Wrykesham* (both **2** 39); f.ns. *Channyns worcke* (1590) **1** 258, †*Hullewerk* **2** 250, *Workelond* **4** 12, †*Ydelwerk* **4** 160 [*cf.* **works**]

wer, **wær** OE 'weir'. Fishweir (Lane), *Horsewere* **3** 368, Maidenford Weir, Sluice Weir, Stony Weir, Sturt's Weir, Wareham, Weir (× 6), Weir Lane, *la Weremull* **3** 368; f.ns. *Fysshewer* **2** 8, *le Were* **3** 82, *la Weres* **3** 35

werg ModE dial. 'willow', f.n. *Werge Close* **4** 70

***wering** OE 'weir', f.n. ?*Weryn* **4** 387. In AS charters, ?*weritun* **2** 58

wērloga OE 'traitor, demon, warlock'. *Warlocke ford* **4** 395

west OE 'west'. Buckhorn Weston, Langton West Wood, Stalbridge Weston, West Barn, West Bay (× 3), West Bottom, Westbridge Fm, West Brook (*v.* RNs), West Burton, Westbury **3** 363, Westbury Fm, West Chelborough, Westcliff, West Cliff (× 2), Westcombe Coppice, West Compton (× 2), West Coppice (× 2), West Creech, West Down (Fm), West End (× 3), West End Barn, Westend Ctgs, West End Ho, West Fm (freq), West Field Coppice, West Fleet, Westfields, Westford Fm, West Gate, West Hall, Westham, West Hanging, Westhay (Fm), West Hemsworth, West Hill (freq), West Holme, West Ho, West Lane (freq), West Lears Fm, Westleaze, Westley Fm & Wood, West Lodge (freq), West Lulworth, Westmead Ho, West Melbury, West Mill(s) (× 5), West Moors, *Westmyngton*, Weston (× 2), *Weston*, ?Weston Fm, West Orchard (× 2), Westover Bridge & Fm, Westown Estate, West Parley, West Plant., West Point, West Road, Westrow Ctgs, West Spring Lake, West Stour, West Wood (freq), Westwood Coppice (× 4), Westwood Fm (× 2), West Woods Fm, Westworth Fm, Whistle Barn, Whistley Fm; f.ns *Westburie* **1** 264, West Md **2** 70, *Westwode* **4** 35. In AS charters, *olde wested'* **3** 201, *westgeat*, *west geate* **2** 136, *west mæd* **3** 289, *west Sæ* **1** 272

westan OE 'west, west of'. Westbrook Dairy/Fm/Ho, ?Weston Fm, Westport; f.ns. West Brook **3** 98, *Westebrouke* **3** 315, *Biwestewode* **2** 54

westerne OE 'western'. Western Ledges, ~ Street; f.n. *Westernecrosse* **3** 315. In AS charters, *westrene cumbe* **1** 33

westerra OE 'more westerly'. Litchfield Copse, *Westretun* **3** 131; f.n. *la Westrelinche* **2** 54. In AS charters, *corfgetes westran cotan* **4** 27

westmest OE 'most westerly', f.ns. *Westmosteforlang* **4** 19, *Westmostrawland* **2** 11

wēt OE (**wǣt** WSax) 'wet'. ?Whatcombe Down, ?Whatcombe Fm & Ho; f.ns. *Watfurlong* **4** 35, *la Wetelonde* **2** 54

weðer, ***wiðer** OE 'a wether'. ?†Weatherby Castle, Wether Hill, ?Wytherstone Fm; f.ns. Weather Leys **4** 25, *Wetherfurlang* **3** 208, *Wethermore* **2** 38

wevehous ME 'building for weaving'. *le Wevehowse* **1** 33

Wey pre-English r.n. of uncertain origin. R. Wey (*v.* RNs), giving name to Broadwey, Upwey and Weymouth

weyour ME 'fishpond', f.n. ?*le vayre* **1** 365

wheel-house ModE 'building in which cart-wheels are stored'. Wheel House Ctgs

whim ModE 'windlass (used for lowering stone to boats)'. Tilly Whim Caves

whin ModE dial. 'gorse' or 'hard rock'. ?Whin Croft, ?Whinthorpe, ?Winspit **1** 68

whirligig eModE 'merry-go-round'. *Le Hurlegogge* **3** 149, Whirlegog **4** 491

whirry, **wherry** eModE 'a rapid or sudden movement'. ?Wherry Pit

whistle ModE. Kittwhistle, Kit Whistle Barn, Kitwhistle Gorse

whitten eModE 'water-elder, mountain ash', f.n. *Whitting Ash* **2** 54

wīc OE 'dwelling, (dairy) farm, specialised farm'. Butterwick Fm, *Cokdenswych* **2** 39, Dodding's Fm, *Grayswick* **1** 357, ?Huntick Fm, Shapwick, *Stroudwitch Lane* **3** 182, Swanage, ?Weekley Coppice & Lane, Week Street Down, The Wick, †Witchampton, ?Wych (Fm) **4** 473, Wyke, Wyke Fm (× 3), Wyke Oliver Ho, Wyke Regis, Wyke Wood; f.ns. *Chadenwich* **2** 10, Chiswick Md **3** 155, *Cokdeneswych* **2** 39 (here 'salt-works'), Commerwick **3** 185, *le Courtewyke* **1** 313, Smithick Fd **3** 98, Wick **4** 188, *Wika* **1** 329, *Wikele* **2** 58. In AS charters, *twam wicon* **4** 454, *wic herepaþes* **2** 144, *wicweie* **2** 82.

wice OE 'wych-elm'. Witcham Fm, Witch Fm, Witch Lane (× 2), ?Wych (Fm) **4** 473, ?Wych Channel & Lake (*v.* RNs, giving name to Wytch Fm in Corfe Castle), Wytch Green; f.ns. (*La*) *Horewich* **2** 259, Winywick **3** 315, Witchcraft **3** 133, *atte Wyche* **4** 331, *Wyccheworthe* **1** 169, *Wychegrave* **2** 54

***wīc-hām** OE 'village associated with a Romano-British settlement or *vicus*'. †Witchampton; f.n. ?*Wickham Mead* **4** 302

wīd OE 'wide'. Whitcombe (Bottom), ?Whitecombe, ?Whitegate Fm, *widemere* **2** 212, Wide St.; f.ns. *Wydecumbe* **4** 12, *Wydefoldessherde* **2** 175

***wierde** OE 'watch, beacon'. ?Woodsford [but now *v.* Addenda **1** 186], ?Worbarrow; f.n. ?*Wyreberge* **2** 11

wīf OE 'woman', f.n. *Wyvenwolle* **4** 22

wīf-mann OE 'woman', f.n. ?*Wymynschete* **1** 239

wifel OE 'beetle'. ?Wavering Lane; f.n. *Wefulschote* **4** 344. In AS charter, ?*hean wifeles hylle* **3** 288

wigga OE 'beetle'. Wigmore Coppices; f.n. Wig Marsh **4** 476

wīl OE 'a wile, a trick, a snare for animals'. Wyld Fm; f.n. Wyld **1** 271

wilde OE 'wild, uncultivated'. Wild Church Bottom; f.ns. ?*Wild Crosse* **1** 243, *Wildeham* **4** 411, *Wildlese* **2** 217, *Wyldemore* **4** 74

wildernesse ME 'desolate spot' or 'wild garden in parkland'. (The) Wilderness (freq); f.ns. (The) Wilderness **3** 136, **4** 193, Wilderness Plant. **4** 464

***wilig** OE 'willow'. Highbridge Willow Bed, Mythe Mead Willow Bed, Williford, *Willow beare* **4** 327, Willow Bed (× 2), Willow Tree Fm, Willow Way; f.n. Whealey Cl **4** 308

willow-flower eModE 'willow-herb', f.n. *Willy flowers* **4** 153

wince OE 'winch'. Winspit [*cf.* **wink**]

wind[1] OE 'wind'. ?Windsbatch, ?Woodstreet Fm; f.n. *Wyndcroftes* **4** 35. In AS charters, *windberghes* **3** 259

(ge)wind[2] OE 'winding ascent, winding path or stream'. High Stoy, ?Windsbatch, ?Woodstreet Fm; f.ns. ?*Brechy Wyne* **4** 380, *Wynbroke* **3** 35

***windels** OE 'a windlass'. Broadwindsor, Littlewindsor, ?Wyndlam Fm

***wind-geat** OE 'wind-swept gap or pass', f.n. *Wyngeates* **2** 38

windig OE 'windy', in AS charters, ?*windee bergh* **2** 58

winding eModE 'twisted'. *Winding Stoye* **4** 163; f.n. winding oke **4** 157

windmill ModE. Windmill (on Portland), *Windmill Ashes* **3** 224, Windmill Barn & Knap, Windmill Barrow, ~ Clump, ~ Hill (× 2), ~ Plant.; f.ns. Wind Mill Fd **2** 45, Windmill Fd **1** 258

windy ModE. Windyridge Withy Bed

wine ModE. *Wine or New Street* **2** 42

wīn-geard OE 'vineyard'. Wennet Hill, ?Winyard's Gap; f.n. *Wynard* **4** 12

wink Do dial. 'winch, crank', f.ns. *Wink* **4** 500, Winkhorn **1** 134, Wink Orchd **4** 283 [*cf.* **wince**]

***winn**[1], ***wynn** OE 'a meadow'. *Wimborne* (*winburnan*) (*v*. RNs), Wimborne Minster and St Giles, Woolland

***wïnn**[3] PrWelsh 'white, bright'. Winfrith Newburgh, R. Winniford (*v*. RNs), Wynford Eagle

winnowing ModE 'separating chaff from grain'. Winnowing Knap

winter OE 'winter'. Winter Field, Winterhays; f.ns. Winterhead **3** 392, Winterwell **3** 315

winter-burna OE 'winter stream'. R. North Winterborne (*v*. RNs, giving name to Winterborne Clenston, ~ Kingston, ~ Muston, ~ Stickland, ~ Tomson, ~ Whitchurch, ~ Zelstone), R. South Winterborne (*v*. RNs, giving name to Winterborne Abbas, ~ Came, ~ Herringston, ~ Monkton, ~ St Martin, ~ Steepleton); f.n. *Winterbourne* **3** 98

wīr OE 'bog myrtle'. ?Whyr Coppice, ?Wire Drove, ?Wreden Plant.; f.ns. Whir **4** 209, Wire Shard **2** 119

wire-thorn ModE dial. 'yew', f.ns. Wirethorn **2** 224, *le Wirthorne* **2** 54

wisc OE 'marshy meadow'. ?Fishey Lane, *Hollish*, *Wysselay*; f.n. ?*Vissmor* **2** 143. In AS charters, *alle þiscan/discan*, *holen wicken*, *olle discan* **1** 14

***wiscel** OE 'marshy place', f.n. ?Whistles **1** 61

***wise** OE 'river, swamp'. ?Wizard Bridge; f.n. ?*Wisefurses* **1** 239

wishing ModE. Wishing Well

witch ModE, f.n. Witches' Corner **4** 208

wīðig OE 'willow'. Crooked Withies, Elwibber, The Withies, Withies Croft Quarries, (The) Withybed(s) (freq), Withybed Plant., *Withybere* **4** 13, Withyhook Mill, *Wytihoug* **2** 240. Freq in modern minor names as ~ Withy Bed, e.g. Blackmanston Withy Bed, Came Withy Bed, Causeway Withy Beds, East Morden Withy Bed, Froghole Withy Bed, Langford Withy Bed, Lutton Withy Bed, Moormill Withy Bed, New Withy Bed, Nyland Withy Bed, Rodford Withy Bed, Seven Acre Withy Bed, Stroud's Moor Withy Bed, Tynehams Withy Bed (freq), Warren Withy Bed, West Withy Bed, Whiteway Withy Bed, Windyridge Withy Bed. Freq in f.ns, e.g. Withy Hams **3** 8, Withy Pit Fd **2** 172, Withy Plot **2** 200, *le Wytheber* **1** 63. In AS charters, *haran wiðig* **2** 161, *wiðgas* **2** 162, *withi* **3** 259, *withi begh*, *wiþibedde*, *wiþig þefele* **1** 33, *wiþiȝheȝe* (*v*. Addenda **4** 193), *Wiþig þeuel* **2** 69

wiððe OE 'withe, osier', f.n. Little With **4** 58

with(y)wind eModE dial. 'bindweed'. Withywind Coppice

wodecok ME 'woodcock', f.ns. *Wodecoklane* **3** 346, Woodcock Mdw **3** 166, Woodcocks Walk **1** 207

wodehewer ME 'wood-cutter', f.n. *Wodehewerescroft* **3** 35
wōh OE 'crooked'. Oborne. Also in f.ns., e.g. *Wogheland* **4** 19, Woo Greys Cl **3** 204, *Wowacre* **3** 35, (*L*)*a Wo*(*u*)*welonde* **1** 42, **2** 133, 251, **4** 133. In AS charters, *wonalre* **3** 137
woodbine ModE. Woodbine Ctg; f.n. Woodbine Cl **4** 425
***wōr** OE 'wood-grouse', f.n. ?Woor Md **2** 127
workhouse ModE. Beaminster Union Workhouse, *le Werke house* **2** 186, Workhouse Fm
works ModE 'factory, workshop'. Riverside Works, Works (× 2), Works Wall [*cf.* **(ge)weorc**]
worm-stall ModE dial. 'cattle shelter'. Wormstall Fm; f.ns. (*le*) *Wormestall*(*e*) **1** 239, **4** 12, 271
worð OE 'enclosure'. Bloxworth, ?Charisworth, Dibberford, Eastworth Fm, Elworth, E & W Hemsworth, Holworth, *Kentlesworth*, Keysworth, *Kinsworth* **3** 54, E & W Lulworth, Piddles Wood, Tillworth, Turnworth, Westworth Fm, Widworthy, Wilksworth Fm, Wortheal, Worth Matravers. [Some ME *-worth* forms found for Hamworthy]. Also in f.ns., e.g. Benford **3** 235, *Blakkeworthe* **4** 411, Chatsworth **4** 221, *Denneworthe* **1** 168, Lillwood Mdw **3** 186, Patchward **3** 253, *Pendeworthe* **2** 53, *Waseys* **4** 411, *Welteswurthe* **3** 376, *Woorthe*, *Woorthayes* **3** 166, (*la/le/the*) *Worth*(*e*) **2** 72, 193, **3** 393, **4** 35, 117, *Wyccheworthe* **1** 169 [*cf.* **wyrð**]
worðig OE 'enclosure'. ?Bugworthy, Hamworthy, Widworthy, Wortheal. [Occasional ME *-worthy* forms found for Bloxworth, Holworth, East Lulworth, Turnworth, Wilksworth Fm]; f.ns. *Densworthy* **4** 222, *Gyleworthy* **3** 231, *Lodewurthy* **3** 257, Worthy **3** 45
worðign OE 'enclosure'. Occasional *-warden* forms found for Holworth
wōþ OE 'sound, melody' or ***wōþ** OE 'sweet, pleasant'. *Woth* (*v.* RNs) giving name to Camesworth and Wooth Grange in Netherbury and other names in Bradpole and Symondsbury
wrǣna OE 'stallion' or **wrenna** OE 'wren', f.n. *Wernehull* **2** 14
***wrǣsel** OE or ***wrǣst** OE 'something twisted', f.n. Gt & Lt Wrestle **4** 425
wrang OE 'crooked', f.n. Ranklands **4** 343
wreath Do dial. 'rod used in hurdle-making', f.n. Wreath Cl **1** 81
wrecca OE 'fugitive, outlaw', f.n. *Wrechebergh* **4** 70
***wrēo** OE 'crooked', f.n. *ByWroghmede* **4** 184
wrīd OE 'bush' (Do dial **wride** 'bush'). Right Bottom

wring ModE 'a press for cider', Wringhouse **4** 218, *la Wryngehous* **4** 272

wring ModE 'to hurt, to squeeze', Ring-Bum Gdns

***wrocc** OE 'buzzard or other bird of prey'. Wraxall; f.n. Wraxhill **4** 437

wudig OE 'wooded'. East & West Woodyates

wudu, **widu** OE 'wood'. Abbey Wood, Abbott's Wootton Fm, Ailwood, Afflington Wood, Alder Wood, Gt & Lt Alderwood Coppice, Gt & Lt Almer Wood, Bagber Wood, Bagwood Close, Bailey Ridge Wood, Baker's Wood, Balfour's Wood, Barewood Copse, Hr & Lr Barrowland Wood, Bartleys Wood, Bear Wood, Beechwood, Beerwood Ctg, Benecke Wood, Bere Wood, Bexington Wood, Bickham Wood, Big Wood, Bishop's Caundle Wood, Bishop's Wood, Black Wood Bottom, Blagdon Hill Wood, Blindman's Wood, Bolland's Wood, Bothenwood, Bothewood, Bottom Wood, Bowles's Wood, N & S Bowood, Boyden Wood, Boys Wood, Boywood Fm, Breach Wood, Briar's Wood, Brickles Wood, Bridge Wood, Brierhill Wood, Brimland Wood, Buffet's Wood, Bully Wood, Burcombe Wood, Burngate Wood, Burton Hill Wood, Burwood, Bussey Stool Woods, Butterwick Wood, Cadford Wood, Calcots Wood, Came Wood, Captain's Wood, Castle Wood Fm, Castlewoods Fm, Catemwood Coppice, Caundle Marsh Wood, Chapel Wood, Charity Wood, Claywood Coppice, Lt Cleeve Wood, Clenston Wood, Cliff Wood, Clifton Wood, Coach Hill Wood, Cole Hill Wood, Cold Park Wood, Cole Wood, Coll Wood Ctg/Gt Coll Wood, Coltwood Coppice, Combe Wood, Common Chalk Pit Wood, Common Wood (× 2), Coneygar Wood, Coombe Wood (× 3), Courtwood, Cowpound Wood, Crate Wood, Cuckoo Lane Wood, Daggons Wood, Dalwood, Delcombe Wood, Denbose Wood, Den Wood, Devils Wood Lane, Dewlands Woods, Dorset Wood, Down Wood, Dry Wood, Duncliffe Wood, Eastcombe Wood, Eastward, East Wood, Eastwood, Elmwood, Everett's Wood, Fallow Wood, Farm Wood, Farnham Common Wood, Farnham Woods, Ferney Down Wood, Fern Wood, Fiddle Wood, Fifehead Wood, The Folly Wood, Fontmell Wood, Foxbrake Wood, Foxholes Wood, Furzey Close Wood, Garden Wood, Garston Wood, Goathill Wood, Gorewood, Grange Woods, Gravel Pits Wood, Great High Wood, Great Wood, Grey's Wood, Grove Wood, Hail Wood, *Hakeryge Wode* **4** 403, Ham Wood Fm, Harley Wood, Harvey's Wood Cottage, Hawk Wood, Haydon (Hill) Wood, Hay Wood, Henroost Wood, Henwood (Hill), Hey Wood, Higher Wood,

Highfields Wood, High Wood (freq), Highwood, Highwood Wood, Hill Wood (× 3), Hilton Wood, Holcombe Wood, Holloway's Wood, Holmwood, Holt Wood(s), Homer's Wood, Home Wood, Honeycomb Wood, Hookswood Fm, Horton Wood, Houghton Wood, Hove Wood, Humber Wood (× 2), Incombe Wood, Ivest Wood, Jubilee Wood, Killwood, Kimerley Wood, Kingsettle Wood, King's Wood (× 2), Knowle Wood, Lamb House Woods, Langton West Wood, Launceston Wood, Lay Wood, Lee Wood, Leigh Wood, Lewcombe Woods, Little Wood (freq), Littlewood, Lode Wood, Lodge Wood, Longthorns Wood, Long Wood, Look Wood, Loscombe Wood, Lousley Wood, Lower Wood, Lyon's Wood (× 2), Manor Coppice & Hill, Manswood, Marl Pits Wood, Marshwood, Marsh Wood, Middle Wood, Milborne Wood, Mill Hill Wood, Misle Wood, Moigne Wood, Monk's Wood, Hr & Lr Monkwood, Monkwood Hill Fm, Morgan's Wood, Netherwood Mead, New Leaze Wood, Noddy Wood, North Wood (× 2), Northwood (Coppice & Fm), Norwood (Coppice & Park), Oakers Wood, Oakley Wood, Oatclose Wood, Oborne Wood, Ochill Wood, Old Park Wood, Old Rectory Wood, Old Road Wood, Old Wood (× 2), Over Wood, Oxleaze Wood, Paradise Wood, Park Wood (× 3), Parley Wood, Patley Wood, Penn Wood, Penny Farthing Wood, Piddle Wood (× 2), Pimperne Fox Wood, Plumley Wood, Pollards Wood, Potterne Wood, Povington Wood, Pound Wood, Pressham Wood, Presswood Copse, Preston Wood, Prestwood Moor, Pribdean Wood, Prince's Wood, Puncknowle Wood, *Purley Wood*, Ratcombe Wood, Ratleigh Wood, Rickett's Wood, Ridge Wood, Riveridge Wood, Robin Wood, Rolf's Wood, Rookery Wood, Roundbush Wood, Rowditch Wood, Sares Wood (× 2), Shitterton Wood, Short Wood, *Shortwood Common*, Shot Lake Wood, Six Acre Wood, Sleech Wood, Snagharbour Wood, Stavordale Wood, Stickland Wood, Stockbridge Wood, Stockwood, Stock Wood (× 2), Stone Hill Wood, Strawberry Wood, Studland Wood, Sutcombe Wood, Swilletts Wood, Sydling Wood(s), Target Wood, Tatcombe Wood, Tenantry Wood, Thickthorn Wood, Thorncombe Wood, Trent Wood, Turnpike Wood, Turnworth Wood, Twinwood Coppice, Tyneham Great Wood, Underwood Fm, Up Cerne Wood, Upwood, Vanity Hill Wood, Verwood, Warmwell Wood, Warre Wood, Watcombe Wood, Waterley Wood, Well Wood, Westley Wood, Weston's Wood, West Wood (freq), Westwood Coppice (× 4), Westwood Fm (× 2), West Woods Fm, Whitefield Wood, White Hall Wood, White Hill Wood, White's Wood, Whitfields Woods,

?Widworthy, Wilcocks' Wood, Wilkswood Fm, Wills Wood, Wiltshire Wood, *Wodehous* **2** 87, **3** 366, Wood Bar Looe, Wood Barn, Woodbridge (× 2), Woodbury Coppice/Copse, †Woodbury Hill, Woodclose (× 2), Wood Close, Wood Coppice, Woodcote, Wood Ctg, Wood Covert, Woodcroft, Woodcutts, Woodcutts Fm, Wood Fm, *Wood Fm*, Woodhays Fm, Wood Hill, ?Woodhill Coppice, Wood Ho, Woodhouse, Wood House Covert, Woodhouse Cross, ~ Fm, Woodhouse Hill, Woodhouse Lane, Woodlake, Woodlands (freq), Wood Lane (× 3), Woodleaze, Woodlynch Plant., *Woodmill* **4** 405, Wood Mire Copse, Woodmoor Copse, Woodrow (Fm) (× 3), Wood's Coppice, Woodsend Fm, Woodside (freq), ?Woodstreet Fm, Wood Town Fm, Woodville (Lane), Woodwalls, *Woodwater* (*v*. RNs), Woodwater Fm, East & West Woodyates, ?Woolford's Water, Wootton Fitzpaine, ~ Glanville, North Wootton, Gt Wootton Wood, Wraxall Woods, Wyke Wood, Wynford Wood, Wytherstone Wood. Freq in f.ns., e.g. *Hallewode* **2** 216, *Westwode* **4** 35, *Woodhey* **4** 411, Woodlakes Cl **2** 10. In AS charters, *littlen wde* **2** 291, *sudwde* **3** 163, *wde* **1** 33, **2** 121, **3** 106, *wde bricge*, *-brigthe* **3** 105, *wdes* **2** 82, *widesgete* **2** 82, *wþe* **2** 57

wuduwe OE 'widow'. Woodyhyde Ctgs & Fm; f.n. *Wydowes Close* **2** 143

wuduweard lOE 'the keeper of a wood, forester'. ?Woodward's Way

wulf OE 'wolf'. ?Wolvercroft, Wood Fold Hill; f.ns. Wolfern Hill **4** 348, Wolve Md **3** 263, ?Woolley Down **2** 78, ?*Wulrygg* **4** 12

wulf-pytt OE 'a wolf-pit or trap', f.ns. *Wolfputte* **4** 133, *Wulpit* **4** 157

wull OE 'wool'. *le Wol*(*le*)*house* **2** 43, ?Woolhouse Fm; f.n. ?Woolyards **1** 61

wunung OE 'dwelling', f.n. *Vyncentyswonyenge* **2** 251

wyrm OE 'snake, dragon', f.ns. Wormhill Pond **2** 32, ?Worms Hill **4** 476

wyrð OE 'enclosure', in AS charters, *beteswirþe sled* **3** 104, *bilian wyrþe* **3** 289, *ealdmanes þyerðe*, *-wyrde*, *eldmannes wrthe*, *heldmannes wrthe* **3** 111, 136, 178, *wænecan wyrð* **2** 162 [*cf.* **worð**]

wyrt OE 'plant'. ?*Vrtebrok'* (*v*. RNs); f.n. *le Woortcloses* **1** 294

wyrttruma OE 'root', in AS charters, *wertrumen* **1** 32, *wirtrone* **3** 259, *wirtrume* **3** 102, *wirtrune* **3** 113, *wurttrune* **3** 113, *wyrtruman* **3** 238, Addenda **3** 267, **4** 305

wyrt-tūn OE 'vegetable garden', f.n. ?Wattinghams **2** 105

yallery Do dial. 'yellow'. Shit Yallery Hole

yatehouse ME 'gatehouse', *le Yatehouse* **1** 329

yeaning ModE dial. 'lambing', f.ns. Yeaning Close(s) **1** 264, **4** 481, Yeaning Plot **4** 125, Yeanning Cl **4** 302

yearling ModE 'animal in its second year'. ?Yearlings' Bottom & Drove

yfer OE 'edge or brow of a hill'. Ivers; f.ns. *le Iver* **3** 273, *Iver Mead* **3** 375, Ivory Hill **4** 221. In AS charters, *yfre* **3** 259

yonder, **yender** ME 'yonder'. Yonder Coppice (× 2), Yonder Hill Creamery; f.ns. *Yenderclose* **3** 80, *le yonder house* **4** 466

zariba ModE (from Arabic) 'fenced enclosure or camp in the Sudan'. Zareba Clump

zigzag ModE. The Zigzag

PERSONAL NAMES
IN DORSET PLACE-NAMES AND FIELD-NAMES

These lists includes personal names incorporated in place-names and field-names, page references being given for the latter. In some cases it is not certain that the examples contain the personal name, but they are listed here if there is a possibility that they contain it. The discussions of individual place-names and field-names in Parts **1-4** should be consulted in all cases. Some place-names and field-names are entered more than once in the lists, as alternative etymologies involving different personal names are given for them. Some personal names which are cited in their OE forms continued in use in ME (especially as surnames) and it is often impossible to say whether a place-name or field-name originated in the earlier or later period. Names which are not independently attested are marked *.

A. OLD ENGLISH

Abba Ebblake
*__Æbbi__ Ebblake
Æddi Attisham
Æffa/Ælffrið Affpuddle
Ægen Ensbury
Ælfflǣd fem. f.n. ?*Alfledemers* **3** 164, ?Allweston
Ælfgifu fem. ?Allweston
Ælfhēah ?Allweston
Ælfhild fem. f.n. *Alphillicomme* **1** 198
Ælfrǣd *Albretesberge Hundred*
Ælfrūn fem. Afflington Fm
Ælfsige (WSax ***I(e)lfsige**) Ilsington, Ilsington Wood
Ælfweald ?Als(t)on
Ælfwīg ?Allweston
Æðel(a) ?Allington
Æðelflǣd fem. f.n. ?*Alfledemers* **3** 164
*__Æðelgēat__ f.n. *Eiletzbergh* **3** 256
Æðelgifu fem. Ailwood
Æðelhelm Athelhampton
Æðelnoð ? f.n. *Alnodesmed* **4** 270
Æðelweald ?Als(t)on
Æðelweard f.n. *Alwardescum* **3** 254

***Ætta** or **Ætte** fem. *aettandene*, *aettan wylle*, *aettan wylles heafde* **3** 310 (AS charter)
Aldhelm St Alban's or St Aldhelm's Head, St Aldhelm's Chapel
Atta f.n. ?*Atleye* **2** 173
Babba *babbanmede* **3** 254 (AS charter)
Bacca Bagber (× 2)
***Baccela** Battleford Fm
Bacga *bacging berghe* **2** 122 (AS charter)
Badda Badbury Rings
Bæde ?Bedchester
***Bæll**, ***Bælli** ?Bazon Hill
***Bætti** f.n. *Bat*(*t*)*escumb* **3** 216
***Ban(n)a** ?Banbury Hill
***Bār** ?Bowerswain Fm, f.n. ?*Boreswill* **3** 31
Bassa f.n. *Bassehull* **1** 326
Bata Batcombe
Bēage Beaminster
Bēaghild fem. Belhuish Fm
B(e)aldheard ?Baltington
B(e)aldhere ?Baltington
B(e)aldhūn ?Baltington, *bealtunes ersc* **3** 288–9 (AS charter)
B(e)aldðryð fem. ?Baltington
***Beassa** f.n. ?Beachy **4** 498
Bebbe Beaminster
Becca f.n. *Beckeberwʒe* **1** 137
Beonna ?Binnegar
Beorht *Bertes welle* **3** 205 (AS charter)
***Be(o)rhtel** ?*breþling made* **3** 178 (AS charter)
Beorhtsige ?Buzbury Rings, f.n. *Bryxieswode* **3** 79
Beorn Barnston Fm, f.ns. *Bernesham* **1** 78, ?Burnham **1** 78
Beornheard Barnsley Fm
Beornrǣd *beorreding mæd* **3** 286 (AS charter)
Beornweald f.n. *Bernoldesclif* **2** 125
Betti Bettiscombe
Bica ?Bickham Fm, f.n. ?*Bikenhurste* **3** 199, ?*biken settle* (AS charter for Sixpenny Handley **3** 113, *v.* Addenda)
***Bida** ?Bedmill
***Bīedi**, ***Bēdi** ?Bedchester
Bil Bilshay Fm
***Bil(l)a** Bilcombe Coppice, *bilian wyrþe* **3** 289 (AS charter)
***Bit(t)a** ?Bidcombe Mill, Bidlake Fm, *Bitteworth* **1** 63

Blaca f.n. *Blakkeworthe* **4** 411
Blæcmann Blackmanston Fm, *blacmanne bergh* **3** 205 (AS charter)
***Blǣdhild** fem. *blehildewelle* **3** 255 (AS charter)
***Blocc** Bloxworth
***Bodd(a)** *boddingc weg* **2** 136 (AS charter)
Bōfa Bovington
Boia ?Boywood Fm
Bota ?*Boteridge* **1** 85–6
***Brant** ?*brandes hricg* **3** 156 (AS charter)
Breoweald *br(e)owoldes ham*, *-heal* **4** 302 (AS charters)
Brūn ?Brownsall Hundred/Brunsell's Fm & Knap
***Brūnoc** Brownsea Island
Brȳni Brenscombe Fm, ?Brimscombe Fm, Brinsham Fm
Brȳnic ?Brenscombe Fm
Bubba Bookham, Bucknowle Ho
Bucca ?Hr & Lr Buckham
Bucge fem. ?Buckshaw Ho, ?Bugley
***Bula** ?Bulbarrow Hill
Buna ?Bonsley Common
Burgwine ?Burstock
Burgwynn fem. ?Burstock
***But(t)el** Biddlesgate Fm
***Byrra** *byrran broc* **4** 302 (AS charter)
***Byt(t)el** *Bittlesmore*, *bytelesmor* **3** 153 (AS charter)
Cada f.ns. Cadenham **3** 176, *Cadenham* **3** 374
***Cana** ?Cancer Drove, *Canendone Hundred*, Canford, ?Canon Hill
***Catt** ?Cattistock, f.n. ?Chatsworth **4** 221
***Cattuc** f.n. ?Chatsworth **4** 221
***Ceacca** ?Checkridge Fm
Ceadda f.n. *Chadenwich* **2** 10
***Ceatt** f.n. *Chattoshale* **4** 284
Ceatta Chetnole
***Ceatwa** ?Chebbard Fm, *ceatwanberge* **1** 303 (AS charter)
Cedd, **Cedda** ?Chedington, f.n. Chetenhill **4** 233
Cēna f.n. *Ken(e)ley(e)* **2** 53
***Centel** *Kentlesworth,* f.n. *Kenteliscombe* **2** 216
Cēol Chalbury **2** 134, *chelesbergh'* **2** 122 (AS charter)
Cēola E & W Chelborough, ?Chilbridge
Cēolbeorht *chelbrichtes dich* **3** 163 (AS charter)
Cēolhelm ?(Hr) Chalmington
Cēolmǣr ?(Hr) Chalmington

Cēolmund ?(Hr) Chalmington
Ceorl f.n. ?Charles Down **3** 108
***Cidd** *ciddesbeara* **2** 161 (AS charter)
***Cippa** f.n. ?*Chypylthorn* **2** 52
Cnihtwine Knitson
***Cobba** ?Cobley Fm
***Cocc** ?Coaxdon Hall, ?*cockes þorne* **2** 122 (AS charter)
***Cocca** ?Cocknowle, f.n. ?*Cockenell* **1** 128
***Cōda** f.n. *Codemor* **2** 215
Cola Colber Crib Ho, ?Cole Combe, Colly Fm
***Cossa** f.n. *Cossymour* **3** 294
Cotta Coltleigh Fm, Cotley, *cotta dene* **4** 306 (AS charter), ?*Cot*(*t*)*cote* **4** 146
***Cucel** Cuckoo's Hill, *cuceles hylle* **4** 28 (AS charter)
Cuda ?Cuddle, ?Cudnell
Cumbra ?Common Mead **3** 19, f.n. ?Commerwick **3** 185
Cūðbeald f.n. ?*Cutebolstre* **1** 345
Cūðburh Cuthburga Rd **2** 184
Cydda Chitcombe Fm, Chitmoor
***Cylferð** ?Culliford Tree (Hundred)
***Cylfweard** ?Culliford Tree (Hundred)
Cȳma Kimbert's End, ?Kimmeridge
Cynestān Kinson
***Cyra** f.n. ?*Kyrton* **2** 10
Dene Densham Fm, f.n. ?*Densworth*(*e*)*y* **4** 222
Denna f.n. *Denneworthe* **1** 168
Dēorwine Durweston
Diga ?Dymor
Dod(d)a ?Ditton's Still, ?Dodding's Brick Yard & Fm
Dol(l)a *dollen beorge*, *-berch* **3** 102 (AS charter)
Drēama Drimpton
Dudd Duds Bury, f.n. ?Dudgemoor **1** 180
Dud(d)el ?Dudley Moor, f.n. ?*Duddeleston* **4** 182
Dunn ?Little Dunster Fm
Dycgbeorht Dibberford
***Dydda** f.n. *Dud*(*e*)*hull* **3** 237
***Dydel** Didlington Fm
Ēadbeorht Ibberton
Ēadhelm *eadelmes melne* **3** 138 (AS charter)
***Ēadmōd** ?Edmondsham
Ēadmund ?Edmondsham, *Edmondesthorp* **3** 353

Ēadrǣd f.n. *Addredeston* **2** 172
Ēadweard St Edward's Bridge
Ēadwine *adwines imare* **3** 163 (AS charter)
Eafa f.n. *Evelegh* **4** 270
Ealdhelm ?*Audelyngton* **2** 134
Ealdmann *ealdmannes þyerðe*, *-wyrde*, *eldmannes wrthe*, *heldmannes wrthe* **3** 111, 136, 178 (AS charters)
Ealdwine *Aldewinesbrigge* **2** 187
***Earm** f.n. *Yearmesborowge* **1** 294
Ēata Adber, Yetminster
Ecga Eccliffe, Egg Moor
Ecgel ?*Eglysham*
Ecgwulf *ecgulfes treow* **3** 310 (AS charter)
Ella ?Elwell Fm, Elworth, ?Hillcombe Bottom & Coppice
El(l)i ?Ellston Hill
Elta f.n. ?*Eltecumbe* **1** 113
Enna *ennan pol* **3** 298 (AS charter)
***Eohhere** Eggardon Hill & Hundred
***Eorma**, ***Eorming** ?Armswell Fm
Fitela Fiddleford, f.n. ?Vittleford **1** 81
***Focga** *foczan-*, *fozza crundel* **4** 303 (AS charters)
Forst ?West Fossil Dairy & Fm
Franca Frankham Fm, f.ns. *Frankemere* **2** 174, *Frankham* **2** 250
Friðurīc ?*fideriches dene* **3** 128 (AS charter)
Frōda f.n. ?*Froudemere* **2** 10
Gāra f.n. ?Coreans **1** 79
Gōda ?God Nore
Godmann Godmanstone
***Godrūn** fem. Godderthorn Hd, *Godrenelonde* **4** 467
Godwine f.n. *Atgodewynesthorne* **2** 85
***Goldgifu** fem. f.n. ?*Guldevelond* **1** 260
Goldrūn fem. f.n. *Guldrunethon* **2** 242
***Guppa** Guppy
***Gūðmǣr** ?Gummershay Fm
***Gupp** f.n. ?Gobson **3** 185
***Gylla** Gillingham
Hæcga ?*hacggen hamme* **3** 237 (AS charter)
***Hafoc** f.n. ?*Haukeshulle* **1** 137
Hagona ?*Hawneferne*
***Hāmweald** ?*ham woldes þorn* **2** 158 (AS charter)
***H(e)afoc** Hawkchurch, f.n. ?*Haukeshulle* **1** 137

Hemele E & W Hemsworth
***Hem(m)ede** E & W Hemsworth
Hengest ?*Hængestes ricg*, *hengestes rigc* **4** 374 (AS charter)
Heorstān Hursey
Here Herston
Herebeorht *herebrittescombe* **3** 256 (AS charter)
***Hippa** *hippe pad* **3** 206 (AS charter)
Hnott(a) Nottington
***Hogg** ?Hogchester, ?*Hogford*, ?Hogstock
Horsa f.n. ?Horsingtons **3** 27
Hræfn *Ramesham*, ?Rampisham, f.n. *Ravenestorp* **3** 280
***Hremn** ?*Ramsbury*, ?Ransom Coppice, ?Renscombe Fm, f.n. ?*Remeswelle* **1** 70
Hrēðel ?*Ruddlesford*
***Hucca** f.n. ?*Huckelake* **2** 221
***Hudela** f.n. ?*Hudeleforundele* **2** 130
Hūn(a) *Hunesberge Hundred*
Hund ?Hounstout Cliff
***Hūngār** f.n. *Hungresbergh* **2** 136
Hwīta ?Whithay Lane & Moor
Ifa, **Iofa**, **Eofa** ?*Eveley* **1** 115
Ine Innsacre (Fm)
***Irma** ?Heron Grove Coppice
***Lǣsta** ?Lastridge
Landbeorht *la-*, *lobirihtes gete* **4** 454 (AS charter)
***Leaxa** f.ns. ?*Laxted forlong* **2** 53, ?*Lexdon* **1** 121, *leaxan oc* **2** 82 (AS charter)
***Len(d)ca** f.n. ?*Lenkethorne* **2** 70
Lēofa or **Lēofe** fem. *leouen imare* **3** 136 (AS charter), f.n. ?*Leveridge* **2** 130
Lēofgēat *Leftisford*
Lēofsige Leeson Ho
Lēofwaru fem. Loverley Fm
Lēofwīg ?Lewesdon Hill, Leweston
Lēofwine Tarrant Launceston, f.n. *Luuenesburg* **2** 216
Lill f.n. Lillwood **3** 186
Lilla *lillanwelle* **3** 257 (AS charter)
Lippa *lippen stagen* **3** 102 (AS charter)
***Lod(d)a** ?Lodmoor, f.n. *Lodewurthy* **3** 257
***Lolla** ?Lowbrook Fm
Ludmann *ludmannes putte* **3** 111 (AS charter)

Lufa ?*Lovard*, Lowness Coppice, Luton Fm, f.n. ?*Louecote* **1** 70
***Lugga** ?Luckford Coppice
Lulla ?Lowbrook Fm, East & West Lulworth, f.n. *Lullebrouk* **4** 454
***Lutta** ?Lutton
***Lylla** Lillington
***Lytig** *litiges heuede* **3** 163, 3 199 (AS charters)
Mac(c)a f.n. *Makedoune* **2** 10
Man(n) Manston, f.ns. Mans Croft **2** 171, *Mansham* **1** 293
Mann(a) Mannington
***M(e)arna** ?Marnhull
***Mocc** f.n. ?*Muxerigge* **4** 465
***Mogga** f.n. ?*Mogwelle* **1** 327
***Mort(a)** ?Morcombe Fm, Morcombelake, Mosterton
Mucel, ***Mucela** ?Muckleford
Mula ?Moulham
***Murt** ?*murter* [?*murtes*] *wyll* **4** 304 (AS charter)
Nunna f.n. ?Nunbarrow **1** 69
Od(d)a *Odenham*, f.n. ?*Odeslade* **3** 6
Olla *?Hollish*
***Ord** f.n. *Ordesham* **2** 10
Ōsa *Osanstoke* **4** 338, f.n. *Oselake* **2** 11
Ōsgār Askerswell
Ōsmund Osmington
Ōsweald Osehill Green
***Pac(c)a** f.n. *Pakkeberewe* **1** 365
***Pæcc** f.n. ?Patchward **3** 253
***Pæcca** f.n.?*Pacchorn* **1** 314
Pǣga *pegan beorh* **3** 122 (AS charter)
Pælli Pallington
***Pampa**, ***Pempa** ?Pamphill
Penda f.n. *Pendeworthe* **2** 53
Peof(a) Povington
***Peohtrīc** Petersham Fm
Pēol ?Pelsham Fm
Pīc f.n. ?*Pyxden* **1** 328
***Pinna** ?Pinford
***Pipp** Pipsford Fm
***Poc** ?Pixon Barn, ?Poxwell
***Poca** ?Poxwell
***Prym** ?*Primesley*
Puda Putton, f.n. *Pudecombe* **3** 257

***Puma** Puncknowle
***Pūna** Poundbury
Punt Poyntington
***Pyd(d)a** ?Pidney, ?Pitcombe Dairy, f.n. ?*Pyd(de)mylle* **3** 368
Pyttel ?Piddles Wood
Rædel, **Rædla** Rollington Fm
***Rēðel** ?*Ruddlesford*
***Rodda** *roddan pol* (AS charter)
Sǣlāf ?*Selavestune*
Sǣweard or **Sigeweard** ?*Seawardeston* **4** 50
***Sceaft** ?Shaftesbury
Sc(e)ortmann ?Chapman's Pool
***Sceota** ??Shotover Moor
Selegifu fem. f.n. ?*Selegyevessetle* **3** 34
***Sigelāf** ?*Selavestune*
Sigemund Symondsbury
Snell ?Snelling Fm, *snelles hamme weghe* **3** 112 (AS charter)
***Stott** ?Stottingway
***Stotting** ?Stottingway
Strutheard Coombe Bottom **1** 10
Stunta f.n. ?*Stuntecnoll* **1** 185
***Sutta** f.n. ?*Suttewey* **1** 199
***Tāda** ?Tadnoll Dairy & Mill
***Tæcca**, ***Tacca** ?Acton
***Tāda** f.ns. ?*Taddehull* **3** 205, ?Tatcombe **1** 305
Tāta ?Tadnoll Dairy & Mill, Tatton Fm & Ho, f.n. ?Tatcombe **1** 305, *tatanbeorge* **2** 122 (AS charter)
Tila ?Tillworth
***Tilluc** *til luches lege* **3** 123 (AS charter)
***Tippa** f.n. ?*Tippelynch* **1** 329
***Toppa** f.n. *Top(e)hull* **3** 59
Tot(t)a Tatcombe Wood, Tatnam Fm, ?Todber, Totcombe Hundred, ?Totnell, ?*toten berg'* **3** 123 (AS charter)
Tud(d)a f.n. ?*Tudelane* **3** 35
Udd, ***Udda** Uddens Ho
***Ucg** Uggescombe Hundred
***Waba** Wabmill Fm
Wāda f.n. ?Watchet **4** 311
***Wæneca**, ***Weneca** *wænecan wyrð* **2** 162 (AS charters)
***Wærla** f.n. ?*Warledenne* **1** 230
Wǣrmund f.n. *Weremundysden* **4** 133

Wulfsige ?*Wulsieslande* **4** 261
Wulfðrȳð fem. f.n. *Wolderdam* **4** 228
Wyrta f.n. ?Wattinghams **2** 105

B. CORNISH
Probus *Lanprobi*, *Propeschirche* (Sherborne) **3** 358

C. MIDDLE ENGLISH (SCANDINAVIAN)
Grímr Grimstone
Gunnhildr fem. f.n. *acram Gunilde* **2** 127
Krókr f.ns. *crockesdich* **2** 161, *Crokescumba* **1** 326
Rualet *Rowaldsway*
Þorsteinn *Thurstanesley*, f.n. *Thursteyneswell* **1** 238
Tóla Tolpuddle

D. MIDDLE ENGLISH AND CONTINENTAL
Alverad Okeford Fitzpaine
Andrew Andrewshays Fm
Brian Bradford Fm, Bryanston, Bryants Puddle
Burde ?*Bush-head Gdn*
Burdel Burleston
Catherine St Catherine's Cross
Clare Clare Towers
Col(l), **Col(l)yn** Colliton Ho
Dodde ?Dodseye
Dru, **Drew** Druce Fm, ~ Hangings
Eggelin Egliston
Elen, **Helen** Green Island
Erlewin Cross (Chaldon Herring)
Gerin, **Jerin** Gerrard's Hill
Giles Corfe Mullen
Godelin ?Godlingston
***Godling** ?Godlingston
Gummar ?Gummershay Fm
Guy Guy's Marsh
Haket Beer Hackett
Herman ?Harmshay
Huberd, **Hubert** Corfe Mullen

E. SURNAMES AND BY-NAMES (INCLUDING ALL THOSE USED IN MANORIAL AFFIXES)

Chideock Chideock Fm, ?Stourton Caundle
Christmas Christmas Close
Clap Clapps Mead
Clapcot(e) *Clapcotts*
(**de**) **Claville** Clavel Tower, Clavell's Hard, ?Cleavel Point
Clench Winterborne Clenston
Cocc, **Cock** Cox's Cross
Coilard Stock Gaylard Ho
Cok Cook's Plant.
Coker Coker's Frome
Cokyn Cookney
Cole Coles Fm
Coleman Colman's Fm
Col(l), **Col(l)yn** Colliton Ho
Colmer Colmer's Hill
Colt ?Colston Fm, ?Colthay Coppice, ?Coltsclose Corner, Coltwood Coppice
de Columbar(iis) Purse Caundle
Colyer Piddletrenthide
Colyn(g)don ?Cullington
Cosin Stoke Wake
Co(u)ch Scouse Fm
(**le**) **Craneson** *Frome Cranchen*
Crei (Little) Waddon
(**de**) **Cricket**, (**de**) **Cruc(k)et** *Creketway*
Croc Cruxton
Crubbe Kingston (Maurward)
Culpeper Cull-peppers Dish
Cunne ?Cumheigh or Cymeheigh
Curse ?Kershay Fm
de Curtenay Iwerne Courtney or Shroton
Cusin East Stour
Cutt(e) Cothays Fm, Cuthay Fm
Daggeuill Daggles Drove
Dagon Daggons
Dawtry ?Dottery
Deneys Danes Hill, Wraxall
de Deverel(l) *Combe Deverell* **1** 311, Deverel Fm, Farnham
Digon Dickens's Marsh
de Dina(u)nt Gussage St Michael
Dole Dole's Ash Fm

Grael, **Grele** Griddleshay Fm
Grant Grant's Barn, Grant's Ctg & Plot
Gray, **Grey** ?*Grayswick* **1** 357, Grey's Bridge, Grey's Wood
Grigge Grighay Fm
Guah(e)bon *Winterburn Guah(e)bon* **2** 84
le Guldene Langton Long Blandford
Gumbald Gummershay Fm
Gum(m)er ?Gummershay Fm
de Gundeville Eastbury Ho, Farnham, Tarrant Gunville, Tarrant Hinton
Guppy Guppy's Fm
Gurewambe *Winterburne Gurewambe* **2** 83
Gyffard *Gyffordsparke* **2** 130
de Haddon Stourton Caundle
Hallyng Hallings or Hallam Manor
Hancok Handcock Coppice
Harang, **Hareng** Chaldon Herring, Herrison, Langton Herring, Winterborne Herringston
Harbin Harbin's Park
Harman Harman's Copse & Cross
Harpere ?Harper's Fm
de Hawey Over Compton
Hayward Hayward's Fm
Heidbury Headbury Big Cove
Henning Cruxton
Herman ?Harmshay
Herring Heron Grove
Hodder Hodder's Fm
Hogg(e) ?Hodge Ditch, ?*Hogford*
Holles *Holles Frome* **1** 369
Honeybourne Honeybins
de Horsenden Barford Dairy & Fm
Horsey Melcombe Horsey
Howe Howe's Eype
Huntingdon *Winterborne Hundyngton*
Hyde Hydes, Hyde's Quay
Iryssh Colber Crib Ho
Jacklin Jacolines Crock
Jan Jan's Hill
Jenkins Jenkins's Fm
Jobbe Job's Plain

Mayer Mayers Pond
Michel *Milborne Michelston*
Moger Mogers Leaze
Moigne Buckhorn Weston, Moigne Comb & Wood, Moigns Down (Barn), Owermoigne
de Moion, **Mohun** Hammoon, *Mohun Castle* **4** 333, Gt & Lt Mohun's Bridges, Winterborne Houghton
Monteacute *Newton Montacute*
Moris Morey's Orchard
Moure Moorcourt Fm
Mowlam Mowlam's Barrow
de Musters Muston Fm, Winterborne Muston
Mydnell Midlands Fm
Myle Miles's Lane
de Neuburgh Winfrith Newburgh
de Neville Fifehead Neville
Newman Newman's Coppice & Ctg
Norris Norris Mill
Oliver Wyke Oliver Ho
Orescuilz Sandford Orcas, Stourpaine
Osmond, **Osmund** Melbury Osmond, *Osmond's Manor*
Oven ?Gt Ovens Hill
Paen E & W Chelborough
Panter Painter's Barn
Patchin Patchins Point
Payne Edmondsham, Payne's Coppice, Stourpaine
Paynel *Southcombe* **1** 116
Penn(e) Penns Bath & Weare, Pennsylvania Castle, ?Pentsome Coppice, Toller Whelme
de Percy Piercy Fm
Perham Pergin's Island
Petre Peter's Gore
Peverel(l) Barford Dairy & Fm, Bradford Peverell, Newton Peveril, Peveril Point
Philip *Phil(l)ips*
Pickard Pickard's Coppice
Pigace Upwey, Winterborne Monkton
Pike Stanton St Gabriel
de la Pile ?Stourton Caundle
Piterych Woolcombe Fm

Skyllyng ?Skilling Fm
Skynner Skinner's Coppice
Smewyne ?Sminhay Fm
Smith Smith's Kilns
Snel(l) Snailscroft Fm
Snelling ?Snelling Fm
Speke Speaks
Spere Spearhay, Spear's Coppice & Lane
Stickland Stickland's Ctg
Stourton Stourton Caundle
Strangways Woodsford
(de) **Strode** Stroud's Moor Withy Bed, Wooth Grange
Sturt Sturt's Weir
Symes Symses
Syward Syward Lodge
Talbot Tabbit's Copse, Tabbit's Hill Fm, Talbothays, Talbot's Wd
de (la) Tilli, **(de) Latylle**, **Tilly** Langton Long Blandford, Tillicum, ?Tilly Whim Caves
Tirel *Tirel's*
Tollard Tollard Farnham
Tonere Turners Puddle
Tregonwell Milton Abbas
Trump Trimshayes Fm
Tudde Tudhay
de Turberville Bryants Puddle, Melbury Sampford, *Turberville Stoke*, West Stafford, Winterborne Muston
Turges Melcombe Horsey
Turner Turner's Cove
Tynham Tynehams Withy Bed
de Valencia Compton Valence
de Ver Vearse Fm
de Vilers Tarrant Rushton
le Vineter *Vinter's Fee*
Vye Vyse Barn
Vyne Vine's Down Bldgs
Wadham *Wadham's*
Wake Caundle Wake, Stoke Wake, East & West Stour
Waket ?Melbury Sampford
(le) **Waleys** Langton Matravers, Mandeville Stoke Fm
Weld Weld Arms
de la Weye, **atte Weye** ?Waytown

de W(h)ytewell *Whitwell*
de Witefeld Frome Whitfield
Wood Wooden Cabbage Fm
Woodman Woodman's Cross
Woolgar Oakers Wood
(de) Worthe *Woors or Worths*
Young Young's Fm

SOME CATEGORIES OF FIELD-NAMES AND MINOR NAMES

(a) **Names for productive land or pleasant places (complimentary names)**: Best Buttery **2** 241; Best Clothes **4** 462; Bounty **4** 347; Dripping Pan **1** 79; Easy Go **1** 112; Fancy **3** 347; Fat Ox **1** 305, **1** 312, **2** 104; Goodluck **3** 347; Heaven **3** 378, *Heaven down* **4** 435; Jacobs Ladder **3** 64; Joy Mead **4** 424; *Merghedene* **2** 216, Merry Fd **3** 40, **4** 391, ~ Mead **2** 237; *Merveile* **2** 10; Mount Pleasant **1** 38, **1** 126, **1** 156, **4** 34 (freq); Paradis(e) **2** 82, **2** 139, **2** 157, **3** 57, **3** 279, **4** 360 (sometimes applied ironically as at **3** 57); Pleasure House Plant. **2** 139; Purse **4** 80; Turnpenny Fd **3** 198.

(b) **Names for unproductive land or unpleasant places (derogatory names)**: *Armemede* **4** 384; Barley Bread **1** 292; Bedlam **4** 39; Break Heart **2** 25, Breakheart Hill **4** 433; Break Neck **4** 328, Brickneck **1** 27; Catcham **1** 361; Cutt Purse **3** 321; Droy **4** 286; Envy Barn **4** 347; *la Forsakengore* **3** 199; Garbage Md **2** 199; Giddy Lake Plot **2** 191, Giddy Md **2** 8; Grief **4** 463; Hard Labor **1** 119; *Hell Close* **3** 226, *Helle* **3** 256, **4** 10, **4** 88, *Helle doughter* **4** 10, Hell Grd **1** 286, ~ Meadows **4** 88, *Helpytt* **4** 167; Hunger Hill(s) **1** 28, **2** 42, **2** 143; Hungry Hill **2** 37, **3** 300; Iron Box **1** 41; *Judas Garden* **1** 169; Gt & Lt Jug **4** 424; Little Content **3** 26; Little Gains **2** 200; Long Friday **1** 312; *Lortemerforlange* **2** 72; *le Louseslappe* **2** 289, Lousy Grd **4** 435, Lowsy Bank **2** 45, Lucehills **2** 105, *Lusarne* **1** 121; Mount Hunger **3** 394; Mount Misery **1** 38, **4** 59; Muchado **4** 337; Murren Plot **1** 142; ?*Nytheclose* **3** 99, Pick Pocket **4** 424; Pick Purse **2** 26; Pinch Croft **4** 464; Piss Lake **4** 268; Poison Cl **4** 464; Purgatory **4** 453; Sharn Md **3** 253; Shitten Can **4** 40, ~ Shard **3** 253, *Shitten Stile* **1** 185; *Shiteroke* **4** 117, *Shyteport* **1** 185; *Skitterlinch furlong* **2** 181; *Slutes hole* **1** 32; Small Gains **2** 200; Sodom **4** 92; *Sorrow bargaine* **4** 11, Starve Acre **3** 216, 310, **4** 249, **4** 331, **4** 436; Starveall **3** 213; Starve Farthing **4** 436; Starve Land(s) **3** 88, **4** 331; Stinking Hill **4** 436; *Stynkbargh'wyswey* **3** 227; Stumbledown **1** 113; Titchy Md **4** 425; Too Bad **4** 383; Torment, Tormenting Plague **4** 425; Trouble Fd **2** 234; Tumble Down **1** 152; *The wast grownd* **4** 387; Weary Hill **2** 238; (The) Wilderness **3** 136, **4** 193, ~ Plant. **4** 464; Worse Again **4** 383.

(c) **Names referring to shape, size or appearance (often figurative)**: Beer Hill Tail **4** 233; Bomb Hill **4** 297, Bombs Lane **1** 226; Bowhays **4** 380; Butter Print **3** 55, **3** 97; Buddick **4** 297, *Buttockespytt* **4** 222; Chopping Knife **2** 119; *Cloveneburgh* **3** 255; Coppert Hill **4** 381; Crinkle Crankle **4** 267; Croked Close **2** 192, Crooked Coppice **2** 32; Crumpley Md **2** 32; Dutch Cap **1** 336, **4** 91; Elbow **4** 499, Elbow Colbourne **3** 196; Forked Bridge Fd **3** 101, Forked Cl **2** 41; *le Fryingpanne* **2** 14; Girdle **4** 192, Girdle Ring **4** 451; Half Moon Plot **4** 423; Handkerchief half **2** 119; Harp **3** 314, Harp acre **4** 56, Harpeland **4** 92; *Hokedwale* **1** 345; *Hornymore* **1** 81; Hundred Acres (often ironical for a small field) **1** 249, **3** 287, **3** 314; Huntinghorn Flg **1** 362; Knappy Cl **4** 151; *Knappydehalle* **1** 184; *Krykeldehull* **4** 182; Leg Plot **2** 171; Long Pannier **4** 330; Marabone Park **4** 486, Marrowbone Barn **4** 474, Marrow Bone(s) **4** 329, **4** 491; *the Narrowe Cont* **1** 259; Navvy Shovel **4** 124; *Nelbowe* **1** 327; Ord Cl **1** 80; Pancake (Close) **2** 9, **3** 78; Peaked Close **3** 57, ~ Fd **2** 85; Peaky Fd **2** 85; Petty Cl **2** 78; Pickaxe **2** 119; Picked Close **3** 57, ~ Jaws **2** 26; Pincushion **2** 205; *Pynteldirne* **3** 375; *Quarentaine* **2** 125, Quarrentine **4** 330; Rimple **3** 187; Roundabout **4** 414, ~ Cl **4** 283; Runnaway **4** 424; *Ryngham* **1** 152, *Ryngwell* **4** 205; Saddle **4** 383, Saddle Back **2** 9, Saddle Cl **4** 330; *Sherpecrofte* **3** 41; Shoulder of Mutton **1** 168, **2** 105, **3** 198; *Smelecombesheved* **1** 42; Stickelewell Md **3** 57, Sticklands **3** 340, Sticklinch **3** 88; Strings **3** 30, **3** 57; *Taylepytte* **1** 122; Tea Grd **4** 249; Three Horned Hay **3** 136; Trendles **4** 92; Two Scondlings **4** 109; Wager **3** 30; Wedgeditch **4** 391; *Wogheland* **4** 19, Woollens Md **3** 31, *Wowacre* **3** 35, *Wowelond*(*e*) **1** 42, **2** 251, **4** 133, *Woylond* **4** 472. See further the section on names referring to landscape features in Introduction *supra*.

(d) **Transferred names**: Boyne **3** 55; Buyenos Ayres **4** 347; Canaan **1** 69; Dungeness **4** 359; Isle of Wight **3** 197; Londonderry **4** 45; Mount Ararat **2** 258; New England **1** 167; Newfoundland **3** 272; New Zealand **2** 28; Pimlico **4** 410; Russia Fds **3** 253; St Vincent **3** 72; Van Diemans (Land) **2** 161, **3** 328, **4** 269, **4** 368; Waterloo **3** 72. See further the section on transferred names in Introduction *supra*.

(e) **Names referring to sport and recreation**: Bowling Green **2** 90, **4** 248, 328, 462, *The bowling greene* **4** 384, Bowling Md **1** 286; (The) Cockpit **1** 283, **2** 139, **2** 234; Dancing Cross Rd **3** 204, *Dawnsinge Plotte* **4** 465; Decoy Pond **2** 214; Games Plot **4** 329; *Huntemede* **3** 157, Huntwell Moor **3** 252; *le hurlegogge* **3** 149; *Joustyng croft* **3** 157; Kings Standing **1** 301; *le Maypole* **1** 258; Nine Pins **4** 471; (The) Play Close

2 90, **4** 330, Play Cross **2** 105, **2** 241, **3** 187, *la pleycrosse* **3** 200, Playford **2** 229, Play Street **4** 499, *Pleystrete* **3** 34, *le Pleymersch* **1** 287; *Playing closse* **3** 188; *Quyntine* **3** 382; Round Robbin **4** 109; *Schetyngstokk* **3** 34, *Showtinge Crofte* **3** 324; Swimming Mdw **2** 9; *tenys cowrt* **4** 360; Troy Town **3** 373; Whirlegog **4** 491. See further the section on names referring to sports and pastimes in Introduction *supra*.

(f) **Names referring to soil type or surface characteristic**: Biscuit Cl **1** 343; *browne grove* **4** 437; Burnbake **2** 85, **2** 109, Burnt Beat **1** 264; Catsbrain **3** 161, *Catsbrayne* **2** 13; *Cheslane* **3** 187; Chessel Fd **3** 132, *Chesyll hyll* **2** 173; *Chelkhulle* **1** 260, *Chelkweye* **1** 345, Chilkhill **2** 288; Clan **1** 198, *Clanebourgh* **4** 476, *Clanfield* **3** 231; Dry Md **3** 55; Girt **4** 188, **4** 451, *le Grette* **1** 237; *Grodbury* **1** 327; *Grotehill* **3** 392, *La Grotene* **2** 85, Grotten Cl **3** 235; Grublands **4** 423; *le Haryleche* **1** 62; *Hethenstrete* **1** 172; Malm Grd **3** 28, ~ Pitts **1** 151; *la Marl* **1** 313; *Marledelond* **1** 327; *Mirremeade* **3** 375; *Mowry Close* **1** 203; *Papilhill* **2** 78, Popple Street **3** 78; Pliddy Moor **4** 391; Pudding Cl **3** 198, ~ Fd **2** 51; Quaggy Brakes **1** 41; Rotten Md **4** 25; Sedgy Grd **1** 100; *Schortmarle* **2** 81; Smythe Hill **1** 363; Soggy Bottom **4** 391, ~ Breach **4** 425; Sogs **4** 453; *Sourehams* **3** 376, *Sourlond* **2** 242, **3** 208; Stone Croft **4** 156; *Stoning Close* **1** 271; *Stonschulye* **1** 230, *Stonschyly* **3** 227, *Overstonschulye* **1** 328; *Stonybrigg* **3** 227, *Stonyham* **2** 175; *Stubbed Close* **4** 223; *le Stubble feild* **1** 134; *le Stubbymerssh* **1** 287; *Swetenham* **2** 175, *la Sweteplek* **1** 329, *Sweytacre* **3** 258; Teoffee Mdw **3** 5; Thirtoop **1** 228; Water Md **3** 178; *Wetfurlang* **4** 472; Whey and Curds **3** 162; *Whitbarrowe* **2** 78, White Flgs **2** 172, Whitmoore **3** 58; *Wild Crosse* **1** 243.

(g) **Names referring to church land, religious practice or charitable use**: *the Abbie Litton* **3** 359, *Abbeyweye* **2** 250; Anchor Md **2** 191; *le Belhous* **3** 150, Belfland **3** 216, *the Bell Acre* **2** 10, Bell Drong **1** 193; Chantry Md **3** 271, *le Chauntrey strete* **2** 186; *le Charterhous* **3** 365; *Cristenecros* **3** 32; *le crosse tree* **4** 10; *Crouchehey* **3** 32; Curfew Grd **3** 216; Freary **3** 26; *Frerene lane* **1** 351, *Fryarnebridge* **2** 78; Genesis **4** 390; *Glebeland* **2** 215; *Gosspell Bushe* **2** 174; *Halgewal* **3** 33; *hermyteswode* **4** 204; Lady Hole **2** 124, ~ *Mead* **3** 219, *levidiforhurde* **2** 85; *Lampeacre* **3** 226, *the lampe halfe acre* **2** 53, Lamp Ground **4** 68, *Lamphayes* **3** 64, Lamplands **3** 287, *the Lamp Light Close* **1** 81, **1** 121; Lenthills **1** 106; *Magdalen* **3** 150; *Mynchinweyslow* **4** 183; *Monkynwell* **4** 374; *Morrow Masse Landes* **3** 33; *Pater Noster Crosse* **2** 174; Path Crouch **4** 343; *Preslye Copes* **2** 175, Press Mdw **2** 191, ~ Street **3** 177,

Prestill Mead **3** 236; *Priors Grove* **2** 175; Sanctuary **4** 370; *Templelond* **4** 500; *Vicar's Breade* **3** 299; *Ye Vycary Leez* **2** 54; Zebedee **4** 193. See further the section on names referring to the church and religious matters in Introduction *supra*.

(h) **Names that are whimsical, fanciful or jocular**: Adam and Eve **2** 109; *Bareleggudthorn* **3** 226; *Bellipot Grene* **2** 106; Best Clothes **4** 462; Carpenters Square **4** 280; *Chyme Silver* **1** 51; Cock and Breeches Cl **4** 328; Collar Maker **2** 214; Cuckolds Md **2** 220, ~ Parlor **1** 27, ~ Pit Moor **4** 423; Cuckoo Pound **3** 287; Duck's Nest **3** 251; Dungeon Close **1** 211, ~ Copse **3** 55, ~ Mdw **2** 157; Eel Fishes **4** 490; Eunucks Md 2 220; *Flaxepoke* **1** 229; *Harvest Bower* **1** 95; Jackalints **1** 60; *Jolly Robins* **2** 201; Kid Whistle **4** 253; *Kill Devil* **4** 398; Kitcats cl **1** 80; *Loveland* **3** 188; Mermaid (Orchd) **4** 330, **4** 424; Nowhere **3** 385; Paddle Dock **1** 29; *Palpitro Hill* **3** 34; Pot Belly and Hollow Belly **1** 94; Resolution **1** 143; Scarcrow Cl **2** 180; Shear and Go **4** 88; Spitting Box **4** 249; Sword and Dagger Breach **4** 425; Trim Trams **3** 30; Weathercock **4** 487; What do you propose to do to this one **4** 465; White Feather **4** 308. See further the section on whimsical and humorous names in Introduction *supra*.

(i) **Names referring to superstition, legend or folklore**: *Alvysch'thornys* **3** 225; Battle Mead **1** 53; Bones **4** 450; Bugbarrow **4** 33, **4** 221, *Bugdowne* **4** 227; Dead Maids' Grd **2** 157; *Dedefolkefurlonge* **2** 13; *Dedman' feld* **2** 52; *Dewelepole* **4** 250; *Elfhede* **1** 313; *Gobelynshole* **4** 35; Grimberry **1** 28, *Grymbargh'* **3** 226; Packmoor **3** 57, *Pokelane* **3** 149, *Pokeput* **1** 328, *Pokershe* **3** 207, *pook acre* **4** 144, Pook (Marsh) **1** 80, Puckhay **3** 101, *Pucklake* **4** 205, Puckmoor **3** 98, *Puckysbarry* **1** 185, Pug Pit **3** 78, *Pukefurland* **1** 243, *Puxeye Wood* **4** 157, Puxy Gdn **1** 203; *Pokelmour* **4** 35, *Pokulchurche* **1** 272; *Sc(h)ochleham* **2** 283; *Thursdyche* **1** 238, *Valgirimede* **4** 12; Witches' Corner **4** 208; Worms Hill **4** 476. See further the section on names referring to superstitious beliefs in Introduction *supra*.

(j) **Names referring to wrongdoing and punishment**: Bannell Mdw **2** 66; Bridewell **4** 115, *Bridewell* **1** 254; Catcham **1** 361; Challenge **4** 462; Friday Street Plot **4** 55; Gallett Rams **4** 192 , Gallytrams **4** 233; Gallows Acre **2** 234, ~ Close **4** 248, *Gallowes Hill* **1** 352, *Gallyhey* **1** 62; Gayhole **4** 337; Gibbett Cl **3** 97; Hangmans Close **4** 434, ~ Corner **4** 248; *Pillory Street* **2** 185, *le Pullery* **4** 11; *Theven'howe* **2** 230, *Thevenwell* **4** 250; *Wrechebergh* **4** 70. See further the section on names referring to punishment and wrongdoing in Introduction *supra*.

(k) **Names referring to administrative or legal matters**: *the Baleys Walke* **3** 31; Beacon Chine **2** 41, Becken Md **1** 215, *Beckyn* **2** 192, *the bekon* **2** 10; *Bordelande* **2** 180, *Bordelond* **1** 120; *Burse Hunte Cross* **2** 173; *le Calenge* **1** 120, *Calengleyn* **1** 345; Contraversy **1** 281; Cosloe **1** 79, Costlow **2** 157; the Court Close **2** 214, *Courte Close* **2** 173, *-Koll* **2** 78; Cozenage Md **2** 37, Cuzenage Coppice **2** 30; *Dawynlonde* **2** 173; *Dowerclos* **3** 32, *the Dowerfield* **4** 471, *la Dowre Oke* **3** 156; Encroachment **2** 157; *Escheated Peece* **1** 229; *Flitegore* **4** 182; *La Forengore* **3** 273; *Hustyngesmersshe* **3** 33; *lansshoredych* **1** 31, *threlansshorys* **3** 227; *Laufulham* **1** 184; *the lawe gutter* **3** 372, *Lawthorne* **4** 182; Look Out **1** 252; *Merkewey* **1** 31; *Modberghfurlang* **1** 327, Moot Hill Pce **3** 5, *Motcomb* **4** 304, *Motweie* **4** 284; *le Moitie* **4** 465; *Morweyefehey* **3** 273; Noad Cl **2** 205; *Nonemanneslond* **3** 232; Propriation Mdw **1** 181; Purlieu **2** 8; Reeve Md **1** 271, *the Rieve Holdes* **2** 175; Roguery **4** 348; Sacmore **1** 169; *Le Sanctarie* **4** 70; Tithing Md **4** 476, ~ Plat **3** 5, Tything Plot **4** 344; Toothill **3** 109, *Totemede* **3** 274, Totter Hills **4** 487, Tout Hill **3** 147, **4** 253; Turnpike Md **1** 100; Tythingman's Land **2** 114; Watchhouse Fd **2** 172, Watch Oak Coppice **2** 238. See further the section on names referring to administrative and legal matters in Introduction *supra*.

(l) **Names referring to rents, tenure and festivals**: Candlemas Md **4** 450; Farthinghold **1** 362, Farthingsland **3** 97, *Ferthyng et Sciling* **3** 266; *Feodary lands* **3** 273; *la Fremede* **1** 333, *la Fryelandshor* **1** 326; *Gawle acr'* **4** 19; Half Crown Grd **3** 186; *Halfepeny Hay* **4** 218, Half pence **3** 101; *le Lentonheys* **1** 203; Long Monday **1** 286; Michaelmas **4** 152; *Pennie Meade* **3** 6, *ye penny hay* **4** 437, Penny Lands **3** 388, Penny Plot **4** 393, Pennys Plot **1** 134, Pennyworth **3** 8, *Penyham* **3** 328; Scotland **2** 294, **4** 343, **4** 499, Scotley Park **4** 209; Shilling Plot **1** 182; *Tolhampmede* **4** 12, Toll Goose **4** 436.

The above names are only a selection, and the Introduction and the Elements section both *supra*, the Index *infra*, and the parish field-name lists in Parts **1–4** should be consulted for further examples of the various name-types. In addition to the above, Thomas Hollis's numerous arbitrary and commemorative field-names in the parishes of Corscombe **4** 287–302 and Halstock **4** 193–204 should be particularly noted.

THE ANGLO-SAXON CHARTER BOUNDARIES OF DORSET

There are over 60 Anglo-Saxon charters (mostly surviving in later copies) for Dorset, and of these 37 have full descriptions in Old English of the boundaries of the grants of land involved (some charters have more than one set of bounds, so 42 sets in all). The individual boundary marks have been fully listed and treated under the respective parishes where they belong, as was considered most appropriate. Here however each set of bounds is given in full so as to show the order in which the boundary marks appear, each one of which is given a reference to the page in volumes **1–4** where it is treated.

The Old English words occurring in these bounds, as well as the personal names found in them, have been included in the general list of Elements and the list of Personal Names respectively, both *supra*.

CORFE CASTLE (parish), 948 (15) *ShaftR* (S 534 (1))
PARS TELLURIS PURBICINGA (**1** 1) …
Þese sanden þe land imare. Arest fram wicanforð (**1** 18) on beam broc. an ðonnen an lang broke (**1** 31) on anne wiþig þefele (**1** 33) þiers ouer þane merse (**1** 31) þare weilaite (**1** 32) þanne or anne ston. of þanne stane (**1** 32) on alle þiscan. of alle discan (**1** 14) on anne dich. þannen suth at lang dich (**1** 31) on ðone here paþ of þen here paþe (**1** 31) suth an lang dich (**1** 31) on þanne cumb. þonne an lang cumbe (**1** 10) on þen hirnen (**1** 31) onne strut heardes paþe of þane paþe (**1** 10) forðe an lange welles (**1** 32) one þane westrene cumbe þanne a dun anlang cumbes (**1** 33) on þane broc þonne anlang stremes (**1** 32) on þe schort mannes pol (**1** 64) þerth ut on sce (**1** 32).
Þane is þis seo westere land sceare. Of sa (**1** 32) an anne stan tor. of þan tore (**1** 32) on þane clif upward (**1** 31) on anne dich (**1** 31). þanen north anlang safundune (**1** 32) on þene richte herepath (**1** 31) þanen north an lang dich (**1** 31). oþ þane ealden stodfald estward (**1** 30). þane forþ on ða ealden hege rewe anlang þare hege rewe (**1** 31) oþ þa dich. of þare dich (**1** 31) on anne linc reawe. of þane linche (**1** 15) on bl[….] þanne anlang broces oþ olle discan. of olle discan (**1** 14) north ouer þane [….]

on scyleford (**1** 94). þane an lang streame (**1** 32) þat it cometh eft to Wikenforde (**1** 18).

CORFE CASTLE (parish), 955 (14) *ShaftR* (S 573)
AT CORF (**1** 5) …. AT BLECHENHAMWELLE (**1** 8) …
Þis sand þare xvi. hyde land i mare at Corf….Arest of sa (**1** 32) on anne dich. an lang dich (**1** 31) uppe on swuren. þwert ouer swuran (**1** 19) on þe þwers dich. of þare diche (**1** 32) on ecge ford be ecge (**1** 31) on Alfricheswelle. of þane welle (**1** 30) on anne weg. of þane wige (**1** 32) on an dich. anlang dich (**1** 31) north richte on wicun (*v.* RNs) niþer anland wicum streames. of þanne streame on anne þorn (**1** 32) north anlang dich (**1** 30) on anne walle. on lang walles (**1** 32) on stanwei (**1** 32) of þane dich. on anne dich. on lang diche (**1** 31) north irihte on wicum (*v.* RNs) niþer anlang wicunstreames. of þat streame on anne þorn. of þane þorn (**1** 32) on þa holendich. an lang dich (**1** 31) uppen iricht þurch þane wde (**1** 33) on þare dich (**1** 31). on þane rupemor. of þane more (**1** 32) on þone crundel of þane crundel (**1** 31) up irichte on þat holenbedde. of þat holnebedde (**1** 31) on þare holne stoke. of þane stocke (**1** 31) on þare stream anlang streames (**1** 32) on auenes broc. of þane broke (**1** 30) on þane ealde weg and lang weies (**1** 31) on hecgan sled. of þane slede (**1** 31) on þare streame. an lang streames (**1** 32). on þat stanene bregge. of þare brigge (**1** 32) on þare wei anlang weies (**1** 32) of irichte on þane richt wei. of þane weie (**1** 31) on anne stan wal. of þane walle (**1** 32) nither irichte on Wickenford. on lang Wikenford' (**1** 18) uppen irichte on beambroc (**1** 31) swo up anlang streames. of þane streame on þat withi begh. of þanne wiþibedde (**1** 33) on anne stan. of þane stane on þane oþerne. of þane stane (**1** 32) on holen wicken (**1** 14) eft on þare stan of þane stane (**1** 32) on þa ealdene dich. anlang dich (**1** 31) on þane wal. an lang welles (**1** 32) on struthgeardes cum (**1** 10) anlang welles (**1** 32) on struthherdes wege. of þane wege (**1** 10) eft on anne wal on lang welles (**1** 32) on seuen willes þry (**1** 32) anlang streames (**1** 32) ut a irichte on seortmannes pol (**1** 64).

CORFE CASTLE (parish), 956 (14) *ShaftR* (S 632)
BLAKENWELLE [*rubric*] (**1** 8) … AD CORF (**1** 5) … AT BLACHENWELLE …
dis sand þa land imare to Corf and to Blechenenwelle þare vii hide; Arest of Wikenforde (**1** 18) an land wiken of Scylenford, of Scylenforde (**1** 94) on richt wege, of þane weie (**1** 31) on olle discan (**1** 14), þanen on blechene of þanne welle (**1** 8) on þane hlinc, anlang hlinkes (**1** 15) on anne dich (**1** 31), þanen one þo ealde rode, onlang rode (**1** 31) onne

þo alde stodfald, of þane falde (**1** 30) on anne dich, suth anlang dic (**1** 31) on þare herepaþ, of þanne herepaþ (**1** 31) suth anlang saþendune (**1** 32) on anne dich onlang dich (**1** 31) oþe clif (**1** 31), þanen ut on Se (**1** 32).
þanen sant þis þat westrene landimare. of Se one þe stod dic (**1** 30) þanen forð be wertrumen (**1** 32) on anne stanweal, of þanne walle (**1** 32) þweres over Smalencumbe (**1** 32) of þa þorn' upwarde (**1** 32) þanen forð be euisc (**1** 31) one þat northene stod dich, of þare dich (**1** 30) on anne stanen wal nord on lang walles (**1** 32) on stan wege, an lang weies (**1** 32) on anne dic, þan' norþ anlang dich of þare diche (**1** 31) on wicean of wichen (*v*. RNs) on anne þorn (**1** 32) þanen on anne diche of þare diche (**1** 31) on anne þorne (**1** 32) þanen north on iricht wege, of þane i wege (**1** 31) on Alfstanes paþ (**1** 30), þan forð be eficlif (**1** 31) on aueres broc adune anlang brokes (**1** 30) oþ þane bige, of þane bige (**1** 31) on anne þorn (**1** 32), þanne suth on irichte on anne mor a dune, anlang mores (**1** 31), on þicean, up anlang þicean (*v*. RNs.) eft on Wichenford (**1** 18).

WYKE REGIS (parish), 978 × 984 (14) *Add* (S 938)
Uuike (**1** 267) … prope uidelicet insulam de Portland (**1** 217) …
Þis synd þa land gemere to Wike. Ærest fram þare west Sæ (**1** 272) in Saggelorð. And fram Saggelorð (**1** 272) inne muledich (**1** 272). And swa andlang þare dich to blake ston. fram þane stone (**1** 271) to Soredich. fram Soredich (**1** 238) to lodomor (**1** 232) and swo bi richte merke (**1** 238). to ellenstubbe. of ellestubbe (**1** 237) in þare æst Sæ.

LITTLE CHESELBOURNE (lost, in Puddletown parish **1** 316), 965 (10) *Ilch* (S 736) [NOTE almost certainly to be located here rather than in Cheselbourne parish itself, both places taking their name from the same stream: the forms from these bounds included under Cheselbourne parish **3** 202–8, *v*. especially the comment made under the *TA* f.n. Head 3 Yards **3** 205, should now be included under Puddletown parish. For this identification of the bounds with Little Cheselbourne, *v*. Hutch[3] **4** 348 and Keynes, AbbotsC 16 fn. 6 who notes that Abbotsbury abbey had some interest here whereas the other Cheselbourne was a Shaftesbury estate, cf. also the DB entry for this link with Abbotsbury]
æt CEOSOL BURNAN (now **1** 316 rather than **3** 202)
Ærest on cyselburnan (*v*. RNs) on ðone ham of ðam hamme (cf. *TA* f.n. Ham **1** 324) andlang streames. to hwitan wyllon of hwitan willon (cf. lost f.n. *Whithewell* **1** 329) ondlang streames to forda of ðam forda (cf.

Chiselburneford **1** 316) to herpoðe of herpoðe to þære hege ræwe (cf. *TA* f.n. Hedge Row **1** 324) þonone to herpoðe anlang herpoðes an furlang.
[*Endorsed*] þis is þara þreora gyrda land boc æt Ceosol burnan.

MAPPERTON (in Almer parish, now in Sturminster M. parish), 943 (15) *ShaftR* (S 490)
AT MAPELDERTUNE … to Mapeldertune … (**2** 56)
Arest of elmere (**2** 55) up anlang streames on Winterburne (*v*. RNs.) on þa dich (**2** 57). and þanne an lang riscemeres (**2** 67) oþ hit cumd to þere alde dich (**2** 57) þanne and lang diche (**2** 57) foren ongen Wiþig þeuel (**2** 69) þanen on Ruþanþorn. of þane þorne (**2** 58) on sex þorn. of sex þorn (**2** 58) on anne þorn þiuel (**2** 58) with Winterburne. and þanne andland Winterburne (*v*. RNs.) on anne water pet. of þane pitte (**2** 69) on stanhecheres ande. of stanhecheres ande (**2** 58) on grenenhille (**2** 57) of þa hege reawe onlang heie reawe (**2** 57) on hþete cumb. anlang cumbes (**2** 58) to wþe. innen wde (**2** 57) on anne crundel. of þane crundele (**2** 57) on þe elþen stret. of þare streate (**2** 57) on anne weie (**2** 58) to horgate. of horgate (**2** 57) on weritun. of wertune (**2** 58) on limbenlee. of limbenlee (**2** 58) on cinninces dic (**2** 71) anlang standene (**2** 58) to molenhame. of molenhame (**2** 58) to cellor. of cellor (**2** 57) on windee bergh. of þane berghe (**2** 58) an lang weies (**2** 58) on winterburne ford anlang streames (*v*. RNs.) eft on elmere (**2** 55).

TARRANT HINTON (parish), 935 (15) *ShaftR* (S 429)
AD TARENTAM [*rubric*] … ad Terentam … (**2** 119)
Arest of pimpernwelle (**2** 110). and þanen to þare rede hane (**2** 122). and þanen to þare hwitendich (**2** 122). and þanen and lang dich to dungete (**2** 121) ut þurch þane wde (**2** 121) and þanen to þan stanegan crundel (**2** 122) and þanen to fildene lane uppende (**2** 122) and þanen to terrente (*v*. RNs) and þanen anlang streames oð þanne ford (**2** 122) and þanen eft anlang herepaþes (**2** 122) od littlen wde (**2** 291) and þanne an lang þere fures (**2** 122) one dat dich (**2** 122) and þanen to tatanbeorge and of þane berge (**2** 122) an lang wic herepaþes (**2** 144) to chirchelford and of þanne forde (**2** 274–5) to bacging berghe (**2** 122) and þanen on þane oþerne beoit. and þannen to horsedich (**2** 122) and þannen on worres berg of þane beorge (**2** 122). est on þone med ham suðe wardne (**2** 122) and þanen on þat beorhlem (**2** 122) be westen cockes þorne (**2** 122). and þanen þiyres ouer chelesbergh' (**2** 122) and þanen wið norþen þanen graetem beorge (**2** 122) and þanne of dune to pimpern (**2** 110).

on þa burnestowe middewardde (**2** 122) and þanen eft to pimpernwelle (**2** 110).

DIDLINGTON (in Chalbury parish), 946 (14) *Harl* (S 519)
DUDELINGTONE ... ÆT DYDELINGTUNE ... ÆT DIDELINGTUNE …
into dydelingtune (**2** 134) Arest on lindune eastesweardre (**2** 151) to þara hina gemære (**2** 146) on cheoles burge eastgeat (**2** 134). þonon ut on þæt westgeat of þam westgeate (ib) on dydelingdune. of dydelingdune (**2** 135) on winburnan. up andlang winburnan (*v*. RNs.) oþ hore tuning gemære. andlang hore tuning gemæres (**2** 159) oþ stigla pæþ of stigela paþe (**2** 136) ofer þone readan weg (**2** 135) þæt eft on lindune (**2** 151).
[*note added*] and þa men þe on þaen londee standeþ and seo mæd liþ on ige þe þær to gebyred (**2** 136).

DIDLINGTON (in Chalbury parish) and **UDDENS** (in Holt parish), 956 (14) *Harl* (S 609)
ÆT DYDYLINGETUNE … ÆT UDDINGC [*rubric*]
ÆT DYDYLINGTUNE … ÆT UDDINGC …
to dydylingtune (**2** 134) … æt udding (**2** 152) …
Ærest on roddan pol nioþeweardne. and of roddan pole (**2** 136) on þane langan hlinc. and of þan langan hlince (**2** 136) on fugelan pæþ. from fugelan paþe (**2** 136) on þa twegen ærcas (**2** 136) þær boddingc weg ut gæð. and from bodding wege (**2** 136) ofer þane redan weg. and of þam readan wege (**2** 135) on lindune estawearde (**2** 151) þær to þam ealdan wege. and swa andlang weges (**2** 136) to mearc broces heafde. from mearc broces heafde (**2** 136). and lang weges in an cheoles byris east gete. and of þam est gete eft ut æt þam west geate. and of þam west geate (**2** 134) on dydeling dune middewearde. and of dydeling dune (**2** 135) on þa haran apuldran. and of þære haran apuldran (**2** 136) þære adune innan winburnan oð midne stream. and swa and lang streames (*v*. RNs.) oð roddan pol (**2** 136).
[*note added*] and at crawan forda (**2** 136) synd. vi. mæd æcras þe þær to gebyriað and on ige twegen (**2** 136).

And þis sint þæs hywisces land gemære æt udding (**2** 152). Ærest on cucan dene. and of cucan dene (**2** 158) on cran mor middeweardne. and of cran more (**2** 158) on þæne heal. and of þam heale (**2** 158) on þa fulan lake. and of þare fulen lake (**2** 158) on secg broc. and up þanan and lang brokes (**2** 158) oþ þa brydenan bricgge. and of þare brycgge (**2** 158) and lang weges (**2** 158) on ham woldes þorn. and of ham woldes þorne (**2** 158) on cucan (**2** 158). and on ige. II. mæd æcres þær to (**2** 136).

HORTON (parish), 1033 (12) *SherC* (S 969)
Hortun (**2** 159)
Ærest cealdan broc scyt (**2** 161) on Winburnan, of Winburnan (*v.* RNs.) on Mapoldorlea middeweardne, of Mapoldorlea (**2** 286) innan eastlea, ut of eastlea (**2** 161) on ðæt slæget [linlæge mor *written above*] of ðæm slægete [in achirste *written above*] (**2** 162) on suðbeara suðeweardne [in ða wiðgas *written above*] of suðbeara (**2** 162) on linleaga mor, of linleage more (**2** 162) on ða gemearcodan lindan, of ðære gemearcodan lindan (**2** 162) on biricgaran (**2** 161) be ðære norðeage (**2** 161) of biricgaran on ðone norðheal, of ðam norðheale (**2** 162) on blacan dune, of ðære dune (**2** 161) on ciddesbeara, of ciddesbeara (**2** 161) on ða haran stanas, of ðam haran stanon (**2** 161) andlang blacan dune ecge norð (**2** 161) on ða haran apoldran, of ðære apoldran (**2** 161) suð on ðone mere (**2** 162) on ðona haran wiðig, of ðam wiðige (**2** 161) west on wænecan wyrð (**2** 162) on ða beorgas (**2** 161) of wænecan wyrðe on suðheal suðeweardne, of ðan heale (**2** 162) on holan broc, of ðan broce (**2** 161) to ruwan leges gete, of ðan gete (**2** 162) on hyrn, up andlang hyrn (**2** 161) on ðone mearchagan, of ðam hagan (**2** 162) on ðone readan weg, of ðam wege (**2** 135) on heara Wulfrices gemære, of ðam landgemære (**2** 161) andlang streames eft on cealdan broc (**2** 161).

COMPTON ABBAS (parish), **SIXPENNY HANDLEY** (parish) and **IWERNE MINSTER** (parish), 956 (14) *ShaftR* (S 630)
to CUMTUNE [*rubric*] … Cumtune (**3** 99)
þis x. hide hereð. in to þanne fiftigen hiden atte Cumtune
þis sent þare x. hide lande gemare. arest on tor scylget (**3** 102). þanen be wirtrume (**3** 102) oð dollen beorge (**3** 102). þanen on þane imeren hole weg (**3** 102) þanen on holenwylle. of hollenwelle (**3** 102) on holencumb (**3** 99). þanen on þa lake (**3** 102) to smale broke (**3** 109) þanen adun one þat lake to smale broke þanen adune one þat lake (**3** 102) at her weg (**3** 102) on stirchel (*v.* RNs.) anlang streames on þat litlen lake (**3** 102) at Meleburge imare (**3** 129). þanen on fern helle (**3** 102) forð þanen on anne hus (**3** 102) þanen on meleberig dune duneward (**3** 131) on þat ealde ad (**3** 102, 131) þonne þanen on lippen stagen (**3** 102) swo be wirtrume (**3** 102) eft on torchil gat (**3** 102).

at heanlegen (**3** 113) … at Iwern (**3** 123) …
Hanlee [*rubric*]

Arest on litlen ac lee estward of aclee (**2** 266) on pegan beorh (**3** 122). þanen on berendes beorh (**3** 122). þanen on þere herepað (**3** 122) at mes delle (**3** 117). þanne on þat get at seuen diche suð ende (**3** 123) anlang diche o þes sledes Northecge (**3** 123) þan' on þane an lipien þorn (**3** 122). þanne be þane on heueden (**3** 122) adun to wege. andland weges (**3** 123) on þannen mylen stede (**3** 117). þan' on þane hege reawe (**3** 122) oð ða furuh. on land furuh (**3** 122). on toten berg' (**3** 123) þan' on ac hylle (**3** 119) on til luches lege (**3** 123). þanen on biken settle (*v.* Addenda **3** 113) þanen on mealeburg norþewarde (**3** 116). þan' eft to anclee [*sic*] estward.

þanne sent þis landimare at Iwern (**3** 123)
Iwern [*rubric*]
Arest on terente dene (**3** 129). þanen to fideriches dene uueward (**3** 128) þanne on þane crundel (**3** 128). at þare weie itwislen (3 129) on þane crundel. þanen on budencumbe hracan (**3** 128) þan' on wung wylle (**3** 40, 129) þane on þane ealden forð (**3** 128) þanen on cranmere (**3** 128). þanen on cing hille (**3** 128) þanen on ne smal þornes (**3** 41, 129) þanen on swylles (**3** 129) þanen on lacmere (**3** 41, 128). þanen an lang streames oð sandford (**3** 129) þanen on ears mores heaued (**3** 128) þanen on styb (**3** 129). þanen on foxlee nortwarde (**3** 128). þanen on agen þorn (**3** 128). þanen on þere crundel northwarde (**3** 128) þannen on waddene uuewarde (**3** 129). on hlinc reawe (**3** 128) þanen on þone grenwai (**3** 128) at merewege uue (**3** 129) þanen on þa here apeldren (**3** 128) þ' eft on terente dene (**3** 129).

FONTMELL MAGNA (parish), 932 (15) *ShaftR* (S 419)
JUXTA FFUNTEMEL [*rubric*] … FUNTEMEL (**3** 103)
Arest on wde brigthe (**3** 105). and þanen up on beteswirþe sled (**3** 104). þanen to snelles hamme weghe (**3** 112). þanne on þurch þo aelres (**3** 106) to holencumbe. of þanne cumbe (**3** 102, 111) to holenwelle. of þan welle (**3** 102, 111) on hamelendune north ecge (**3** 106). þanen to langen cumbes hauede (**3** 104). þanen on dollen berch (**3** 102). an þanne to wde (**3** 106) on done hagen (**3** 110) oð þas soces seað (**3** 112) ðanen on holewei. of holleweie (**3** 111) on sledwich. of sledweie (**3** 112) on hriycsþeg. þannen anlang hricg þeges. of rig wei (**3** 111) ut to þe wines weie (**3** 112) þanen to burch linken (**3** 110) dannen on þa gereþrenc (**3** 110) withinne da chealc seðas (**3** 110) þanen to holebroke (**3** 109). þanen on ðæs lutlen searpennes suðeke (**3** 104) þane on sandhellesled (**3** 111). þanne on ða riþe (**3** 111) to styrd (*v.* RNs.) þanen anlang streames eft to wde bricge (**3** 105). þanne on halgan weies lake (**3** 110).

þanen to þanen welle siluen (**3** 110). þanen on þa hege reawe (**3** 111) to þane shamelen (**3** 112). an to þan herepaþe (**3** 111). þonne west be Winteintune (**3** 113) ut þurch sulangraf (**3** 112). þanne one þe hegen (**3** 110). be weste hegen paðe (**3** 110) and þar of dune to þare stigele (**3** 112) þanne west anlang hegen (**3** 110) to heldmannes wrthe suthward (**3** 111, 136) þanne north. to ludmannes putte (**3** 111). þanne be wirtrune (**3** 113) on blinches broc. of þanne broke (**3** 91–2) to Wigheardes stapele. fram Wigheardes stapele (**3** 112) to cludesleghe. of þare lege (**3** 110) to þanen ealden herepaþe þ' schet to blinchesfelde (**3** 90–1, 111). þanen forð be wurttrune (**3** 113) to scearpenhame (**3** 112). þanen one þat lake (**3** 111) to stirchel anlang streames (*v*. RNs) eft to wdebrige (**3** 105).
[*note added*] and at suttune (**3** 80) ligð xxiiii akeres meade þat hirð in to funtmel (**3** 81, 103). on ðes horderes land oþerhalf hewisse of suttune (**3** 80) erþe in to þis land.

EAST ORCHARD (parish), 963 (15) *ShaftR* (S 710)
at ARCHET (**3** 133)
Ðis sant þe landimare to Archet
of eldmannes wrthe (**3** 136) on irichte anlang hina imares (**3** 136) to þan scamelen fram þanne scamelen (**3** 136) anlang hine imares oð wdebrige. of wudebricge (**3** 105) a dune anlang streames (**3** 136) oð land scorforð. of þane forde (**3** 136) anlang leouen imare (**3** 136) oð kinghes imare and gemare (**3** 136) oð funtemel forð of þanne forð (**3** 136, *v*. RNs) adune mid streame oþes bissopes imare ut sceoþaþ (**3** 136). þanen on irichte to þan lipgete. from þane gete (**3** 136) on irichte to anne wonalre. of þane alre (**3** 137) on irichte to sucgimade hauede. of þane hauede (**3** 136) on irichte þurch ðene holt (**3** 136) to Wlgares imaren. of þan imaren (**3** 136) eft on eldemanneswyrde (**3** 136).

WEST ORCHARD (parish), 939 (15) *ShaftR* (S445)
AD ARCHET [*rubric*] … AT ARCHET … (**3** 133)
Þis sent þa land imare to archet þare westere fif hide (**3** 133–4).
Arest suthward of ealdmannes wyrde (**3** 136) at langhecgan (**3** 138) to sugging made (**3** 138) þane on i richte to stirthel (*v*. RNs) on þane alr (**3** 138) beneþan scealden forð (**3** 139) and þanen anlang higweges þ' of higeweges ande (**3** 137) to funtemel (*v*. RNs) þry akeres wid eastem þane broc (**3** 139) þonne anlang funtmeales oþ eadelmes melne (**3** 138) of funtemel to þane twam þornen (**3** 139) þat forð on irichte to stigel hege (**3** 139) ouer stirtel (*v*. RNs) and þonun on irichte to ginum hocum (**3** 138) to þrem land sharen (**3** 139) þat on archet hamm (**3** 138) and

þanen cagbroc up anlang cagbroges (*v*. RNs) to land share hegen þonne anlang heges (**3** 138) to ealdmannes wyrþe (**3** 136).

SHAFTESBURY (parish), small area in, 958 (15) *ShaftR* (S 655)
SHAFTESBURY [*title*] … AT SCEAFTESBERI … (**3** 139)
Đis sant þe landimare at Scaftesberi. Arest of bogen wylle (**3** 148) on brandes hricg andlang hrichtes (**3** 156) on bytelesmor. of bitelesmore (**3** 153) on wyndrede dic. of þare diche (**3** 159) on bokenwelle (**3** 148).

HINTON ST MARY (parish), 944 (15) *ShaftR* (S 502)
HAMTUN' [*rubric*] … AT HAMTUNE … (**3** 159)
Þis sand þe landes imare at hamtune. Arest of hefdeswelle (**3** 162) on nunnen linc. of nunnene linche (**3** 163) on litiges heuede. of litiges heuede (**3** 163) anlang diche (**3** 163) on sture (*v*. RNs.) at oxene bricge (**3** 163) unen up anlang sture (*v*. RNs.) at sudwde (**3** 163) þannen of þane hegen. of þane hegen (**3** 163) to dich. þanen anlang dich (**3** 163) ouer cinerget mor (*v*. RNs.) on þane mapelder. of þane mapeldere (**3** 163) be þane akere heueden (**3** 163) on chelbrichtes dich (**3** 163) anlang cuterget mores (*v*. RNs.) on þane stream anlang streames (**3** 163) on adwines imare up anlang imares (**3** 163) þat eft on efdeswelle suthward (**3** 162).

THORTON (in Marnhull parish) and part of **IWERNE COURTNEY** (parish), 958 (15) *ShaftR* (S 656)
ÆT ÞORNTUNE (**3** 171)
Þis sant ða land i mare at þortune. þe ierð in to Hamtune (**3** 159)
arest on arcetham. of archethamme (**3** 138) on Rumanhelle (**3** 171) þanen of breþling made (**3** 178) þ' on þigerðes stapel (**3** 178). þanen to tudesleghe of þere lege (**3** 178) be kings imare on blinnesfeld (**3** 91). þanen be wde on þere hinc imare anlang mares þ' on ealdmannes þyerðe (**3** 136, 178) þanan on þa seales. of þan sealen (**3** 138) þat eft on arcetham (**3** 138).
And þis sent þe tweire hide land imare þe cyna achte þe ealde (**3** 37)
arest on iþern broc of þane broke (*v*. RNs.) on þe lang dich (**3** 41). þan on þene weie. anlang weies (**3** 41) to late mere (**3** 41). þane onlang sledes (**3** 41) on ða smale þornes (**3** 41). þanne on ða dich. þanne swo be diche (**3** 41) on cranemere. of cranemere (**3** 41) eft. on ipern broc (*v*. RNs.).

STURMINSTER NEWTON (parish), 968 (14) BCS 1214
NYWETONE … STOURE [title] … AT STOURE (**3** 188)

Þis beth þe landmere to stoure þare thritti hide. Erest op of stoure (*v.* RNs) on trildoune (**3** 191–2) vppe tril (*v.* RNs) on landscher' ford ouer þane ford (**3** 199) bi merehale (**3** 200) on Conibrok' of þan brok' (**3** 191) bi merehawe (**3** 200) on wadeleighe northward (**3** 200) on stanwey. of stanweie (**3** 200) bi merehawe (**3** 200) on deuelisch. of defelich (*v.* RNs) on stut þare on vppe stut (**3** 200) on þar dich southward be dich (**3** 199) ouer sanddoune (**3** 200) on filithleighe (**3** 199) þanen on Holambrok' of þan brok' (**3** 348) on hedem'chil (*v.* Addenda **3** 189) þane' on Lideman bi lidenan (*v.* RNs) on stoure (*v.* RNs) on þan olde wested' (**3** 201) vp on stoure (*v.* RNs) on þer dich be dich (**3** 199) on litegesheued. of lutegesheued (**3** 199) bi groue (**3** 199) on minnanlinche (**3** 200) þanen north bi heuede (**3** 199) þanen est to hedeswelle (**3** 162) þanen bi sixe acres north an heued (**3** 200) on þare þorn rewe (**3** 200) þanen on sand mor of þan more (**3** 198) on smeryate by seuerȝate (*v.* RNs) eft on stoure (*v.* RNs).

CHESELBOURNE (parish), 859 ?for 870 (15) *ShaftR* (S 334)
… II. CASSATOS JUXTA CHESELBURNEAM … [*title*] … duos cassatos in loco qui dicitur be CHISELBURNE (**3** 202)
Ðis is þes landes imare þus uten ynisald [?for winfald = Wenfled, *cf.* S 485(i) *infra*, a grant of lands to Wenfled, i.e. *Wynflǣd*].
Arest on land scorhlinc (**3** 207) to chiselburne (**3** 202). þannen anlang streames þannen up of streame on anne hlinchesheaved (**3** 207) þanen be suþe Scaftesbury (**1** 303) on þane hlinc. of þat ihlinche (**3** 207) on anne castel (**3** 206) at swindune (**3** 208). upward of dy castele (**3** 206) to burnstowe (**3** 205) þanen on anne linkesheaved (**3** 207). þanen to anne castel. of ði castele (**3** 206) on anne herepaþ westward over herepaþ (**3** 206) on hlinc reawe (**3** 207) þanen on anne crundel (**3** 206) to burstowe (**3** 205) on anne cnap (**3** 206) on anne Ierðlonde Northward (**3** 207) þanen wið anne crundeles. of þane crundele (**3** 206) on þes heges hirne (**3** 206). þan' est to herepaðe. of herepaðe (**3** 206) on heandene beorg (**3** 218) on deuelisc (*v.* RNs) of dune anlang streames. up of streame on anne furch (**3** 206) on anne stan castel. of ða icastele (**3** 208) on bleomannes berge. of þa iberge (**3** 205) on land scarlinc (**3** 207).
[*note added*] an sex made eres be frome (*v.* RNs. and Frome Mead Addenda **1** 317) be deuelisc made (Addenda **1** 304–5). þe ierþ on þise lande.

CHESELBOURNE (parish), ?870 (15) *ShaftR* (S 342)
… ad Cheselburniam [*rubric*] … be CHISELBURNE fif hide … (**3** 202)
Ðis land is þisen imare. þus ute þinsald (*v*. S 334) arest on landscar hlinc (**3** 207) to cheselburne (*v*. RNs) þannen an lang streames. þan' up of streame on anne linkes haued (**3** 207) þanen be suðe ceatwanberge (**1** 303) on þanne hlinc. of þanne i hlinche (**3** 207) on anne castel (**3** 206) at swindone uuewearde (**3** 208). of þe castele (**3** 206) to burnstowe. þanen up anlang burnstowe (**3** 205) þan' on anne linkes heued (**3** 207) þanen to anne castel of þo icastel (**3** 206) on þere herepaþ westward op ðe herepað (**3** 206) on þat hlinc reawe (**3** 207) þanen on anne crundel (**3** 206) to burnstowe (**3** 205) on anne cinep (**3** 206) on ðan zerðe nordewarde (**3** 207) þanen wið anne crundeles of þi crundele (**3** 206) on des heges hirnen (**3** 206) þanen eft to herepað of þe ereðe (**3** 206) on hendune beorch (**3** 218) on deflisch (*v*. RNs) of dune anlang streames. up on streame on anne furch (**3** 206) on anne stancastel. of þi castele (**3** 208) on blieqz mannes beorg of þa iberge (**3** 205) on land scare hlinc (**3** 207).
[*note added*] and sex made eres be frume (*v*. RNs. and Frome Mead Addenda **1** 317) wið deuliscmad (Addenda **1** 304–5) ierð to þise land.

CHESELBOURNE (parish, east part) and **WINTERBORNE TOMSON** (parish), 942 (15) *ShaftR* (S 485(1))
AT CHESELBURNE (**3** 202) … to cheselburne
of lisebroke (**3** 203) on þane fearngaren. of þane fearngaren (**3** 206) on þone ston istel. of þan istelle (**3** 208) anlang ricges (**3** 207) on þa iheafde. of þane theafde (**3** 206) on þane forð erthe acre. of þane forerð akere (**3** 206) on cheselburne stream (*v*. RNs) of þane streame on þe twifelde dich of þare dic (**3** 208) onne flescumbe (**3** 206) on þane crundel. of þane crundel (**3** 206) on deulisc stream. of þane streame (*v*. RNs) on langhelee northward on langhelee (**3** 207) on þe blaken þornes northward. of þane blake þornen (**3** 205) anlang hricge weges (**3** 207) to þe beorge. of þe berge (**3** 205) on ðone crundel. of þane crundele (**3** 206) on ða holu suthward (**3** 207) to deulisc streame (*v*. RNs).

at WINTERBURNE … at Winterburne (**2** 81)
Arest of Winterburne (*v*. RNs) to chelfgraue. of chelfgraue (**2** 82) one wicweie neþerward (**2** 82) anlang broke (**2** 82). one meairweie anlang weies (**2** 82) to widesgete (**2** 82) anlang cinincces [*for* cunucces] dich (**2** 71) on þe sherd. of þene sherde. of þane shearde [*sic*] (**2** 82) anlang þides (**2** 82) to nearn þan anstigan. of þan anstigan (**2** 82) to midelgete

of midelgate (**2** 82) þurh þane garen (**2** 82) anlang herepaþes (**2** 82) on hornget. of horngetes hirne (**2** 82). lang land share (**2** 82) to leaxan oc (**2** 82) an lang wdes (**2** 82) to þane ealde seale (**2** 82). þanen one þo berges astward (**2** 82) to Winterburne. of Winterburne (*v.* RNs) tou þares túnes hirne (**2** 82) on lang þane herepaþe (**2** 82) on þene depe crundel (**2** 82) of dune an lang land share (**2** 82) eft on Winterburne (*v.* RNs).

CHESELBOURNE (parish, except Lyscombe), 1019 (15) *ShaftR* (S 955(1))
de Cheleburna [*rubric*] … Cheselburne (**3** 202)
Arest of Bertes welle (**3** 205) on þare ealde diches heued; of þare dich (**3** 206) on flexcumbes heuede (**3** 206); on þane ealde paþe; of þane paþe (**3** 206) on þa ellen þirnen; of þare þirnen (**3** 206) on deuelisc stream; andlang streames (*v.* RNs) on þone hwiten welle; of þane welle (**3** 207) on þane bergh (**3** 205) uppen morhelle (**3** 207); of þane berghe (**3** 205) on þa ealde berig; of þane Bery (**3** 206) on þane rugen bergh; of þane bergh (**3** 207) on sebergh; of sebergh (**3** 207) on þorþivel; of horþiuele (**3** 208) on neþan pol; of þane pol (**3** 207) inne þa þornen; of þare þyrnen (**3** 208) on þane þorn up an Gretindune; of þane þorne (**3** 206) on þo stancysten on holencumbe; of þane stancyste (**3** 208) on blacmanne bergh; of þanne berghe (**3** 205) be anne hefden (**3** 206) on hippe pad; of þane pade (**3** 206) on Cheselburne (**3** 202) on shete bergh; of þane berghe (**3** 207) on braden bergh; of brade berghe (**3** 205) on þare diches hirne; of þere hirne (**3** 206) on þorndunes cnep; of þane sneppe (**3** 208) on þane greate hlinc; of þare linke (**3** 206) on burestowe (**3** 205) on þat ealden reshbed; of þane bedde (**3** 206) bi streame in þane miliere; of þane miliere (**3** 207) on þo apeltren; of þare apeltren (**3** 205) on þar haren torre (**3** 206) on hiþisc bergh; of þane berghe (**3** 206–7) on liscbroc; uppe be broke (**3** 203) on þa sticelen lane; of þare lane (**3** 208) on þo tweie pettes of þone petten (**3** 208) on þane hordþiuel; of þane þiuel (**3** 207) on þane bergh; of þane bergh (**3** 205) on anne þornstub; of þane stubbe (**3** 208) on þane snieren fyrsgaran; of þan garan (**3** 207) eft on Bertes welle (**3** 205).

WOOLLAND (parish), 833 (14) *ShaftR* (S 277)
Wennland [*rubric*] … wennland (**3** 232)
Primitus of alor riðe (**3** 233) uppe on anne þoure (**3** 238) þanen on anne Walle dich (**3** 238). þann’ þurh reowe (**3** 238) on delesburg middenwearde (**3** 237). þanen to cylberge (**3** 237). þan’ on wyrtruman (**3** 238) þane on burg (**3** 233) þanne on anne dich (**3** 237) one þane

ealdene hage. þanen anlang ðies hagen (**3** 237) oð anne þorn (**3** 238) on hacggen hamme (**3** 237). þane' on anne hege ðat on irichte (**3** 237) on anne oc (**3** 237) ðonne est ouer melenbroc (**3** 236) on anne þorn on anne stapel (**3** 238) eft to anne aler riðe (**3** 233).

PLUSH (detached part of Buckland Newton parish), 891 (14) *Glast* (S 347(2))
PLYSSCH … plussh … plyssch' (**3** 245)
Erest on Ruanberghe (**3** 257) þanne on stanwey (**3** 258) þanne on stanbrok (**3** 258) þanne on radenwey (**3** 257) þanne on lillanwelle (**3** 257) þanne op on hagen (**3** 256) þanne on an hul (**3** 257) to hamme (**3** 256) þanne on an welle (**3** 259) up on brugh doune (**3** 255) on warrode (**3** 258) þanen on windberghes (**3** 259). þanen endlang herepathes (**3** 256) on þar weilete (**3** 258) þanen on widdisȝate (**3** 259) þanen over thorncombe (**3** 246) on anne crundil (**3** 255) þanen in on þe weilete (**3** 258) þanen west over plissh' (*v*. RNs) on an Sticholnelinche (**3** 246) þanen on an White Wey (**3** 259) þanen on schilleburghe (**3** 258) þanen on ellene crundil (**3** 256) þanen on cliveleghe (**3** 255) þanen over whetecombe (**4** 123) on alden doune (**3** 254) op bi wirtrone (**3** 259) on an ruwe dich (**3** 257). þannen adoun over yfre (**3** 259) þanen out on bromhulle of bromhulle (**3** 255) eft on Ruanberghe (**3** 257).

BUCKLAND NEWTON and **PLUSH**, 941 (14) *Glast* (S 474)
BOCLONDE … BOCLANDE … PLYSSH' [*heading*] (**3** 239, 245)
AT BOKLONDE … ad PLUSSH'
Erest on Scherdenberwe (**3** 258) endlangweies (**3** 258) on an burghen (**3** 255) þanen on ane firesrewe (**3** 256) þanen on langengroue suthward (**3** 257) adun on anne rig' (**3** 253) to streme endelangstremes (**3** 253) on doggeneford (**3** 255) þanen up on doggeneberwe (**4** 161) þanen on amesheal (**3** 254) þanen on þar elde lidyate (**3** 256) a doun on herebrittescombe (**3** 256) on þane streme endlang stremes (**3** 258) on babbanmede (**3** 254). þanen on hawen endlang hawen (**3** 256) on þan here path (**3** 256) þanen on ane dich endlang dich (**3** 255) on birninge (**3** 254) þanen on bradeleighe (**3** 255) þanen on stiþeleighe southward (**3** 258) þanen on sic of þan sice (**3** 258) on bournandign' (**3** 255) þanen up on dounen tit (**3** 242) on anne heued Wriȝt (**3** 256) þanen on þar Elde Waldich (**3** 256). þanen a doune on here path (**3** 256) þanen on þe merhawin. endlang hawin (**3** 257) on risshdene (**3** 257) þanen on neodbournen endlang stremes (*v*. RNs) þanen on þan rissbed (**3** 257) þanen on apildore (**3** 254) þanen riȝt on cristemeleighe (**3** 255) þanen on þar thre landschere of þan (**3** 258) on þe merhawe enlang hawin (**3**

257) on hengersten (**3** 252) þanen on an withi (**3** 259) þanen riȝtwey (**3** 257) þanen on igland (**3** 257) on an Wirnen (**3** 258) þanen on streme endlang stremes (**3** 258) on þan merehawe endlang hawin (**3** 257) on blehildewelle (**3** 255) þanen houndesete (**3** 257). þanen on þan southerne Wistonwey (**3** 259) þanen on anne þorthn (**3** 258). þanen on Schepleghʼe (**3** 258). þanen endelang here pathes (**3** 256) eft on Scherdanbourgh (**3** 258).

WOOTTON GLANVILLE (GLANVILLES WOOTTON) (parish), 956 *Add* 15350 (BCS 969, S 608) [NOTE Identification of these bounds once considered very uncertain, hence not included under this parish **3** 267–274, but now thought quite likely to belong here]
æt Wudatune…to Wudetune (**3** 267)
Ærest on þam ge middu' (*v.* **(ge)mȳðe**) and lang þære weal dic (*v.* **weall**, **dīc**, cf. *elde waldich* **3** 256 and the *TA* f.n. Wall Orchard **3** 273) on canuc (*v.* **canc**, cf. Cank **3** 322) ufeweardæ swa and lang þære weal dic on canu rih (*v.* **hrycg**) ufæ weardne swa and lang hrigges on ðone weg þæ scyt on stænæ wæg swa adun on stan wæg of stan wæge (*v.* **stān**, **weg**, cf. High Stoy **4** 163) east and lang hærpaðæs (*v.* **here-pæð**). swa forð bæ clyfæ (*v.* **clif**, cf. the f.n. *lez Clyfes* **3** 273) oð ða coc rodæ. swa of þære coc rode (*v.* Cockroad **3** 270) adun east and lang weges and [*sic*] deoh holes hyllæ of deoh holes hyllæ (*v.* **hol**[1], **hyll**, first el. uncertain, cf. perhaps the *TA* f.n. Darsells **3** 272) on þonæ cealc pyt (*v.* **cealc**, **pytt**) swa ford [*sic*] wyð þonne weg westan oðða ða dunæ (*v.* **dūn**) ufe wearde on þa æþenam byrigelsas (*v.* **hǣðen**, **byrgels**). swa adun be wyrtruman (*v.* **wyrt-truma**) æft to ge miðum.

STALBRIDGE WESTON (in Stalbridge parish), 933 (12) *SherC* (S 423)
APUD WESTONAM (title) … æt westtune (**3** 283)
Ærest of ciric hylle (**3** 281) on horgan sloh (**3** 289) thanon on beorhleage (**3** 281), thanon on west mæd ufewearde (**3** 289), thanon on bydeburnan (*v.* RNs), thanon on anna mapulder (**3** 289) thanon on bilian wyrþe (**3** 289), swa forþ on hagan (**3** 289), thæt on beorreding mæd norþewearde (**3** 286), thanon on þa greatan ac (**3** 289), thæt forþ on hagan (**3** 289) oth stanbroc (**3** 289), thonon on hean wifeles hylle (**3** 288) oth þone ealdan weg andlang weges (**3** 289) oth fyldena wylle (**3** 289) thanon on þone ealdan hagan (**3** 289), þæt swa forth ut on bealtunes ersc nyþeweardne (**3** 288–9) thonon on æsc leage of þam æsce (**3** 288) forth on hagan (**3** 289) on gerihte ut on feld (**3** 289) on anne stan (**3** 289) thæt swa forth east andlang weges (**3** 289) eft on ciric hylle (**3** 281).

BRADFORD ABBAS (parish), including Wyke in Castleton parish, 933 (12) *SherC* (S 422)
BRADEFORD … Æt BRADAN FORDA (**3** 294)
Ærest on east healfe, of aettan wylles heafde (**3** 310), and lang lace (**3** 298) to mylenburnan (**3** 298) thonne and lang streames oth gifle (*v.* RNs) thonne eft and lang streames oth ennan pol (**3** 298) thanon and lang lace (**3** 298) oththa ealdan dic, thonne and lang dic (**3** 298) oth hloscumbes heafud (**3** 328), thanon on ecge, and lang ecge (**3** 310) othha dic thanon and lang dic (**3** 310), oth ecgulfes treow (**3** 310), thanon east to aettandene middeweardre and lang dene (**3** 310) eft to aettan wylle (**3** 310).

THORNFORD (parish), 903 for 946–951 (12) *SherC* (S 516)
APUD TORNEFORD (title) … ðornford (**3** 376)
This synd tha land ge mæru. On east healfe on Cumtun bricgge (**3** 325) upon tha lace (**3** 380) oth ran wylle, of ran wylle (**3** 380) upon thone eastemestan holan weg (**3** 380) up to hricgge, and lang hrigcges (**3** 379) west be ecge (**3** 380) on mere (3 377) thonon on lyngærstun eastewerdne (**3** 380), thonon on slaep, thone and lang streames (**3** 380) on gifle (*v.* RNs), oth mylenburnna ut scyt (**3** 380) up on stream eft to Cumtun bricgge (*3* 325).

PORTISHAM (parish), 1024 *Ilch* (KCD 741, S 961)
æt Porteshamme (**4** 26)
Ærest of mylebroce (**4** 32) on cuceles hylle. of cuceles [hylle] (**4** 28) [on holun. of] holun (**4** 35) on motbeorh. of motbeorge (**4** 35) on þa stræt of þære stræt (**4** 35) on þone ealdan weg (**4** 35) þurh þone graf (**4** 35) and swa on þone Smalan pæþ (**4** 35) ofer þa stræt (**4** 35) on þone ealdan weg and swa andlang weges (**4** 35) on atange (**4** 34) […….]eorh and swa on æt steallesbeorh of þam beorge (**4** 29) Suð on þa ealdan wale (**4** 34) Swa on corfgetes westran cotan [MS totan]. of þam cotan (**4** 27) Suð be wale (**4** 34) on þære dice hyrnan (**4** 34). and swa west be dic on oðre hyrnan (**4** 34) swa suð be dic on þære mæ[de] hyrnan (**4** 35) and þær west be dic on oðre mæde hyrnan (**4** 35) and Swa suð andlang dic (**4** 34) on halgan wyl. of halgan wylle (**4** 35) west be dic on þa clæian lane. of þære clæian lane (**4** 34) on myle broc adun be streame. eft of mylebroce (**4** 32) on cuceles hylle (**4** 28).

PIDDLETRENTHIDE (parish, northernmost of the three tithings), 966 (15) *ShaftR* (S 744)
AT VPPIDELEN (**4** 167)

Đis sanden þe land imaren at uppidele (**4** 167). of pidelen streame (*v.* RNs) on hlosstedes crundles suð ecge of þane crundle (**4** 173) on þat mere sled. of þat mere slede (**4** 173) on ðeo herepaþe. anlang herepaþes (**4** 173) on mearhhilde mere. of mearhhilde mere (**4** 173) on þane haþene berielese (**4** 173) on midde þane punfald of þanne punfalde (**4** 171) on pidelenstream. of pidelenstreme (*v.* RNs) anlang burnstowe (**4** 173) on gretenlinkes. suth ecge of þane gretenlinke (**4** 173) on chellenberghe (**4** 173) þ' eft on pidelen streame (*v.* RNs).
[*note added*] And se mad be frome (*v.* RNs) þat to þanne tune ibereth.

HALSTOCK (parish), 841 (12) *SherC* (S 290)
aet halʒastoke (this predates the spellings given **4** 193, v. NOTE)
Aerest up on þone esc (*v.* **æsc**) ðonne ut on preosta leʒe (*v.* **prēost**, **lēah**) on ane dic (*v.* **dīc**) ðonene on þone ealdan herpað (*v.* **eald**, **herepæð**) aet niþerstoce (**4** 194, predating the spellings given, *v.* NOTE) ðonne on þonne ford (*v.* **ford**) on live dic ðonne up of [*sic*] live dic (**4** 304, *v.* **dīc**, first el. uncertain, *v.* Forsberg 217) oð holan broc ðonne up an holan broc (*v.* **hol**², **brōc**) ðonne on hlorbroc of hlorbroce [Finberg (1964) hlosbroc] (v. ?**hlōse**, **brōc**, cf. the f.ns. *Loose Land* and *Lostick* **4** 205) on wiþiʒheʒe (*v.* **wīðig**, **hege**, cf. *TA* f.n. Withy Hay **4** 204) ðonne on horsford ðonne up on hors ford (*v.* **hors**, **ford**) on fyres hylleford (**4** 303) ðonne up on hlidan (**4** 299) ðonne ut on hunaleʒe (*v.* ?**hund**, **lēah**, cf. f.n. Humley **4** 304) ðonne up on cruc of cruc (**4** 290) on hunda troh (*v.* **hund**, **trog**) ðonne on tymbercumb (*v.* **timber**, **cumb**) þonne on aescleʒe of aescleʒe (*v.* **æsc**, **lēah**) on fald heʒe (*v.* **fald**, **hege**) ðonne on readan weʒ (*v.* **rēad**, **weg**) þonne up on stiʒele (*v.* **stigel**) ðonne an wibbelescumb (*v.* **cumb**, first el. an OE pers.n. **Wibbel*, cf. f.n. *Weblers* **4** 305) up on aesc eft…
[*added*] in oriente fossam antiqua de amne lyfdic usq' ad aliam lyfdic (**4** 304) et in meridie browoldes ham (**4** 302) et in occidente cruc (**4** 290) et in aquilonali parte aescleaʒe deinde in radicem on ðone aesc et sic iterum in antiquam fossam.
[NOTE: most of the forms from this set of the Anglo-Saxon bounds of Halstock were unintentionally omitted from the entry for Halstock parish **4** 193–205, hence the extra annotations given here. It will be seen that the spellings for Halstock itself and for Netherstoke predate those given **4** 193–4].

CORSCOMBE (parish), 1014 (12) *SherC* S 933)
Corigescumb (**4** 287)

Ærest on aerne dene (**4** 302) ðanone on hornes beorh (**4** 303) ðanone on leas ende (**4** 304) ðanone on miclan corf (**4** 302) on miclan cruc middewearne [*sic*] (**4** 290) ðanone andlanȝes hrucȝes (**4** 303) on hlydan pol (**4** 299) ðanone on fyres forid [*sic*] (**4** 303) ðanone on þone hagan (**4** 303) ðanone on breowoldes ham (**4** 302) ðanone on byrran broc (**4** 302) ðanone on lyf dic (**4** 304) ðanone on lyc haȝan (**4** 304) on ȝrenan weȝ (**4** 303) ðanone on focȝan crundel (**4** 303) ðanone on tryndlea weȝe (**4** 305, also *v.* Addenda) ðanone on earna lea weȝe (**4** 290) ðanone forð andlanȝes herpaðes (**4** 303) eft on aerne dene (**4** 302).

CORSCOMBE (parish), 1035 (12) *SherC* (S 975)
Corigescumb (**4** 287)
Ærest on hornes beorȝ of þam beorȝe (**4** 303) on leas ende (**4** 304) on corfȝet of þam ȝete (**4** 302) on cruc middeweardne ðonne of cruc (**4** 290) on ðone smalan hrucȝ andlanȝ hrucȝes (**4** 303) on hlyda pol of þam pole (**4** 299) anlanȝ streames on fyrs ford of þam forda (**4** 303) andlanȝ readan hyrstes (**4** 304) to hlype ȝete fram hlupe ȝete (**4** 303) to þam col pytte fram col pette (**4** 302) to þam healfan treowe (**4** 303) ðanone forðrihtes on browoldes heal (**4** 302) et ðanon on staford (**4** 304) uppon ȝean stream of þam streame (**4** 305) on stocchylle of stocchylle (**4** 301) on foȝȝa crundel eastweardne of þam crundele (**4** 303) forð be wyrtruman (**4** 305) on þone fulan slo of þam slo (**4** 303) forð be wyrtruman (**4** 305) on brumdun middewearne of brumdune (**4** 302) on murter wyll of murter wylle (**4** 304) on ȝean stream (**4** 305) on tollor æwylman of þam æwylman (**4** 291) on cotte dene of cotta dene (**4** 303) on ruwan beorȝ of þam beorȝe (**4** 304) on riȝ cumb middeweardne of riȝcumbe (**4** 304) on ȝyrdlea middeweardne of ȝyrdlea (**4** 292) on þone staniȝan weȝ of þam weȝe (**4** 304) on þone wæter pytt of þam pytte (**4** 305) on coccelmæres dene of þære dene (**4** 302) on earne dene on [sic] earna dene (**4** 302) eft on hornes beorȝ (**4** 303).

CHARMOUTH (parish), 737 (12) BCS 158 (S 254 MS 1), 737 (12) *Add* 15350 (S 254 MS 2)
CEARN [Cearnæl MS 2] (**4** 370)
Ærest of sæ upp on Hængestes ricg [hengster hricg MS 2] (**4** 374) þonan norð on gyrd weg [gyrð MS 2] weg (**4** 374) swa norð andlang rihges [hrigces MS 2] (**4** 374)] on middeweardna sealtera cumb þ' bæ cumbæ (**4** 374) ingon holan broc þonne andlang streames (**4** 374) in on hlosmoc (**4** 374) adune on stream on seofan æceras [syfon eceras MS 2] (**4** 374) þær of streame æft on butan seofan æceras [sceofan eceras MS 2] æft suð on hlosmoc ðæt a dune on stream of ða suran apældran [apeldran

MS 2] (**4** 375) ðæt suð on þa norð rewe (**4** 374) ðanan eft ofer Cærn [on cern MS 2] (*v*. RNs) on gata forð (**4** 374) ðet suð bufan litlan [lytlan MS 2] graf (**4** 374) swa suð on earnes hlinc of earnes hlince (**4** 374) æft utt on sæ.

CHARMOUTH (parish), 938 (12) BCS 727 (S 443)
CEARN (**4** 370)
Ærest of sæ up on hengestes rigc (**4** 374) ðonan norð on gyrdweg (**4** 374) swa norð andlang hricges (**4** 374) on middeweardna sealtera cumb þ' be cumbe (**4** 374) ing [*sic*] on holan broc þ' andlang streames (**4** 374) ing [sic] on hlos moc (**4** 374) adune on stream on seofan æceras (**4** 374) þær of streame eft onbutan seofan æceras (**4** 374) eft suð on hlos moc (**4** 374) ðæt adune on stream oð ða suran apeldran (**4** 375) þ' suð on ða norð ræwe (**4** 374) ðanan eft ofer cern (*v*. RNs) on gata ford (**4** 374) ðæt suð bufan litlan graf (**4** 374) þwa [*sic*] suð on earnes hlingc of earnes hlince (**4** 374) eft ut on sæ.

ABBOTT'S WOOTTON (in Whitchurch Canonicorum parish), 1044 (m11) *Ilch* (S 1004)
æt WUDETUNE … to wydetune … wydtune [*dorse*]
Ærest of stræt forda (**4** 454) up andlang þær grenan weges (**4** 454) to smiceran æc (**4** 454) andlang wegges (**4** 454) to ketan lege ufewerdre (**4** 454) andlang herpaþes (**4** 454) to þam herpaþes on norð healfe lobirihtes gete (**4** 454) of þam herpaðe benorþan labirihtes gete (**4** 454) on ðære dice hyrnan (**4** 454) adune on þære land sceare (**4** 454) to þam twam wicon (**4** 454) adune andlanges drian hyrstes lace fram ðære lace (**4** 454) to þam ligcgendum æc fram ðam licgendum æc (**4** 454) innan sticelan ford of ðam forda (**4** 454) andlang herpaðes (**4** 454) of wulfrices dice hyrnan fram þære dice hyrnan (**4** 454) andlang hæccean (**4** 454) to bridian beara lace (**4** 453) on þone hnottan stybb of þam hnottan stybbe (**4** 454) andlang hæccean (**4** 454) on ðone mycclan mapeldran (**4** 454) andlang hæccean (**4** 454) to þære ræw apoldran (**4** 454) andlang hæccean (**4** 454) to cern hlinces herpaðe andlang herpaðes (**4** 454) to þan holan wege (**4** 454) Andlang dic suð (**4** 454) to þære holan lace andlang ðære lace (**4** 454) vpinnan þone herpað andlang herpaþes (**4** 454) innan hweoul forda (**4** 454).

INDEX OF THE PLACE-NAMES OF DORSET

This index includes all the major names and minor names treated in Parts **1–4**. Parish names are printed in small capitals and lost names in italics. Field-names are not indexed, but page references are given to the more interesting field-names listed in the section on Categories of Field-Names *supra* and also to all those included in the list of Elements *supra*.